Orienting Istanbul: Cultural Capital of Europe?

Planning, History and Environment Series

Published titles

Planning Europe's Capital Cities: aspects of nineteenth century development by Thomas Hall (**paperback 2010**)

Selling Places: the marketing and promotion of towns and cities, 1850–2000 by Stephen V. Ward

Changing Suburbs: foundation, form and function edited by Richard Harris and Peter Larkham

The Australian Metropolis: a planning history edited by Stephen Hamnett and Robert Freestone

Utopian England: community experiments 1900–1945 by Dennis Hardy

Urban Planning in a Changing World: the twentieth century experience edited by Robert Freestone

Twentieth-Century Suburbs: a morphological approach by J.W.R. Whitehand and C.M.H. Carr

Council Housing and Culture: the history of a social experiment by Alison Ravetz

Planning Latin America's Capital Cities, 1850–1950 edited by Arturo Almandoz (**paperback 2010**)

Exporting American Architecture, 1870–2000 by Jeffrey W. Cody

Planning by Consent: the origins and nature of British development control by Philip Booth

The Making and Selling of Post-Mao Beijing by Anne-Marie Broudehoux

Planning Middle Eastern Cities: an urban kaleidoscope in a globalizing world edited by Yasser Elsheshtawy (**paperback 2010**)

Globalizing Taipei: the political economy of spatial development edited by Reginald Yin-Wang Kwok

New Urbanism and American Planning: The conflict of cultures by Emily Talen

Remaking Chinese Urban Form: modernity, scarcity and space. 1949–2005 by Duanfang Lu

Planning Twentieth Century Capital Cities edited by David L.A. Gordon (**paperback 2010**)

Olympic Cities: city agendas, planning, and the world's games, 1896–2012 edited by John R. Gold and Margaret M. Gold

Planning the Megacity: Jakarta in the twentieth century by Christopher Silver

Designing Australia's Cities: culture, commerce and the city beautiful, 1900–1930 by Robert Freestone

Ordinary Places, Extraordinary Events: citizenship, democracy and urban space in Latin America edited by Clara Irazábal

The Evolving Arab City: tradition, modernity and urban development edited by Yasser Elsheshtawy

Stockholm: The making of a metropolis by Thomas Hall

Titles published in 2010

Dubai: behind an urban spectacle by Yasser Elsheshtawy

Capital Cities in the Aftermath of Empires: planning in central and southeastern Europe edited by Emily Gunzburger Makaš and Tanja Damljanović Conley

Orienting Istanbul: cultural capital of Europe? edited by Deniz Göktürk, Levent Soysal and İpek Türeli

Lessons in Post-War Reconstruction: case studies from Lebanon in the aftermath of the 2006 war edited by Howayda Al-Harithy

Olympic Cities: city agendas, planning and the world's games 1896–2016 edited by John R Gold and Margaret M Gold

The Making of Hong Kong: from vertical to volumetric by Barrie Shelton, Justyna Karakiewicz and Thomas Kvan

Orienting Istanbul: Cultural Capital of Europe?

edited by
Deniz Göktürk, Levent Soysal and İpek Türeli

Routledge
Taylor & Francis Group
LONDON AND NEW YORK

First published in 2010
by Routledge
2 Park Square, Milton Park, Abingdon, Oxfordshire OX14 4RN

Simultaneously published in the USA
by Routledge
711 Third Avenue, New York, NY 10017

Routledge is an imprint of the Taylor & Francis Group, an informa business

Typeset in Aldine and Swiss by PNR Design, Didcot
Printed and bound in Great Britain by TJ International Ltd, Padstow, Cornwall

This book was commissioned and edited by Alexandrine Press, Marcham, Oxfordshire

British Library Cataloguing in Publication Data
A catalogue record of this book is available from the British Library

Library of Congress Cataloging in Publication Data

Orienting Istanbul : cultural capital of Europe? / edited by Deniz Göktürk, Levent Soysal, and İpek Türeli.

p. cm.—(Planning, history and environment series)

Includes bibliographic references and index.

1. Istanbul (Turkey)—Civilization. 2. Politics and culture—Turkey—Istanbul. I. Göktürk, Deniz, 1963- II. Soysal, Levent. III. Türeli, İpek.

DR726.O75 2010
949.61'804—dc22
2010008410

ISBN: 978-0415-58010-6 hbk
ISBN: 978-0415-58011-3 pbk
ISBN: 978-0203-84442-7 ebk

Contents

Acknowledgements

This book could not have been brought to fruition without the help of friends and colleagues, to whom we owe deeply. With three editors, twenty-three contributors and nineteen chapters, the process was long and arduous but intellectually satisfying and productive. Along the way many individuals have become part of the project that we named 'Orienting Istanbul'. Arif Aşçı provided us with one of his beautiful photographs of Istanbul, which not only appears on the cover of the book but also inspired us to think about 'orienting Istanbul'. We would like to express our profound appreciation to our contributors for their industry in taking up numerous revisions and for making this interdisciplinary effort possible.

In the process of editing the book, we have received tremendous support, assistance, and encouragement, from friends, partners, and colleagues, some of whom go unnamed but provided ideas, motivation, and helping hands when needed. This book is the culmination of conversations which began in 2005. Then there was the panel at the meetings of the American Anthropological Association in San Jose, California, in 2006. Dikmen Bezmez, Banu Karaca, İlay Romain Örs, Aysim Türkmen, İpek Türeli, Sibel Yardımcı, and Michael Herzfeld, as discussant, participated with their research and perceptive ideas in our early ruminations on Istanbul, in a panel titled 'Remaking Istanbul: Globalizing Desires, Urban Futures, and Cultural Landscape of the World City', organized by Levent Soysal.

Then came the interdisciplinary conference, 'Orienting Istanbul: Cultural Capital of Europe?', which we three organized at the University of California Berkeley, 25–27 September 2008. Nezar AlSayyad, cordially offered us advice, and support through the Center for Middle Eastern Studies, and a generous Al-Falah Grant, which made the conference possible. Our discussants and respondents, Greig Crysler, Richard Reinhart, Cihan Tuğal, and Barbara Wolbert provided much needed critical guidance and intellectual rigour to our discussions and exchanges. All conference participants, among them İpek Akpınar, Marshall Berman, Feride Çiçekoğlu, Aslı Daldal, Sibel Erol, Nilay Kayalp, Zeynep Korkman, Jeremy Pine, Brian Silverstein, Savaş Zafer Şahin, and Şebnem Yücel, expanded our intellectual vistas on Istanbul. Although their work is not part of this book, their contributions have been invaluable. At the Istanbul 2010 European Capital of Culture Agency, Korhan Gümüş and Oğuz Öner were always accommodating and found a way to respond to our requests. At Kadir Has University, Yücel Yılmaz and Deniz Bayrakdar, kindly supported our efforts. From Berkeley and Kadir Has, Nilgün Bayraktar, Muykey Chongtan, Kristin Dickinson, Robin Ellis, Joe Gouig, Khatchadour Khatchadourian, Priscilla Layne, Mejgan Massoumi, Sara McCarthy, Kathryn Schild, Umut Salih Siliman, Kaya Tabanlı and Andy Wand were there whenever we called for them. For their hard work and amity, we are simply thankful. Our sponsors were, at the University of California Berkeley, the Center for Middle Eastern Studies in collaboration with the Doreen B. Townsend Center for the Humanities (especially Anthony J. Cascardi), the Institute of European Studies, Department of German (especially Anton Kaes and Niklaus Largier), Department of Architecture, the Berkeley Center for New Media, and the Pacific Film Archive (especially Susan Oxtoby); in Istanbul, Kadir Has University and the Istanbul 2010 European Capital of Culture Agency, and Turkish Airlines. Their generous support and donations made the conference and continued collaboration possible.

Designed as an interdisciplinary forum, the conference at Berkeley included an exhibition which opened the conference to a wider public. A selection of video works from the Tenth Istanbul Biennial, curated by Hou Hanru and featuring works by Emre Hüner, Wong Hoy-Cheong, and the Map Office, connected Istanbul-Berkeley with moving images. An interactive installation commissioned for 'Orienting Istanbul' and created by Gökçe Kınayoğlu and Wenhua Shi, with Gökhan Kınayoğlu, brought the 'Sounds of the City' to Berkeley. We are grateful to the artists and curators for sights and sounds that enriched our experience of the city. *Arredamento Mimarlık* (December 2008) contributed to this conversation by publishing a review of the conference to make available in Istanbul the lines of thought that emerged in Berkeley.

This book cannot offer an 'end', a final conclusion to ongoing conversations. As proofs were arriving, five of the authors, Çağlar Keyder, Banu Karaca, İpek Türeli, Özlem Ünsal, and Jeremy Walton came together at another Istanbul conference at NYU in April 2010 (organized by Maryam Hariri), and many other scholarly exchanges have already been planned.

We owe intellectual debt to Çağlar Keyder for paving the way, with his edited volume, *Istanbul between Global and Local*, and inspiring us to undertake this project. We thank Michael Herzfeld for being on board and accepting the challenge of bringing his scholarship to Istanbul. We are grateful to the anonymous reviewers, whose comments were most helpful in improving the content and vision of the book. Nicholas Walter Baer and Aleks Göllü read some chapters of this book with a keen eye and offered crucial comments and corrections. Our deeply felt gratitude goes to our editor Ann Rudkin. Under her exacting supervision and editorial dexterity, this process came to fruition and *Orienting Istanbul* is now in your hands.

Deniz Göktürk wishes to thank Minoo Moallem for lively conversations about places here and there, Angela Göktürk for maintaining the home in

Istanbul and sharing her unique 'view' on the city, Aleks Göllü for participating in the conversation, keeping up good spirits, and sharing a home in California, Armen and Jermen Göllü for nourishing our tastes and memories of Istanbul. Levent Soysal expresses his heartfelt thanks to Suat, Mine and Yaşar Soysal, for their unwavering camaraderie, support, and comfort and Yasemin Nuhoğlu for being there, away but close, in London, when needed. İpek Tureli would like to thank Gökçe Kınayoğlu for his love and companionship, and express her profound gratitude to Alptan Türeli and Gülhan Türeli for their continuous support in her academic endeavours and for their good humour, and wisdom.

The Contributors

Elif Akçalı is currently completing her PhD thesis, 'Discontinuity and Contemporary Narrative Cinema', in the Department of Media Arts at Royal Holloway, University of London. Her research interests include film style and narrative, new filmmaking technologies and film history. Prior to her doctoral studies, she completed an MA in Visual Culture, worked as assistant director in commercials and films in Istanbul and as research assistant at the Department of Radio, Television and Cinema at Kadir Has University in Istanbul.

Ayşegül Baykan is Professor of Sociology at the Department of Humanities and Social Sciences at Yıldız Technical University in Istanbul and currently a Visiting Senior Fellow at the Cities Programme in the London School of Economics. She studied Philosophy and Sociology at the University of Pittsburgh. Her recent publications include 'Politics and Culture in the Making of Public Space: Taksim Square, May 1, 1977 Istanbul' with T. Hatuka in *Planning Perspectives* (2010) and 'Visiting the *Harem* of İnci Eviner: A Journey to Utopia' in *Tactics of Invisibility* (Thyssen-Bornemisza Art Contemporary Catalogue, 2010).

Deniz Bayrakdar is the Dean of the Faculty of Communication and Vice Chancellor at Kadir Has University. She has been organizing the annual 'New Directions in Turkish Film Studies' Conference since 1999. She has published eight edited volumes on Turkish cinema, most recently *Cinema and Politics: Turkish Cinema and New Europe* (2009). Other edited works include: *Identity Politics and the Media: Mapping the Margins* (2002); *Mediated Identities* (2001, with K. Ross); *Gender and Media* (1996, with K. Ross and N. Dakovic); *Communications Revolution* (1996, with J. Lotherington).

Nilgün Bayraktar is an actor, director and PhD Candidate in the Department of Theater, Dance, and Performance Studies at the University of California Berkeley. In her dissertation 'Screens Against the Current: The Geopolitics of Mobility in Contemporary Europe', she studies socially engaged video and film works that concern questions of space, place and identity. She has been teaching on migrant cinema in Europe and the aesthetics of mobility at the University of California Berkeley since 2007. Her publications include articles on *coup d'état* novels and migrant cinema. She is also the co-writer of the play *İşte Böyle Güzelim* (2008).

Zerrin İren Boynudelik is Assistant Professor at the Faculty of Art and Design at Yıldız Technical University in Istanbul. She studied History of Art at Mimar Sinan University and Istanbul Techical University (MA and PhD). She was a Fulbright Visiting Scholar at School of the Art Institute of Chicago. Her recent publications are 'Art Practices With

The Public And Contemporary Art: Opportunities And Challenges' in *Cultural Policy and Management Year Book 2009*; 'A Snapshot of Objectivity: Public Reconstruction of TV News in Collaboration with the Artist' (with A. Yurtsever) in *PAJ: A Journal of Performance and Art* (2009).

Marcy Brink-Danan is Assistant Professor of Anthropology and Judaic Studies at Brown University. She holds a PhD from Stanford University. A socio-cultural anthropologist specializing in minority religious communities, her publications include 'Exhibiting Tolerance: Difference & Doubt in a Turkish Museum' in Esther Benbassa (ed.) *Itinéraires sépharades: Complexité et diversité des identités* (2010), '"I Vote, Therefore I am": Rituals of Democracy and the Turkish Chief Rabbi' in *Political and Legal Anthropology Review* (2009) and 'Names that show time: Turkish Jews as "strangers" and the semiotics of reclassification' in *American Anthropologist* (forthcoming). Brink-Danan's monograph about tolerance and the Turkish-Jewish community is forthcoming with Indiana University Press.

Nezih Erdoğan is Professor of Film and Media Studies at Istanbul Bilgi University. He has published essays on melodrama and sound in Turkish popular cinema, and censorship in Turkey in *Screen* and the *Historical Journal of Film, Radio and Television*. He contributed chapters to the *Companion Encyclopedia of Middle Eastern and North African Film* (with D. Göktürk, O. Leaman (ed.) 2001) and *Hollywood Abroad: Audiences and Cultural Exchange* (R. Maltby and M. Stokes (eds.) 2005). He is currently conducting a research project on the early years of cinema in Istanbul.

Jale Erzen has been teaching art and aesthetics since 1974 at the Middle East Technical University in Ankara, Turkey. She is the founder and president of the Sanart Association of Aesthetics, and the editor of the art journal *Boyut* (1980–1984); general secretary of the International Association of Aesthetics (1995–1998) and second vice president of the same organization (2007–2010); recipient of the French Cultural Ministry's Arts and Letters Chevalier Award and the Contribution to Architecture Award 2008. She has published and lectured internationally on Turkish, Ottoman and Islamic art and aesthetics. Her paintings are held in private and state collections.

Deniz Göktürk is Associate Professor of German and Film Studies at the University of California Berkeley. She holds a Dr. phil. from Freie Universität Berlin. Her publications include a book on imaginations of America in early twentieth-century German culture, articles on migration, culture and cinema in edited volumes and journals (*New German Critique, Framework)*, and translations from Turkish literature. She is co-editor of *The German Cinema Book* (2002, with T. Bergfelder and E. Carter), *Germany in Transit: Nation and Migration 1955–2005* (2007, with D. Gramling and A. Kaes) and co-founder of the electronic journal *TRANSIT*. Current projects include *Im/mobilities in a Mediated World*, and *Uniform Identity: Comedy and Community*.

Hou Hanru is Director of Exhibitions and Public Programs and Chair of Exhibition and Museum Studies at the San Francisco Art Institute (www.waltermcbean.com, www.sfai.edu). His curatorial projects include 'The Spectacle of the Everyday', the 10th Biennale de Lyon, 'Not Only Possible, But Also Necessary – Optimism in The Age of Global Wars', and the 10th Istanbul Biennial, Istanbul, 2007. He has contributed to art publications and served as a jury member for international art awards and competitions. His book *On the Mid-Ground* was published by Timezone 8, Beijing-Hong Kong, 2002.

Carola Hein is Professor in the Growth and Structure of Cities Program at Bryn Mawr College in Pennsylvania. She earned her doctorate at the Hochschule für bildende Künste Hamburg in 1995. She has lectured widely on architectural and urban planning, notably in Europe and Japan. In 2007 she received a Guggenheim Fellowship to pursue research on 'The Global Architecture of Oil'. She is currently conducting research on the transformation of Hamburg's waterfront between 1842 and 2008. Her books include *The Capital of Europe* (2004), *Brussels: Perspectives on a European Capital* (2007) and *Cities, Autonomy and Decentralization in Japan* (2006).

Michael Herzfeld is Professor of Anthropology at Harvard University. His most recent authored books include *Cultural Intimacy: Social Poetics in the Nation-State* (1997; 2005), *Portrait of a Greek Imagination: An Ethnographic Biography of Andreas Nenedakis* (1997), *Anthropology: Theoretical Practice in Culture and Society* (2001), *The Body Impolitic: Artisans and Artifice in the Global Hierarchy of Value* (2004), and *Evicted from Eternity: The Restructuring of Modern Rome* (2009). His current research activity includes completion of a book and a film about historic conservation and eviction in Bangkok; current projects include research on Italian-Chinese interactions in Rome and on the profession of town planning in Italy and elsewhere.

Engin F. Işın is Professor of Politics at the Open University. He is the author of *Cities Without Citizens* (1992), *Citizenship and Identity* with Patricia Wood (1999) and *Being Political* (2002). Professor Işın is currently engaged with three different, though related, 'genealogical investigations': concerning 'citizenship after orientalism' with a focus on the Islamic and Ottoman institution, *waqf*; concerning 'acts' especially as it pertains to those acts that constitute subjects as claimants of justice; and, concerning 'governing affects' with a focus on the role of mobilizing emotion in politics.

Banu Karaca is a Visiting Scholar at Sabancı University. She holds a PhD in Anthropology from the Graduate Center of The City University of New York and an MA from Goethe University in Frankfurt/Main. Her dissertation 'Claiming Modernity through Aesthetics: A Comparative Look at Germany and Turkey' examines how divergent claims regarding the civic, socio-political and economic impact of art are mediated among the actors that constitute the artworld in Berlin and Istanbul. Her recent articles interrogate the rhetoric of culture in the making of Europe (*Focaal – European Journal of Anthropology*, 2009), and the politics of intercultural exchange programmes in the EU (*International Journal of Cultural Policy*, May 2010).

Çağlar Keyder teaches at Boğaziçi University and at the State University of New York at Binghamton where he is Professor of Sociology. A native of Istanbul, he completed his primary and secondary schooling in the city and took his university and doctoral degrees in the United States. After a decade of teaching in Ankara during the 1970s he moved to Istanbul. He has written on Turkey's political economy and historical sociology, and on Istanbul's transformation during the global age. His latest book is a collection of essays on the Ottoman Empire and Turkey's bid for candidacy in the European Union.

Introduction: Orienting Istanbul – Cultural Capital of Europe?

Deniz Göktürk, Levent Soysal, İpek Türeli

The (Re-)Discovery of Istanbul

The realization of this book owes much to digital technologies. Email, skype and Google documents enabled collaboration – across continents and time zones – between Providence, Istanbul and Berkeley.[1] Our world is shrinking, while cities are expanding into each other, continuously reshaping our (and their) sense of place within a global horizon. As cities compete for relevance they are challenged to claim both situated difference *and* global connectedness. This book explores such tensions in various spheres of knowledge production such as art, social sciences, market, and governance, which appear increasingly entangled in the everyday practice of the present-day city. Coming from three disciplines – architecture, anthropology and cinema – our interests in cities and in Istanbul in particular converged. This collection offers new multidisciplinary research focusing on Istanbul, but it also speaks to readers curious about cities elsewhere, in Europe and beyond. We are keenly aware that the world is becoming an 'endless city' (Burdett and Sudjic, 2008) – at least for those of us with access to networks of communication.

While we were editing the chapters which follow, the travel sections of international newspapers abounded with articles about Istanbul as a tourist destination. In a list of 'The 31 Places to Go in 2010', the New York Times ranked Istanbul at number 19, highlighting as a major selling-point the 'contemporary art scene … one of the most innovative in the world', rather than the historical sights of the city.[2] In the spirit of participatory digital media, the online edition invited readers to choose their favourite destination and say where they would like to go in 2010 by clicking on an interactive world map.[3] The readers' ranking came out in favour of Istanbul.

> The No. 1 spot, by a fairly wide margin, was Istanbul, with 143 recommendations... This is hardly a surprise. Providing a confluence of cultures and continents, with an innovative

> art scene, creative food offerings and fascinating architecture, Istanbul is now at the top of many travelers' must-visit lists. Its allure in 2010 is enhanced by its designation as a European Capital of Culture.[4]

Publicity images that go along with this discovery no longer focus primarily on historical buildings such as the Hagia Sophia or the Topkapı Palace, but modern features such as the new skyline at night, which are reminiscent of downtown Chicago.

Istanbul's popular appeal internationally should, however, not be taken for granted. It was only in the 1990s that Turkey emerged as a popular tourist destination for package holidays due to the adoption of tourism as economic development policy and the decreasing costs of international air travel. Turkey has invested in major advertising campaigns announcing itself as the world's largest open-air museum, a 'destination museum' (Kirshenblatt-Gimblett, 1998). Internationally high-profile events, such as the Galatasaray soccer team's UEFA championship in 2000, Sertap Erener's victory in the Eurovision song contest in 2003, and, most prominently, Orhan Pamuk's Nobel Prize for literature in 2006, bolstered national pride and amplified Turkey's and Istanbul's image abroad.

Orhan Pamuk is now a global player who has achieved international recognition as *the* writer of Istanbul. In his memoir *Istanbul: Memories and the City*, he introduced himself as an author living in an age of mass migration, rootlessness and exile, whose imagination was nonetheless fuelled by staying in the same city, the same street, the same house, looking at the same view to write seven (by now eight) novels mostly set in Istanbul (Pamuk, 2006, p. 6). In his opening speech at the 2008 Frankfurt Book Fair, which showcased Turkey as a guest of honour on the 80th anniversary of Turkey's change from Arabic to Latin script, he reflected on the imbalances and shifts in cultural traffic with a fine sense of irony:

> during the last century, we Turks have complained so much about the world misunderstanding us that it has become part of our national identity. Most of us believe that our culture and our literature owe their power and their uniqueness to the very fact that no one else knows about them... The political and cultural developments of the last twenty years have made the story of Turkey's two-century-long struggles between tradition and modernity more interesting to world audiences. These days, I almost never hear people complaining about how no one can find Turkey on the map.[5]

Even the proclaimed inside-view is mediated by reflection on how things might look from outside. As Edward Said (1978, 2003) has argued brilliantly in *Orientalism*, the power of discourse is internalized; it does not need to be produced in the West.[6] Pamuk's memoir is exemplary. His autobiographical portrait of Istanbul breathes melancholia (*hüzün*) and longing for the lost glory of the former imperial city (a perspective that Engin Işın counters in Chapter 2 with the concept of *keyif*, pleasure), openly admitting that Flaubert, Nerval and other travelling Europeans informed his own perspective on the city.

Filmmakers based in Istanbul have made a name for themselves at international festivals, mainly with pensive auteur films, which tend to be met with enthusiasm at the Cannes Film Festival. The image of the brooding photographer against the silhouette of the city from *Distant* (2003), which is also found on the cover of a recently published book length study of Turkish cinema in English (Suner, 2010), is reminiscent of Pamuk's melancholic flâneur. Nuri Bilge Ceylan stages in *Distant* the feeling of encroachment on part of the Istanbulite whose space is invaded by the migrant from the country seeking work on freight ships (see İpek Türeli, Chapter 8). Meanwhile, other migrants – engaging in an imaginary return journey from Western Europe – have been discovering Istanbul with more instantaneous energy. For Western eyes, Istanbul has emerged in the last decade as an outpost of 'authenticity'. Hamburg-based film director Fatih Akın, the son of migrants to Germany, projects a utopian vision of polyphonic diversity and East–West amalgamation in his musical portrait of the city, *Crossing the Bridge: The Sound of Istanbul* (2005), which acquires a nostalgic undertone only in its Turkish title *Istanbul Hatırası* (Istanbul Memory) (see Deniz Göktürk, Chapter 10). On the other hand, Yılmaz Erdoğan's *Magic Carpet Ride*, a domestic film production of the same year, also notable for its use of Roma music and aerial perspectives, tackles the expanse and totality of the city, contrasts and convergences between affluence and crime, in an ironic light. Although it had no international distribution, this films speaks with much wit to questions of circulation and Europeanization (see Deniz Bayrakdar and Elif Akçalı, Chapter 9).

The appeal of Istanbul is surely expanding. Domestically produced Turkish television serials showcasing Istanbul, in particular those set in villas on the Bosphorus, such as *Aşk-ı Memnu*, *Gümüş*, *Yaprak Dökümü*, are popular in the Gulf States and the rest of the Middle East (and the Ministry of Culture and Tourism has created special funds for the production of such series). Indian production companies use Istanbul as a location for TV serials and Bollywood films such as *Mission Istaanbul* (2008).

In conjunction with these media representations, Istanbul has also been attracting significant influx of foreign direct investment and international companies. While Turkey's primary trade and investment partnership is with the EU, the growing Asian and Middle Eastern economies are also increasing their engagement in the city. As Saskia Sassen puts it, 'Istanbul is the immutable intersection of vast and diverse mobilities', on the North–South and East–West axes of the world (Urban Age, 2009, pp. 5–6). Social scientists and creative artists compulsively deploy the metaphor of the bridge to locate the city in global maps and cultural imaginaries. However, the extant diversity of flows and mobilities complicate the compartmentalizing geopolitics of 'East' and 'West'.

◆ ◆ ◆

Orienting Istanbul is the first book to capture Istanbul's rise to the world stage set by post-industrial capitalism. It offers new insights into the re-presentation of Istanbul as a city of culture, history, and diversity. Cities around the world adopt global city projects in their competition for a place on the map. However, globalizing desires produce diverse outcomes for civil society, urban politics and mediatized image production in each city. Istanbul's designation as a European Capital of Culture in 2010, while Turkey's membership in the EU is still contested, complicates the logic of area studies separating the Middle East from Europe. This year-long spectacular event aims to foster Turkey's ties to Europe, and via Europe to the world. Taking this event in the making as an occasion, this book analyzes interactions between governmental agencies, NGOs, artists and activists, and publics at large, connected in conspicuous ways.

Divided into five parts (Paths to Globalization; Heritage and Regeneration Debates; The Mediatized City; Art in the City; A European Capital?), with case studies ranging from urban renewal, architecture and heritage preservation to art exhibitions, cinema and literature, *Orienting Istanbul* aims to provide a unique picture of how the course to European integration and globalization is manifested in Istanbul's streetscapes and the lives of its citizens. It includes conversations with practitioners and cultural brokers and thus appeals to audiences beyond an academic readership to public intellectuals and experts in culture industries.

Our introduction to *Orienting Istanbul* proceeds with Istanbul's reign as 2010 European Capital of Culture, an event that focuses the gaze of both foreigners and residents on the city for a year. From the present, we turn to the past, briefly excavating the history of an imperial capital, situated to the east of Europe but with integral connections – material and imaginary – to Europe's economy, politics, wars and culture. Back to the present, we explore how Istanbul's past and present are framed in urban studies – a necessarily concise undertaking, for Istanbul has been seriously neglected in scholarly circles. Istanbul belongs to the 'global cities without privilege', cities which are deemed to be outside the networks that connect 'nerve centres' of the globalized world of the late twentieth century. Even so, we assert, Istanbul, like its counterparts – be they London, New York or Tokyo, the pre-eminent models of the global city, or Mumbai, Sao Paolo or Mexico City, presumed peripheral cities – promises to contribute to and complicate our understanding of the culture, politics and economies of global cities. Istanbul, the city with which we are intimately connected, is not simply an academic concern for us but affects us as citizens. The questions we pose are pressing at a time when culture replaces industry as the economic pillar of the city and a more diffuse politics of difference and identity succeed the nationally organized politics of the earlier decades. We close our introduction with an account of orientation embedded in the cover image of *Orienting Istanbul*, a layered picture taken from the roof of Büyük Valide Han, a building which embodies the presence of the past in the city.

A European Capital of Culture

Spanning two continents, Istanbul is the largest city not only in Turkey but in Europe. Istanbul's selection as one of the three cities (along with Essen/Ruhr and Pécs) celebrated in 2010 as 'European Capitals of Culture' demonstrates that Europe officially acknowledges Istanbul as a key part of its own heritage while remaining ambivalent about Turkey's Europeanness. A paradoxical split imaginary emerges in representations of Istanbul, official or popular: one that separates the city from the rest of Turkey. Some of the contributions to this book examine how this division is internalized and projected onto Istanbul where the national and municipal government, as well as local private capital, stage the city as Turkey's gateway to Europe. Others explore what Istanbul has to offer towards the formation of the 'imagined' identity of a cosmopolitan, post-religious and post-national Europe. Turkey's unresolved bid for membership in the European Union has been employed to legitimate and promote change. European Union membership is expected to bring not only the intensification of relations with Europe but also with the rest of the world. The accession process creates incentives to engage with future-oriented projects to upgrade the city's infrastructure, educational institutions and tourist sites (Keyder, 2008). The city's prospects are implicitly and explicitly connected to Turkey's membership in the EU.

Turkey has been a member of the Council of Europe since 1949 and an associate member of the European Economic Community, the predecessor organization of the European Union, since 1963. Since 1995, the country has been linked with the EU by a Custom's Union Agreement. Turkey's application for full membership in the EU, however, is still pending. The current governments in France and Germany share reservations about admitting Turkey, following the stance of former French President Giscard d'Estaing who had proclaimed in 2002 that accepting Turkey 'would be the end of the European Union'. First and foremost, there are economic reasons for this hesitancy about opening the borders to a country with a mostly young population of over 70 million and a growing economy. Moreover, 'cultural difference' is emphasized as a major obstacle, meaning specifically that a country with a predominantly Muslim population would destabilize Europe's implicit self-definition on the basis of Christian/Western values. The prevailing concerns about terrorism and security following the attacks of 11 September 2001, as well as the ongoing wars in Iraq and Afghanistan, have cast dark shadows over Europe's and the United States' relationship with the Muslim world. The rise of Islamism in the formal political arena has been a contested process in Turkey. Yet, after its victory in the 2002 national election, the Muslim-oriented AKP (Adalet ve Kalkınma Partisi, Justice and Development Party or JDP) emerged a major promoter of liberal market economy and Turkish membership in the European Union. In Turkey, the European accession is perceived as a civilizing process with implications for standardization, human rights, NGOs, civil society and participatory democracy (see Levent Soysal, Chapter 17).

The European City of Culture programme is one of the most acclaimed cultural projects endorsed by the EU (Gold and Gold, 2005, p. 221–245; see also Miles, 2007, p.121–142). The programme was launched in 1985 by the Greek Minister for Culture, Melina Mercouri (well-remembered from her role as the stylish jewel-kleptomaniac in *Topkapi* from 1964). Athens served as the first 'City of Culture'. The programme is designed to showcase the cultural life of the chosen city for one year, encouraging local initiatives to make the city a better environment for its inhabitants and an attractive place for tourists. Cities have to apply with a proposal and are selected by a committee. Some support is granted from the EU, but mostly cities have to raise their own funds to stage their spectacle. For a city like Glasgow in 1990, the festival served as a major catalyst in revitalizing a decaying de-industrialized city (Gold and Gold, 2005). Thus not only 'beautiful' cities with historical heritage could claim the title 'capital of culture', but less showy cities could also participate and benefit from the programme of European integration. Since 2000, more than one city can hold the title 'Capital of Culture' in any one year. The idea of decentralized and mobile capitals is in line with the EU's efforts to engender unity in diversity and engagement with the European project in all corners of the Union. Since 1999, cities in EU-affiliated, but non-member states have been permitted to apply for the title. Hence Istanbul applied and was chosen as one of the designated Capitals of Culture for 2010 (see Oğuz Öner, Chapter 15).

The programme enables interesting insights into the imaginary construction of Europe in different places. The question arises, however, whether the implementation of the programme is really about imagining Europe and seriously engaging with ideas of post-national integration or whether various initiatives – governmental and non-governmental – in the designated cities grab the occasion to pursue their already-existing projects and 'opportunistically' reframe them in European terms (see the argument put forward by Carola Hein, Chapter 14). The programme does not seem to foster much communication and collaboration between cities. There is not a single joint project between Istanbul and Essen, for example, despite the considerable Turkish presence in Germany's Ruhr area.

In Istanbul, the programme opened with fanfare on 15 January 2010. In his speech at the opening ceremony, Prime Minister Tayyip Erdoğan emphasized the flagship importance of the city for his government's goal of joining the European Union:

> Istanbul is a European city. With its history, culture, civilization, people, its past and its future, Istanbul is a city that is facing Europe. As much as Istanbul has absorbed European culture, it has also shaped European culture. Istanbul will only carry the title of a European Capital of Culture for one year, but it will never seize to be a cultural center of Europe. Istanbul alone is proof that Turkey is a European country, that it is a natural member of the European Union.[7]

At the same time, Erdoğan highlighted Istanbul's location at a global crossroads. He stressed the unique confluence of civilizations, races, and colours as well as the coexistence of mosques, churches and synagogues side by side as a source of inspiration for cities across five continents. Despite claiming connections to cities elsewhere, Erdoğan declared that 'Istanbul most of all resembles Istanbul'. This glorification performs a straddling act in multiple directions, gesturing towards Europe, the Middle East and the globe at large, but ultimately looking inward. The paucity of European connections in the 2010 programme suggests that Istanbul is orienting itself to global audiences rather than to a specifically European gaze. One cannot help but ask: What does 'Europe' mean to people on the streets?[8] Who is showing what to whom and who is doing the 'orienting'?

Imperial Capital on the Outskirts of Europe

Not only is the future of the city subject to reorientation but also its past; indeed, the future of the city is imagined via its cosmopolitan past. A recent programme of theatrical performances on Istanbul makes reference to the Greek origins of its name which translates as 'To the City'.[9] Its non-Muslim inhabitants referred to Istanbul as Polis/Bolis (*the* City). Armenians, Greeks and Jews at one time constituted half of the city's population (see Nezih Erdoğan, Chapter 7 on multi-ethnicity and multilingualism in the early days of cinema in Istanbul, also Marcy Brink-Danan on Jews in the city today and their orientation towards Europe, Chapter 16). Following the founding of the Republic in 1923, the colloquial Greek name was adopted as the official Turkish name in 1930, rather than Constantinople, which the Ottomans had revered.

Due to population exchanges with Greece in 1923, 'Turkification' policies, and the exodus following the pogroms in 1955, non-Muslim communities in Istanbul have been dwindling. The city has become much more uniformly Turkish and Muslim today than it ever was before.

For over 1500 years of its history, Bolis was an imperial capital – of the Eastern Roman Empire, Byzantine Empire and the Ottoman Empire. It was established by Emperor Constantine in the fourth century on the site of an older Greek colony as New Rome and in 330 CE renamed Constantinople, the city of Constantine. Conquered in 1453, Constantinople became the capital of the Ottoman Empire which had its prized territories in the Balkans and made continual advances into central Europe.

European views of Istanbul's location and whether it belonged to Europe or not have changed over time (Brummett, forthcoming), the current boundaries hardening in the nineteenth century. By the latter two-thirds of the nineteenth century, under the impact of industrialization, European and self-perceptions of Istanbul started to change (Kafadar, 1997/1998). It was due to uneven developments in the nineteenth century that Istanbul emerged as an 'Oriental city'. When the 24-year-old Charles-Édouard Jeanneret (Le Corbusier) travelled to the 'Orient' in 1911, Istanbul was the farthest point to the east of his itinerary (Le Corbusier, 2007). Le Corbusier arrived in Istanbul by boat in the footsteps of previous travellers, looking for authenticity, but by this time the city was well connected and serviced.

During the nineteenth century Istanbul had undergone a series of street regularizations, following fires, and infrastructure modernization paralleling that in other world cities. Most of the services, however, had benefited the Galata-Pera area, north of the Golden Horn, across from the historic peninsula, and mainly populated by affluent non-Muslims, of Ottoman or European citizenship – so much so that the duality of the historic peninsula and the northern area might have given the impression of a colonial urban design. However, there was no segregation; bridges and boats connected the two sides with regular traffic (Çelik, 1986, p. 160).

By the 1910s, the city had acquired a modern transport network of commuter boats (begun in 1851 under Şirket-i Hayriye, the first Ottoman joint stock company), trams and trains. The Sirkeci Train Station opened in 1890 as the eastern terminus of the Orient Express. In the first years of the Young Turks' rule (1908–1918), the Grand Post Office (1909) and Haydarpaşa Train Station (1908) opened as the western terminus of Istanbul-Baghdad Railway. The first power station, Silahtarağa, was established in 1914 in the deep end of the Golden Horn, to provide electricity to various parts of the city and the new electric-powered tram network.

Modern Istanbul, Back Into Limelight

Istanbul remained the capital of the Ottoman Empire until the Allied occupation from 1918 to 1923. After Ankara was designated the capital of the newly founded Republic of Turkey, Istanbul was cast in official publications in opposition to Ankara. Istanbul was imagined as old and dusty, cosmopolitan and decadent, while Ankara was new and clean – a model for the new Turkey (Bozdoğan, 2001, p. 67).

One of the most important interventions of urban design in the years immediately following the establishment of the Republic was the reorganization of Beyazıt Square, and the opening of the Atatürk Monument in the middle of Taksim Square. After a decade of relative 'neglect' (Gül, 2009), the Republican era city was shaped according to the design of French planner Henri Prost who worked for Istanbul from 1936 to 1950. Prost had experience in the Maghreb and was author of the Master Plan for the Paris Region when recruited for Istanbul. His initial plan for Istanbul targeted the historic peninsula and to its north, the Beyoğlu district (which included the areas formerly called Galata and Pera) and aimed to improve the street network, open new boulevards, preserve monumental buildings and the city's distinct silhouette.

In line with planning paradigms of the time, he sought to reorganize the city into an automobile friendly space with (industrial, commercial, residential and recreational) zones. The Golden Horn served as the principal industrial zone. Prost's plan was gradually and only partially implemented. Atatürk Boulevard cut through the historic peninsula to divide Sultanahmet and Zeyrek neighbourhoods; major demolitions in Eminönü, Karaköy, Beşiktaş, and Üsküdar, on the Anatolian side, opened up 'squares' by the water; the Artillery Barracks were demolished to reorganize the Taksim Square and to make way for the promenade (*Taksim Gezisi*).

In the 1950s, Turkey's new role in the post-war international order turned the government's attention back to Istanbul. After Prost's departure, the General Directorate of Highways played an important role in the development of Istanbul's urban form, continuing along the lines drawn up in Prost's plan, overseeing massive demolitions to open up wide boulevards through the city.

Several new investments such as the Hilton Hotel in Maçka Park, the first garden suburb of Levent to the north, the model town of Ataköy to the west, along with the new boulevards (e.g., Vatan (Country) and Millet (Nation) Avenues in the historic peninsula, Barbaros Boulevard from Beşiktaş to Levent), became showcases for the government.

Post-war governments continued piecemeal urban form interventions and civic improvements, while closing their eyes to increasingly visible squatter settlements which grew into whole neighbourhoods. The city was rebuilt and expanded with speculative housing developments on all sides. Concrete-frame walk-ups rapidly replaced the existing residential fabric.

One of the most important developments was the building of the first Bosphorus Bridge in 1973; together with its connecting highways, this opened new areas for development and facilitated the west–east expansion of the city along the Marmara Sea. Now served by highways, industry gradually moved out of the Golden Horn area and spread to the Anatolian side.

In the 1980s, with economic liberalization a new phase of urban restructuring carved out the Tarlabaşı Boulevard and, via demolition and infill, turned the banks of the Golden Horn into public parks. The new Central Business District between Levent and Maslak was now realized with the addition of glass-clad high-rises on the Büyükdere Asphalt. This is the skyline contemporary Istanbul projects as a counterpart to that of the historical peninsula with its domes and minarets.

All this rapid transformation meant that the traditional fabric almost disappeared. Only in the 1970s did calls emerge to preserve not only monuments but the wooden houses and the neighbourhoods they constituted emerge. Parts of the historic peninsula were eventually designated as world heritage by UNESCO in 1985. Some of the older neighbourhoods, especially those Bosphorus villages formerly inhabited by non-Muslim communities, such as Kuzguncuk, were discovered anew as desirable residential areas (Mills, 2010).

Since the 1950s the city has witnessed not only rapid urbanization but also the rise of a consumer society, the expansion and cultural ascendancy of the

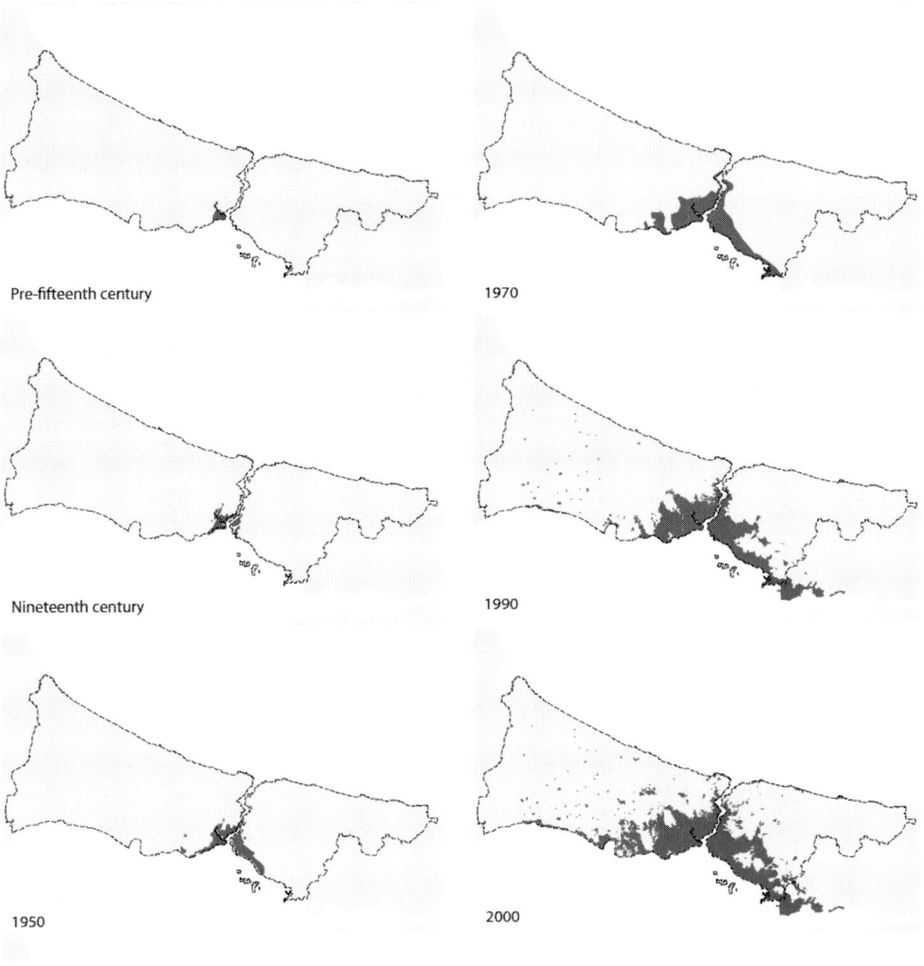

Figure I.1. Historical evolution of Istanbul's footprint. (*Source*: Urban Age, London School of Economics, www.urban-age.net)

middle class, and the development of modern techniques of image production and consumption. The 'opening' up of the city, through its physical expansion and intense population movements, gave way to an obsession with the city in the realm of culture which persists to the present day (Türeli, 2008). The present is heavily burdened with a pervasive feeling of loss in response to Istanbul's unplanned growth and Turkey's inchoate position *vis-à-vis* Europe.

Following worldwide trends, starting in the 1980s, a surge in oral histories, memoirs and exhibitions about the city's past have turned 'Old Istanbul' into a popular site consumed by a larger public (Türeli, 2010). Among the most notable are the restoration and commercial opening of several Ottoman-era houses and mansions by the Touring and Automobile Club of Turkey run by Çelik Gülersoy, and the formation of the Economic and Social History Foundation of Turkey (*Tarih Vakfi*) in the early 1990s by Turkey's leading historians and social scientists with a focus on Istanbul's history. The Foundation's multifarious activities in producing histories of the city, such as the 1996 exhibition 'Istanbul-World City' at the Imperial Mint Building (*Darphane-i Amire*) on the occasion of the UN Habitat meeting in Istanbul, have generated a lively debate, and many other foundations and organizations have turned their attention to the city's past. A relatively recent one is the Suna and İnan Kıraç Foundation's Istanbul Research Institute operating under the Pera Museum (2005). Meanwhile, Islamist groups and organizations have also emerged as important actors re-enacting and invoking the city's Ottoman past (see Jeremy Walton on localized neo-Ottomanist heritage preservation and global aspirations, Chapter 5).

The past remains omnipresent in Istanbul. The construction of Marmaray, a new suburban train line tunnelling under the Bosphorus, was delayed by several years, when in 2005 digging unexpectedly uncovered the site of an ancient port, taking back the time-line to 10,000 years ago. Construction thus turned into archaeology, reminiscent of a scene of the construction of the underground in Frederico Fellini's film *Roma* (1972). The past resurfaces in contentious debates about architectural preservation.

Global Cities without Privilege

Despite increasing popular interest, Istanbul is underrepresented in scholarly publications. This book seeks to fill an obvious gap by assembling recent research on the city conducted in a variety of disciplines. The absence of Istanbul from research agendas has to do with the 'asymmetrical ignorance' (Roy, 2001) that inherits Orientalism.

Even though there have been several plans, never fully realized (Çelik, 1986; Gül, 2009), Istanbul's development in the post-war period with its sprawl, shanty-towns, crooked streets, and unplanned housing development, has been considered explosive, quasi-organic, and dominated by the small-scale 'build and sell' system. For these qualities, Istanbul is readily referred to as a 'Mega City' or 'Third World City'.

Recent urban studies in Turkey have been conducted in two realms: architectural urban history and sociological studies (İçduygu, 2004). Architectural histories of the modern era have focused on Republican nation building and neglected Istanbul. This is especially true for the post-war period, a time of rapid urban expansion outside the control of architects' and planners' visions. Sociological studies have turned to the city early on, but with a focus on migration, squatter settlements and the question of 'integration' (Erman, 2001). It was only in the 1990s that studies of the city started to examine changing 'lifestyles' brought about by neoliberal economic policies and efforts to turn Istanbul into a world city, implemented by governments after the coup in 1980 (Bali, 2002).

Çağlar Keyder's *Istanbul: Between the Global and the Local*, published in 1999, was the first book to bring together this new generation of studies with a cultural perspective and a focus on practices of the middle class (see also Kandiyoti and Saktanber, 2002). The question Keyder asked in the 1990s was whether Istanbul would be able to achieve its potential to become a 'global city' in the sense defined by the concept's theorists such as Sassen or whether it would miss the opportunity which unfolded in the aftermath of the dissolution of the Soviet Union. In the post-Cold War world order, Turkey gained new geopolitical significance vis-à-vis the Middle East and Central Asian Turkic states. It was not clear in the 1990s which path the Islamists in power, with Erdoğan then Istanbul's mayor from Refah (Welfare) Party, would take regarding the global city project. Contributions to Keyder's volume reflect that anxiety about Istanbul's status and trajectory. Since 2002, under a single party, AKP, an off-shoot of the Welfare Party, Istanbul witnessed relative stability and a more formal push (Keyder, 1992; 1999; and Chapter 1 in this volume).

The concern for Istanbul's status demonstrates the power of the 'global cities discourse' (Smith, 2001) around which much urban studies literature converges. The conceptualization of 'the global city' derives from the assumption that certain privileged Euro-American cities have acquired unique roles in the world economic system, as control and command centres, in the aftermath of the post-Fordist organization of production (Friedmann, 1986; Sassen, 1991).

Earlier studies of the global city focused on ranking cities based on economic data. They were criticized for narrow-minded ranking at the expense of historical specificity (King, 1990) and agency (Smith, 2001) and for perpetuating, through the 'regulating fiction' of the global city, the divide between 'First World' global cities as models, and 'Third World' mega-cities as problems (Robinson, 2006). Despite sustained criticism, ranking studies are in no way obsolete.[10] More intriguingly, these scholarly abstractions turn into reality by shaping urban policy and even popular perceptions. The experience of global cities has emerged as a field of aesthetic investigation, influencing even the categorization and curation of art. An exhibition at Tate Modern in London on 'Global Cities' (20 June to 27 August 2007) featured Istanbul along with Shanghai, Cairo, London, and Los Angeles and compared the cities according to size, speed, form, density and diversity.[11]

More recently researchers have focused on the changing role of the state – from a regulator to an agent of the market – and the impact of economic neo-liberalization processes on urban transformations. The shift from manufacturing to services in de-industrializing cities caused class polarization, as cities invest in public-private partnerships, aiming to attract capital, rather than in services for citizens. The relocation of industries to poorer countries and to their special economic zones with lax environmental and labour regulations amplified processes of urbanization and exacerbated problems of access to already limited public services. These shifts were accompanied by important implications for the built environment and architectural culture. Neoliberal restructuring projects have materialized in the built environment of cities in the form of increased residential segregation (fortification), the growth of networked Central Business Districts and unbundled infrastructure provision – what Stephen Graham and Simon Marvin (2001) called 'splintering urbanism'.

Scholars who focus on neoliberalization examine how projects of restructuring and inherited institutional, social, and spatial landscapes intersect and produce diverse results in different localities. Scholarship on Istanbul has been conversant with concurrent debates in urban studies – somewhat dominated by the Anglo-American academy. Accordingly, more recently, scholars working on Istanbul have taken up issues of gated communities (Danış and Pérouse, 2005), gentrification in the historic quarters of the city (Behar and Islam, 2006; Mills, 2010; Uzun, 2003), generalized gentrification of slums (Ünsal and Kuyucu, Chapter 3 in this volume), and the revitalization of de-industrialized zones of the city (Bezmez, 2008).

Benefiting from all of the above, we argue that new studies need to take into account various strands of theory together (as in Jacobs, 1996). Furthermore, in order to interpret the politics of the present, we assert that our analytic vistas should focus not only on how the past informs the present, but also on how the present inspires a rethinking of the past – as elegantly elucidated in Michael Herzfeld's ethnographic explorations of the politics and poetics of historical preservation in cities large and small in Europe, moving from Rethemnos in Crete (1991) to Rome in Italy (2009). We maintain that urban studies need to reveal the role of urban imaginaries in representations of cities. *Orienting Istanbul* follows this path, placing due emphasis on conceptualization of the city, of its past and present, in the arts, visual media and everyday practice.

The experiences of cities like Istanbul can generate a productive theoretical framework for studying new urban formations in a comparative perspective (see Michael Herzfeld's epilogue). Just as Chicago and Los Angeles provided models with which to compare other cities around the world, so can the study of Istanbul (and Cairo, Bombay or Mexico City) enrich the scholarly perspective on contemporary urban life around the world. Thus *Orienting Istanbul* does not seek merely to fill an obvious gap as a case study, but also to promote the widening of our theoretical horizons. It adds to emergent scholarship that enriches our intellectual map of the global city by attending to urban formations that were formerly deemed invisible (see Singerman and Amar, 2006). As Andreas Huyssen (2008, p. 14) proposes, the aim of such studies is 'to open up architectural, urban, and cultural studies to the imaginative geographies of alternative or different modernities that are usually sidelined by the still-dominant focus on the northern transatlantic in much of the Western academy'.

The global city and European integration are usually considered disconnected realms. Hence, one of the key contributions of this book is to examine how these two frameworks are intertwined. The 2010 European Capital of Culture programme causes excitement and dissent among policy-makers, intellectuals, and producers of culture. It seeks to forge new connections and to strengthen existing ones towards a new European identity. Istanbul's version presents itself primarily as a marketing strategy, highlighting new urban regeneration projects while supposedly celebrating cultural diversity. The EU programme evidently employs culture to remap or raise the status of its key cities with forward-looking economies which are not themselves key players in the global city network. Policy and resources are geared towards culture-led development rather than structural economic change, which could potentially benefit a broader demographic in cities participating in the programme. This is where the EU programme and popularized visions of the global city coalesce. *Orienting Istanbul* asks critical questions about the ways in which cities embrace culture as urban development policy to remedy omission from the global map.

Citizens, Governance, and Participation

According to the census of 2009, Istanbul has a population of 12.9 million. This means 17.8 per cent – or close to one-fifth – of total population of Turkey, which officially stands at 72.5 million. Most experts estimate the real population of the city to be nearer 15 million.[12] Istanbul's population remained around the million mark in the first half of the twentieth century; it started increasing thereafter with nation-wide increased life expectancy and rural to urban migration.[13] Migration flows from the country to the city, on to Western Europe and back again have been formative forces. Migration still plays a role in population expansion, although births now account for the major factor in growth.

Geographically the city spreads over an area of 5.5 thousand square kilometres. In this vast city-space, one encounters a very high density, 2.4 thousand persons per square kilometre. There are more than 2.2 million residences, mostly situated on the European side of the city, where about 8 million of the city's population lives.[14] The density poses massive infrastructural challenges. Denizens have been overwhelmed by traffic, pollution, crowds, crime and the rapid transformation of the cityscape around them. The city is located on top of one of the major geological faults and is under constant threat of earthquakes (the earthquake of August 17, 1999 killed 17,000 people). Occasional floods, mainly caused by irregular building in vulnerable riverbeds and flood plains, and occasional snow exacerbates the troubles of living in this vast metropolis. As Istanbulites say – and know – 'when it rains, life in Istanbul stops'.

Istanbul has been serviced, since the 1980s, by a two-tier municipal system and the governorship. The metropolitan municipality (Istanbul Metropolitan Municipality, IMM) concerns itself with citywide services and district municipalities serve populations within their boundaries. Nonetheless, both, in particular their elected mayors, have vast powers of decision-making that came with changes in the governance of municipalities, decreasing the role of centrally appointed governors' offices and increasing the sources of income for municipal governments. EU integration efforts and adoption of EU models definitely had accelerating effects on these transformations.

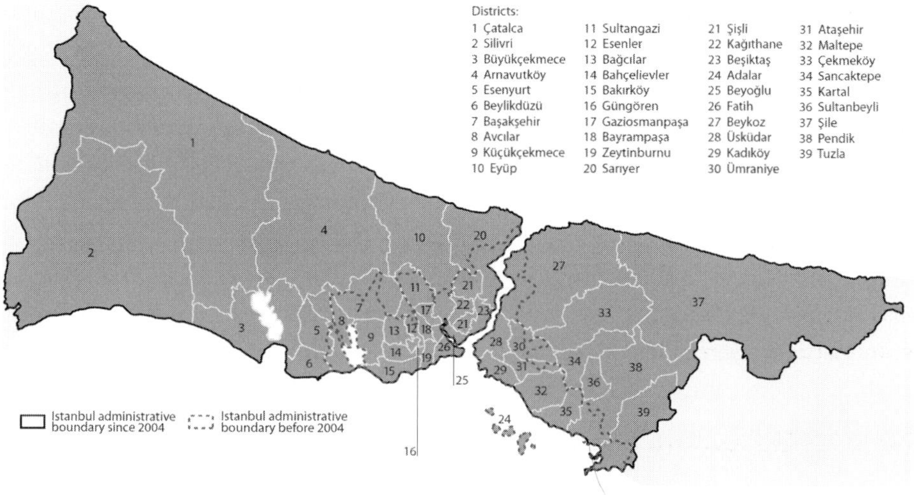

Figure I.2. Municipal boundaries. (*Source*: Urban Age, London School of Economics, www.urban-age.net)

The legal framework, which is at work, does not necessarily facilitate participatory processes in city governance. Working with municipal councils with relatively weak powers, mayors more or less act as CEOs. Citizens learn about decisions through newspaper accounts, mostly after the fact when contracts are assigned and deeds are done. There are citizens' councils (*kent konseyi*) in each municipality, but they operate more like showcases for the mayors than forums for citizens. Although NGOs, citizens' groups and, most effectively, professional organizations of architects and urban planners do raise their voices through the media and courts, it is not wrong to say that there is a lack of accountability to the citizens.

Driven by new powers vested in them, municipalities have been able to privatize the public land and buildings that they owned, found semi-public companies and independently obtain loans and credits, resulting in massive infrastructural investments such as metro-bus lines, subway lines, roads and bridges. For instance, IMM is reported to secure 1.5 billion euros from the European Investment Bank for expansion of the city's metro system.[15] IMM owns twenty-four semi-public companies, ranging from bread factories to ferry transport, landscape design to culture, the revenues or debts of which are not disclosed to the public.

While debts are being incurred, the city's investment and privatization frenzy continues. Two major port areas, Karaköy, where the Istanbul Modern Museum is housed in one of the old cargo warehouses, and Haydarpaşa, where the historic train station that links the city to Anatolia is located, are being opened to construction of prestige residences, hotels and shopping malls. Kartal, one of the heavily industrialized, working-class districts, is set to experience regeneration on a grand scale under a blueprint designed by Zaha Hadid. A third bridge is being planned for the northern end of the Bosphorus as a way to ease cross traffic between the Anatolian and European sides.

The city has a master plan developed by Istanbul Metropolitan Planning and Design Centre (IMP), an agency founded by the metropolitan municipality in 2004, in the words of Istanbul's mayor Kadir Topbaş, 'to give an end to the problems of Istanbul, such as lack of planning, crooked settlement and uneven growth'.[16] IMP has brought together experts, architects, academics and municipal officers and produced a variety of plans and schemes intended to carry Istanbul forward in line with contemporary precepts of urban planning: de-industrialization of the urban landscape; preservation of the green areas in the north of the city; and construction of infrastructure for a city sustained by culture and creative industries. IMP is criticized for the narrowness of its vision for the future of the city, a vision that almost singularly relies on culture industries and gentrification. It was nonetheless a plan. IMP's reign as a city planning office did not last long, however; it is now a defunct organization, whose grand vision and hundreds of plans have been shelved until further notice.

Istanbul is an 'incomplete' city. The population incessantly complains about the unfinished state of their city, its infrastructural insufficiencies, its unplanned growth and inadequacy of services. Municipal bodies lack transparency; participatory mechanisms that involve citizens in their city's business are completely absent. In a city of such size and complexity, completeness can only be a utopian aspiration. Put differently, incompleteness is not a characteristic that categorizes Istanbul as an 'Oriental' and 'Third World' city as some would claim, but remains indispensable in orienting the city to the future.

Industry Goes, Culture Comes

Turkey's central government and the city's local governments have been pursuing the global city project for a long time through various types of regeneration project. Starting in the mid-1980s, urban administrators and local academics began to debate widely the opportunities and challenges of transforming Istanbul into a global city. The initial building programme of this vision included inter-city highways, five star hotels, and the city's first gated communities and shopping malls. Non-governmental organizations and corporations contributed to the effort by initiating international cultural events. By the 2000s, the city turned exclusively to 'culture' and the government to 'generalized gentrification' (Smith, 2002). Cultural institutions are inserted into formerly decaying urban areas while squatter settlements and slums are cleared to make way for state-financed modern housing schemes for the private market.

Istanbul has made serious inroads into becoming a culture city in recent decades, even if in an ad hoc manner. The case of the Golden Horn (Haliç) is illustrative. The Golden Horn was the major industrial zone of Istanbul. In the 1970s, it was home to Istanbul's historical major shipyard and mid-size production plants manufacturing items such as household appliances and engines for agricultural production. Soon the industrial waste literally turned the Golden Horn into a sewer. In the 1980s, Istanbul Municipality embarked on a project to rehabilitate the Golden Horn. After expensive and extensive efforts, the water was cleaned and industry was moved out. Now Golden Horn once again has a green shoreline, which is used as a picnic area by the residents of surrounding neighbourhoods.

What has replaced the plants and factories is a variety of cultural and educational institutions. Istanbul Commerce University, Bilgi University and Kadir Has University occupy the two banks of Golden Horn. Both Bilgi and Kadir Has Universities have their own cultural centres. Rezan Has Museum, housed at Kadir Has University, sits on top of a Byzantian cistern and an Ottoman *hamam*. Santral Istanbul, located in the premises of the Silahtarağa Power Station at the far end of Golden Horn is an important exhibition centre. Rahmi Koç Transportation Museum, Miniatürk, a theme park exhibiting models of architectural heritage (see İpek Türeli, Chapter 6), Sütlüce Congress Centre, built in place of an old slaughterhouse, Feshane, the old *fez* factory turned exposition centre, all line the shore, along with restaurants, boutique hotels, small shops of all kinds. The shipyard is still operational but it is imminent that it will be transformed into a cultural centre. Converted ex-industrial buildings are especially favoured as exhibition spaces.

Most of the recent transformations pertaining to the re-presentation of Istanbul as a cultural site can indeed be observed in other contending global cities; that is, they are not exclusive to Istanbul. International film, dance, theatre, and music festivals, art exhibitions and design weeks crowd the city's calendar. To take the example of the 'biennial fever' that spread across formerly peripheral cities from Sao Paolo to Taipei, economic interests, cultural policy and curatorial and artistic experience converge as city marketing strategies; failure to create difference in content pushes organizers to assert difference through the city (see Banu Karaca for a comparison between the Berlin and Istanbul Biennials, Chapter 13).

The demise of manufacturing has led to a rise in policy emphasis on culture as an economic sector (Zukin, 1995). It is even possible to talk about a 'cultural turn' in urban policy (Miles, 2007). Culture and creative industries appear to policy-makers as a panacea to fuel urban economy and even resolve social and financial problems. While recognizing that arts and cultural events are increasingly deployed as vital urban policy, complementing the more permanent, architectural projects of urban regeneration, we believe it would be reductionist to pass the verdict on arts as a mere instrument of urban redevelopment (see Hou Hanru and Jale Erzen, Chapters 11 and 12, for an important debate on national and global frameworks in thinking about art and society). With their trans-disciplinary scope, the contributions to *Orienting Istanbul* bring concrete and grounded perspectives to debates about spectacle, mediatization, and cultural identity. Contributors consider not only institutional frameworks of arts and cultural events but also the cultural products produced and exhibited on these platforms.

Layered Pictures, Living Monuments

Istanbul's unique topography enables orientation. Throughout the city, hills offer the spectator elevated view points that open onto waterways. This topography connects the dispersed city with its ever expanding outskirts, if only in visual and imaginary terms, and provides a mental map and a sense of place.

We close our introduction with a closer look at the panorama on our book's cover. Taken from the roof of Büyük Valide Han (the living heritage site discussed in Chapter 4 by Baykan *et al.*), this panorama of the city overlooks the silhouette of modern Istanbul from the angle of the historic Peninsula. The photograph contains another picture, a graffiti painting in the foreground on a parapet wall, which mirrors certain structural elements from the visible scene. Photographs tend to be read as direct inscriptions of the real; this quality is referred to as their indexicality. Yet, once overlaid with a highly interpretive 'painting' of the viewed scene, the supposed indexicality is thrown into crisis. Furthermore, this doubling of representation creates the layered effect that renders this photograph symptomatic of the city's experience – simultaneously lived and mediated. The banks of blue that repeat in the graffiti sky, in the waterway between the land pieces, and in the real sky heighten this effect.

Figure I.3. View from the roof of Büyük Valide Han. (*Source*: Reproduced from *Cityrama*, by courtesy of photographer Arif Aşçı)

The graffiti in the foreground highlights the notion of 'orienting' with the arrows – forcefully pointing to an idealized mosque skyline, and geographically to Asia. The photograph lines up for the viewer three domes that constitute a strong axiality: a segment of a dome on the roof of the building where the photographer is standing, multiple domes that lie flat in the graffiti, and thirdly, the domes of Yeni Camii in Eminönü. These multiple domes create a depth of field that stretches the gaze into the horizon towards the Bosphorus and the first Bosphorus Bridge. Competing with this axis is another one, perhaps more powerful and right at the centre, defined by the Galata Bridge. Spanning across the Golden Horn in the middle ground, the bridge invites the viewer to cross to the modern side of the city. Yet, a tension arises where s/he is prevented from doing so by intersecting walls.

Behind the minarets of Yeni Camii lies, in the distance, the Istanbul Modern Museum, where this photograph by Arif Aşçı was displayed in an exhibition, Cityrama. Curated by Engin Özende, Cityrama was organized in collaboration with the International Union of Architects' (UIA's) 22nd World Congress of Architecture (held in Istanbul between 2 and 10 July 2005) which brought thousands of architects from around the world to Istanbul. The introduction to the exhibition catalogue poses important questions about the place of architectural heritage in urban life:

> With what emotions do we look at the grinding sharp-toothed machines of the age and the old buildings that silently surrender without a fight? While Istanbul expands in all directions, carrying its sorrows with it, is something lost or is it a natural process of growth?

According to information from the research team working on Büyük Valide Han (Burak Sevingen and Zerrin İren), the little white hut behind the graffiti houses a textile dyeing workshop. The grass-covered tower to the left is a very old structure, probably from Byzantine times, which is mentioned in sources ranging from Evliya Çelebi and Wolfgang Gurlitt to John Freely and Murat Belge. Various myths about the history of this tower are in circulation among the residents of the Han. Underneath the tower a metal turning workshop run by an old Armenian is in operation. Residents of the Han date the graffiti after 2004 and claim that a group of 'French' students who were never seen again sprayed it in two days. This rooftop, which is locked and accessed with a key kept by the caretaker, serves as both a 'hidden' tourist sight and as a refuge for the people who live and work in the building. As a 'well-known secret', the Han has turned into an attractive site for artists and social scientists alike.[17]

The cover photograph thus emblematically captures some of the processes that we map out in this book, regarding the intersection of contemporary art, social science, lived practices, and urban imaginaries; processes germane not only to Istanbul but pertinent also for our understanding of Europe as a space based on heterogeneity and 'fragmentary unison' (Soysal, 2002).

Notes

1. The seed was planted during a conference at University of California, Berkeley in September 2008. See: http://www.ced.berkeley.edu/istanbulconference/. Accessed 20 February 2010.
2. http://www.nytimes.com/interactive/travel/2010-places-to-go.html. Accessed 20 February 2010.
3. *Ibid*.

4. http://www.mainlinemedianews.com/articles/2010/02/01/main_line_times/life/doc4b6091ed5679 0885937712.txt. Accessed 20 February 2010.

5. http://www.orhanpamuk.net/news.aspx?id=19&lng=eng. Accessed 20 February 2010.

6. See also Herzfeld (1996).

7. http://www.istanbul2010.org/HABER/GP_619547. Accessed 20 February 2010.

8. This is a shared question. See photographer Serkan Taycan's project 'Europe?' http://www.serkantaycan.com. Accessed 20 February 2010.

9. http://www.garajistanbul.org. Accessed 20 February 2010. http://en.istanbul2010.org/BASINODASI/BASINBULTEN/HABER/GP_567197. Accessed 20 February 2010.

10. See, for instance, 'The World According to GaWX 2008' where New York and London are at the top and a city like Istanbul receives at best 'Alpha' status. Available at: http://www.lboro.ac.uk/gawc/world2008.html. Accessed 20 February 2010.

11. http://www.tate.org.uk/modern/exhibitions/globalcities/default.shtm. Accessed 20 February 2010.

12. http://www.tuik.gov.tr/PreTablo.do?tb_id=39&ust_id=11. Accessed 20 February 2010.

13. http://www.ibb.gov.tr/sites/ks/tr-TR/0-Istanbul-Tanitim/konum/Pages/Nufus_ve_Demografik_Yapi.aspx. Accessed 20 February 2010.

14. http://www.ibb.gov.tr/sites/ks/tr-TR/0-Istanbul-Tanitim/konum/Pages/Sayilarla_Istanbul.aspx. Accessed 20 February 2010.

15. http://www.thefreelibrary.com/Turkey%3a+Int+l+firms+lend+1.5+bln+euros+for+%3fstanbul+metro.-a0180205581. Accessed 20 February 2010.

16. http://www.ibb.gov.tr/en-US/Pages/Haber.aspx?NewsID=153. Accessed 20 February 2010.

17. During the 2003 Biennial, the Han was already used as an exhibition site. The British artist Mike Nelson installed a darkroom in one of the Han's cells. *Magazin:* http://www.friezefoundation.org/commissions/detail/mike_nelson. Accessed 20 February 2010.

References

Bali, Rıfat (2002) *Tarz-ı Hayat'tan Life Style'a: Yeni Seçkinler, Yeni Mekânlar, Yeni Yaşamlar* [From 'Way of Life' to 'Lifestyle': New Elites, New Spaces, New Lives]. Istanbul: İletişim Yayınları.

Behar, David and İslam, Tolga (eds.) (2006) *İstanbul'da 'Soylulaştırma': Eski Kentin Yeni Sahipleri* [Gentrification in Istanbul: New Owners of the Old City]. Istanbul: Bilgi Üniversitesi Yayınları.

Bezmez, D. (2008) The politics of urban waterfront regeneration: the case of the Golden Horn, Istanbul. *International Journal of Urban and Regional Research*, **32**(4), pp. 815–840.

Bozdoğan, Sibel (2001) *Modernism and Nation Building: Turkish Architectural Culture in the Early Republic.* Seattle, WA: University of Washington Press.

Brummett, Palmira (forthcoming) Ottoman expansion in Europe, 1453–1600, in Faroqhi, Suraiya N. and Fleet, Kate (eds.) *Cambridge History of Turkey*, Vol. 3. *The Ottoman Empire as a World Power, 1463–1603*. Cambridge: Cambridge University Press.

Burdett, Ricky and Sudjic, Deyan (eds.) (2008) *The Endless City: The Urban Age Project*. London: Phaidon.

Çelik, Zeynep (1986) *The Remaking of Istanbul: Portrait of an Ottoman City in the 19th Century.* Seattle, WA: University of Washington Press (translated into Turkish in the 2000s).

Danış, Didem and Pérouse, Jean-François (2005) Zenginliğin Mekânda Yeni. Yansımaları: İstanbul'da güvenlikli siteler [New Reflections of Wealth on Space: Gated Communities in Istanbul]. *Toplum ve Bilim*, No. 104. pp. 92–103.

Erman, Tahire (2001) The politics of squatter (*gecekondu*) studies in Turkey: the changing representations of rural migrants in the academic discourse. *Urban Studies*, **38**(7), pp. 983–1002.

Friedmann, John (1986) The world city hypothesis. *Development and Change*, **17**(1), pp. 69–83.

Gold, John R. and Gold, Margaret M. (2005) *Cities of Culture: Staging International Festivals and the Urban Agenda 1851–2000*. Farnham: Ashgate.

Gül, Murat (2009) *The Emergence of Modern Istanbul Transformation and Modernisation of a City*. London: I.B. Tauris.
Graham, Stephen and Marvin, Simon (2001) *Splintering Urbanism: Networked Infrastructures, Technological Mobilities and the Urban Condition*. London: Routledge.
Herzfeld, Michael (1991) *A Place in History: Social and Monumental Time in a Cretan Town*. Princeton, NJ: Princeton University Press.
Herzfeld, Michael (1996) *Cultural Intimacy: Social Poetics of the Nation State*. New York: Routledge.
Herzfeld, Michael (2009) *Evicted from Eternity: The Restructuring of Modern Rome*. Chicago, IL: University of Chicago Press.
Huyysen, Andreas (2008) Introduction, in *Other Cities, Other Worlds: Urban Imaginaries in a Globalizing Age*. Durham, NC: Duke University Press.
İçduygu, Ahmet (2004) From nation-building to globalization: an account of the past and present in recent urban studies in Turkey. *International Journal of Urban and Regional Research*, **28**(4), pp. 941–947.
Kafadar, Cemal (1997/1998) The question of Ottoman decline. *Harvard Middle Eastern and Islamic Review*, **4**(1/2), pp. 30–75.
Kandiyoti, Deniz and Saktanber, Ayse (eds.) (2002) *Fragments of Culture: The Everyday of Modern Turkey*. New Brunswick, NJ: Rutgers University Press.
Keyder, Çağlar (1992) How to sell Istanbul? *Istanbul*, No. 3, pp. 80–85.
Keyder, Çağlar (ed.) (1999) *Istanbul Between the Global and the Local*. Lanham, MD: Rowman & Littlefield.
Keyder, Çağlar (2008) A brief history of modern Istanbul, in Kasaba, R. (ed.) *Cambridge History of Modern Turkey*, Vol. 4. *Turkey in the Modern World*, Cambridge University Press.
King, Anthony D. (1990) *Global Cities: Post-Imperialism and the Internationalization of London*. London: Routledge.
Kirshenblatt-Gimblett, Barbara (1998) *Destination Culture: Tourism, Museums, and Heritage*. Berkeley, CA: University of California Press.
Jacobs, Jane M. (1996) *Edge of Empire: Postcolonialism and the City*. London: Routledge.
Le Corbusier (2007) *Journey to the East* (translated and edited by Ivan Zaknic). Cambridge, MA: MIT Press.
Miles, Malcolm (2007) *Cities and Cultures*. London: Routledge.
Mills, Amy (2010) *Streets of Memory: Landscape, Tolerance, and National Identity in Istanbul*. Athens, GA: University of Georgia Press.
Özendeş, Engin (2005) *Cityrama*. Exhibition Catalogue. Istanbul: Istanbul Modern.
Pamuk, Orhan (2006) *Istanbul: Memories and the City*. New York: Vintage
Robinson, Jennifer (2006) *Ordinary Cities: Between Modernity and Development*. London: Routledge.
Roy, Ananya (2001) Traditions of the modern: a corrupt view. *Traditional Dwellings and Settlements Review*, **12**(2), pp. 7–21.
Said, Edward (1978, 2003) *Orientalism: Western Conceptions of the Orient*. London: Penguin.
Sassen, Saskia (1991, 2001) *The Global City: London, New York, Tokyo*. Princeton, NJ: Princeton University Press.
Singerman, Diane and Amar, Paul (eds.) (2006) *Cairo Cosmopolitan: Politics, Culture, and Urban Space in the New Middle East*. Cairo: American University in Cairo Press.
Smith, Michael Peter (2001) *Transnational Urbanism: Locating Globalization*. Oxford: Wiley-Blackwell.
Smith, Neil (2002) New globalism, new urbanism: gentrification as global urban strategy. *Antipode*, **34**(3), pp. 427–450.
Soysal, Yasemin N. (2002) Locating Europe. *European Societies*, **4**(3), pp. 265–284.
Suner, Asuman (2010) *New Turkish Cinema: Belonging, Identity and Memory*. London: I.B. Tauris.
Tanyeli, Uğur (2004) *İstanbul 1900–2000: Konutu ve Modernleşmeyi Metropolden Okumak* [Istanbul 1900–2000: To Read Housing and Modernity from the Metropole]. Istanbul: Akın Nalça.
Türeli, İpek (2008) Istanbul, Open City: Exhibiting Anxieties of Urban Modernity. Unpublished PhD thesis, University of California, Berkeley.
Türeli, İpek (2010) Ara Güler's photography of 'Old Istanbul' and cosmopolitan nostalgia. *History of Photography*, **34**(3).

Urban Age (2009) *Istanbul: City of Intersections*. Available at: www.urban-age.net/publications/newspapers. Accessed 4 March 2010.

Uzun, C. Nil (2003) The impact of urban renewal and gentrification on urban fabric: three cases in Turkey. *Tijdschrift voor Economische en Sociale Geografie*, **94**(3), pp. 363–375.

Zukin, Sharon (1995) *The Cultures of Cities*. Oxford: Blackwell.

Part I. Paths to Globalization

In this opening Part, two seemingly disparate but ultimately complementary chapters locate Istanbul's paths to globalization and set the agenda for the book. Çağlar Keyder, in 'Istanbul into the Twenty-First Century', true to well-established traditions of macro sociology, registers Istanbul's trajectory to becoming a world city. Engin Işın, in 'The Soul of a City: *Hüzün*, *Keyif*, Longing', pursues a different course, a literary and photographic expedition, in search of the sources for Istanbul's vitality at the beginning of the twenty-first century. They seem to be depicting two Istanbuls, worlds apart – for Keyder, a city in the grips of global currents and transformation process, for Işın, a city grappling with its particularities, detached from the world outside. This perfunctory glance is, however, misleading. The Istanbuls narrated in these chapters are not just two sides of the same coin, one socio-economic, the other cultural-aesthetic, but they form a constellation of the world city – in conversation with each other and with other chapters in the book.

Keyder's purview of Istanbul's ascendancy to being a global city proper answers the question he himself posed when prefacing his seminal volume *Istanbul: Between the Global and the Local*: whether Istanbul, then set to typify what he terms 'informal globalization', could become a global city. A decade later, his response is solidly affirmative. By the 'standards' that gauge a city's standing in the world hierarchy of cities, Keyder concludes, Istanbul presents a success story. The city is now incorporated into the coveted real estate markets and financial centres; gentrification is continuing at rapid pace; skyscrapers form a discernible skyline; and the cultural landscape of the city offers an enviable diversity of art exhibitions, fashion shows, and festivals, culminating in the designation as European Capital of Culture 2010. Keyder provides a brief history of this success story, succinctly detailing macro transformations that have prepared the city for its new role as a world city, interlaced with a political history of contemporary Turkey since the 1980s.

Işın's entry point to the discussion is Orhan Pamuk, renowned author of Istanbul, who ascribes the affective label of *hüzün* to the city in his memoir. In what develops at first as a personal account, walking the streets of Istanbul and capturing daily life at work, Işın grows sceptical of the weight Pamuk attributes to *hüzün* and proposes *keyif* (enjoyment of the city) as an alternative that describes the 'soul' of the city. Eventually, he concludes that both *hüzün* and *keyif* are discursive constructions that 'orient' Istanbul, simultaneously 'orientalizing' the city as an 'object of desire' and 'Europeanizing' it by speaking to the Occident through the city. To situate these constructs, Işın turns to the 'local' and searches for the ways social groups that inhabit the city deploy them as instruments of governance of the city commensurate with their tastes, habits, and dispositions.

Both Keyder and Işın are concerned with Istanbul's place and image in a globalized world. Işın finds his objects of study in the representations of Istanbul for audiences in and travellers from the Occident, as well as images of the city that emerge from the everyday practice of the locals. Meanwhile, Keyder is not merely presenting a story of economic transformation; his argument about Istanbul becoming a 'world city' is mediated through cultural representations and paradigms that resonate with the images that Işın deliberates over.

A central point of correspondence between the two chapters is circulation. Whether in pictures of the city tinted in *hüzün* or scenes of *keyif*, Istanbul images travel the world and bring the world back to Istanbul. These images permeate academic research and artistic expression as well as planning policy and city branding to enhance tourism. Istanbul as narrated by Keyder and Işın is not a city of permanent lives but a city of and for migrants or strangers, who come in search of better fortunes, or of uprooted Istanbulites, now living elsewhere, but nonetheless desiring the pasts embodied and left behind in the city. Like these inhabitants, Keyder and Işın, and several other authors contributing to this book, relate to the city as transient citizens, who maintain homes and jobs in Istanbul and elsewhere. These impermanent lives enable a convergence of perspectives, looking at the city simultaneously as insider (experiential of a citizen) and outsider (looking at Istanbul from afar and comparing it to other cities). Empowered with distanced connectedness to the city, the essays in this and the following Parts propose to decipher and complicate the globalizing 'success story' that is Istanbul.

Chapter 1

Istanbul into the Twenty-First Century

Çağlar Keyder

The Globalization Project

During the first decade of the new century Istanbul appeared confident and prospering. There were blocks of newly erected high-rise office buildings, luxury residential compounds and towers, and dozens of shopping malls offering global wares. The city centre was beautified and prepared to anchor global networks. Gentrification of the Beyoğlu area and the historic peninsula, as well as the rebuilding of the waterfront around the Golden Horn, created new spaces of leisure and culture. A feverish pace of projects and plans, all designed to enhance Istanbul's standing in the global sweepstakes for investment, revenue and culture, coincided with the preparations for the coveted status of 2010 European Capital of Culture. By the standards of city marketing worldwide, Istanbul was a success story.

Back in 1980 Istanbul was a fairly typical Third World sprawl. The old city retained most of its glory and the older neighbourhoods some of their charm, but the overwhelming impression was one of dilapidation and crowdedness. Environmental degradation in the industrial periphery of the city was matched by the lack of urban amenities in the shantytowns; constant construction activity produced the ever-present mud and dust that plagued the streets; and there was an unavoidable gloom of air pollution emanating from old cars and cheap coal. But the city attracted investment and migrants from the countryside, and performed as the principal transmission mechanism for the 'modernization' of the peasantry. Istanbul had been responsible for the absorption of a quarter of the new urban population in Turkey during the period from 1960 to 1980. The new migrants could not readily find good employment; but there was a widely shared belief that manufacturing would expand, well-paying jobs would be forthcoming and eventually all households would attain decent comfort. Furthermore, there was provision for the newcomers to access the urban space through informal mechanisms of land occupation and housing construction. As in many cases of Third World urbanization, migrants were permitted to share in the fruits of growth through a politics of populism which suspended the strictures of the law of property. Takings were modest, but their distribution seemed if not fair at least not grossly polarizing.

All this changed when Istanbul, in common with other globalizing cities of the Third World after the 1980s, experienced the shock of rapid integration into transnational markets and witnessed the emergence of a new axis of stratification. Those in the new professions who could participate in global networks gained disproportionately in status and income. Istanbul as Turkey's primate city in the 1980s was already the centre of high-level services oriented to the national economy. Trade and finance were concentrated in the city, as were the culture and media industries catering to the country as a whole. With globalization and liberalization, however, all these sectors – finance, real estate, advertising and media – experienced explosive growth. Along with culture industries and those service sectors that cater to producers, there was also a resurgence of the art scene, well integrated into global networks. All this growth in activity and employment accelerated with the impetus of tourism which expanded rapidly during the most recent period (now reaching eight million visitors a year), boosted by the city being 'discovered' and thus earning a place in the global 'cosmopolitan' consciousness.

The predominance of new service sectors and culture industries, even if these account for a smaller proportion of employment in Istanbul compared to London or New York, follows the pattern of de-industrialized global cities around the world. This thin social layer of a new bourgeois and professional class which adopted the lifestyle and consumption habits of their transnational counterparts seems to define a mission for the city, especially because of the obsessive coverage of their lifestyles in the media. The social and cultural ascendance of the new global class is reinforced through carefully cultivated distinctions in lifestyles, residential choice and material consumption, with all the expected ramifications for social and spatial stratification. It is now clearer than ever where the lines of demarcation lie.

The Triumph of Marketing

There is no doubt that Istanbul's success in capturing a share of the global dazzle is due in large part to the world economy, since the 1980s, favouring the resurgence of the metropolis: this was a period in which the control and management functions of global capital shifted to the great cities of the world and those sectors which are specifically urban gained ground (Scott, 2008). Finance increased its share, real estate development became a leading sector and culture industries expanded. This political and economic shift by itself, however, cannot explain Istanbul's performance: the world economy provides an opportunity but there is also and always resistance against global projects by the potential losers, the 'defenders of old space',[1] those who will be displaced socially and spatially.

In Turkey's case anti-globalization was a banner in the hands of the 'nationalists' suspicious of the claims of the Davos crowd; for Istanbul this meant the deflation of the global-city project by the land-bound state elite in Ankara. In fact, until the 1990s it looked as if Istanbul would miss the opportunity. Caught between a political class committed to populist modernization and a timid bourgeoisie reluctant to alienate Ankara's bureaucracy, actors in the city were unable to mobilize significant resources towards global success. Things changed, however, when the conservative-Islamic party which won the local elections in 1994, receiving the bulk of its support from the peripheral neighbourhoods of migrants, proved to be surprisingly pro-business. Their adoption of the neo-liberal discourse found a perfect fit in projects preparing the city for global exhibit, implying that traditional solidarism would be abandoned in favour of chasing after investment. The new urban coalition – the city government, real estate concerns, the bourgeoisie in its manifold manifestations, and the top echelons of the civil society, including the media and the city-boostering foundations funded by businessmen – strived to consolidate the city around their image of gentility.

The marketing of Istanbul proceeded along expected lines: the historical riches of the city as well as its night-life and culinary diversity were (and are) highlighted, along with dozens of music, art, and film festivals, new museums and exhibits. This, of course, was not a conflict-free process. The archaeological layers of the city's many incarnations were alternative candidates for foregrounding.[2] The Byzantine city, the Ottoman city of many cultures, or the imaginary Islamic city of the devout had to wage a battle with the Turkish city of the Republic. Since this last had too narrow a reference and obviously lacked marketing potential, what won out after a few years of competition was an inclusive Ottomanism, a re-imagined rubric encompassing the multifarious heritage of which the city could boast. The elite were happy to display their mansions and *objets* dating from the Empire; churches and synagogues were carefully restored along with mosques and the architecture of the everyday – apartment buildings and houses of the ordinary people. Ottoman art of the nineteenth century became a staple in the new museums where exhibits helped establish that the Ottoman elite had been very much engaged with European art, music, and literature. This was a new (and post-national) representation of the city in which the peripheral modernity of the Empire seamlessly flowed into an aspired status in contemporary global space. What it achieved was a narrative that could be easily appropriated by the global media, the art world, and taste makers who helped put Istanbul on the map – of investors, discerning tourists, curators of exhibits, real-estate developers, buyers of residences in 'in' cities of the world, and sundry consumers of culture.

I had written in 1999 that the city had embarked on the path to globalization but the process was haphazard and the results could best be termed 'informal globalization' (Keyder, 1999*a*). Istanbul's success in this path, however, has become undeniable in the new century, and it is now grounded on solid institutional foundation. This should be attributed in large measure to the coincidence of political and economic expectations from the city, thus to the coherence of the urban coalitions which strived to upgrade the city's image and marketing potential in the eyes of a footloose global demand – whether for investment, culture, or leisure. Since the mid-1990s, Istanbul has been governed by the same political party, and in fact by its leader Recep Tayyip Erdoğan – first as mayor then as party leader and prime minister, but always hands-on where Istanbul was concerned. This is an unusual continuity given the political instability that characterized earlier periods. Especially after the national election in 2002 when Erdoğan became the prime minister, the central government allocated resources to upgrade the city's infrastructure; the municipality participated in the project to improve urban services and to make city streets look more becoming. The city's bourgeoisie (and the media that they control) were willing partners in this project. Not only did they benefit directly, both in business initiatives and land development, but through a newfound vocation in philanthropy they actively participated in the project. The remarkable spectrum of privately endowed museums which came into existence during the last decade is a case in point. The ongoing series of art exhibits, festivals of jazz, theatre, classical music, and film, at a scale that plausibly rivals any large European city, has been initiated in most part by non-profit private groups (see Yardımcı, 2005).

The success in advertising the city was a perfect counterpart to the claims and aspirations of the liberalizing government in whose conception Istanbul would be the test case for successful integration into the emerging world of global markets. The new urban coalition facilitated the transition from an informal, unstructured and insufficiently institutionalized globalization towards a more formal and deliberate project aiming to capture a desirable position in global networks. This political change was the major impetus to the successful marketing of the city. With the central government committing itself to a global economic orientation, Istanbul's bourgeoisie found itself in a position where they could unequivocally subscribe to the project. They were no longer hesitant due to uncertainty about the central government's intentions. With the municipality, the central government and the urban elite on the same page, the 'marketing' of the city could proceed apace.

Land Development Through Finance

The period during which this urban growth coalition (Logan and Molotch, 1988) came together witnessed the most recent wave of global expansion during which Istanbul participated in the relentless inflation of global fund circuits. In the prelude to the crisis of 2008 there was a disproportionate expansion of the financial sector in the world, translating into speculative investment primarily in financial assets. Real estate was also considered to be subject to speculation, and residential and office space prices increased rapidly in all global cities. In fact, what triggered the bursting of the bubble was the defaulting of debtors on their mortgage payments. The East Asia crisis of 1997 had been triggered in the same manner, as a result of the bubble in real estate prices which had been bid up by easy credit. Developers in Bangkok and Seoul had constructed an excess of office buildings and housing using cheap loans, and had been caught short by the sudden collapse of the market. Istanbul in the 1990s had escaped this dynamic, mostly because the legal conditions for such an expansion were lacking. Turkey did undergo a crisis in 2001, the last of a series of home-made crises stemming from ill-conceived economic policies in Ankara. The governance repairs that were instituted in its aftermath readied Turkey for a much less encumbered economy and banking system as far as the inflow of capital and various kinds of short-term funds was concerned.

As a result the financial expansion of the period 2002–2008 found a counterpart in the flooding of Turkey's 'emerging economy', with an unprecedented volume of foreign funds. There was a huge and increasing volume of money searching for opportunities around the globe, and the Turkish economy was considered to be a surer bet than most. This money found its way through the banking system to new real-estate development corporations which financed both the construction firms and the buyers. Foreign capital also arrived in the form of partnerships with local developers. Inasmuch as financialization, meaning the preference for liquid assets, was the prevailing tenor of the global conjuncture, most of the investment thus occasioned sought to identify speculative opportunities. Istanbul, as a city where population continued to increase and where both office buildings and residential stock were in dire need of upgrading, seemed to provide such opportunity, especially since developments mostly occurred in areas where ground was being freshly broken. The story of Istanbul's physical expansion in this most recent period is the coinciding of the speculative bubble with the newly concocted 'neo-liberal' strategy *vis à vis* urban growth.

An essential component of this strategy lies in normalizing the legal relations in respect of land. Ottoman laws on private ownership were ambivalent at best, even though new and comprehensive legislation regarding land had been issued in 1858 and added to in subsequent years. Private property on land, outside of urban boundaries, was not a clear category, but land that was not actually cultivated was considered to be state land. Although the Republic adopted a version of the Napoleonic Code as the Civil Code in 1926, the Armenian massacres, war deaths, immigration from imperial territories to the land that remained as Turkey, the flight of the Greek population in the aftermath of the World War, and the Exchange of Populations with the Greek state in 1923–1924 had created a highly uncertain situation in terms of ownership of land. The departure of non-Muslims from Istanbul continued, such that in absolute numbers only a few thousand Greeks, around 50,000 Armenians and 25,000 Jews remained in the 1970s (out of a population of 2.5 million in 1975) compared to the early years of the Republic when one-third of the 700,000 population in 1927 was non-Muslim. The property left behind was rarely subject to orderly transfer. With this situation in the background the migrant population filling the city after the 1950s could be met with populist leniency when they squatted on public land, agricultural fields with no residential permit, or 'empty' housing and land recently abandoned. As is well known, their implicit right to squat and engage in the informal construction of *gecekondu* was recognized but the property status was never rectified, with the result that until recently perhaps half of all buildings in Istanbul were to some degree illegal (Keyder, 1999*b*).

The reason to remember this well known history is in order to underline the difficulty it was to pose to land development at the pace Istanbul witnessed in the 2000s. Accordingly, it has been the case during this decade that policies and planning have sought to re-regulate land that was already ceded to the squatters, attempting to create a 'lawful' city out of the chaotic string of villages whose agglomeration Istanbul has become. New legislation gives extraordinary powers to municipalities to clear illegal construction and forcibly move their inhabitants to designated neighbourhoods (see Ünsal and Kuyucu, Chapter 3; Bartu and Kırlı-Kolluoğlu, 2008). In Istanbul this prerogative has been bolstered by 'earthquake legislation' giving the power to the metropolitan municipality to demolish buildings thought to be unprepared for the eventual big tremor. Plans have been announced to demolish 85,000 *gecekondu* houses and move their inhabitants to low-cost high-rise projects in the far reaches of the metropolitan area.

The Mass Housing Administration (TOKİ), which has acquired extraordinary powers, allowing its decision-makers to construct on public land, escaping all planning and zoning scrutiny, has spearheaded the production of space through the extension of the residential area. TOKİ is now responsible for 7–8 per cent of all housing starts and its powerful director justifies its operations as constituting a housing policy consistent with Turkey being a 'social state'.[3] In fact, there has been no housing policy attempting to respond to the vast inflow of migrants into the cities, except for the benign neglect that created the *gecekondu* neighbourhoods. Unlike housing policies in most European cities, the government (TOKİ has a special status as an agency directly under the office of the prime minister and thus escapes bureaucratic scrutiny) since 2002 seems to have decided that ownership rather than tenancy should be the goal for low-income families. TOKİ's low-income housing is sold, in a means-tested manner, cheaply and with small monthly payments (as low as 100 TL or $65 a month) with long-term payment schedules. The development of new land as residential areas on the edges of the urban constellation is part of this project as well. TOKİ creates new residential spaces and contracts private firms to build apartment blocks where they get a share of the flats; thus it is also instrumental in opening new areas as residential districts, around which other developments follow. Where TOKİ goes the municipality has no choice but to follow, with both public and private development oriented to benefit fully from the externalities introduced by infrastructure projects.

Istanbul's sprawl has gravitated in recent years around the 'Second Bridge' over the Bosphorus. The roads connecting to this new bridge are a sort of beltway for the city, but as has been the case for other highways, they have fast become surrounded by business and housing during the last ten years. The second bridge which was opened in 1988 and its connecting roads have added substantially (perhaps one-third as much) to the spatial spread of the city as it used to exist at the end of the 1980s. There are now projects for a third bridge to the north, which together with its connecting roads, would open yet more fresh ground for development. Metropolitan Istanbul is already encroaching into its peripheries, in effect adding smaller cities to its urban area in a serial manner. It has become a sprawl without any clear divide to mark its limits. In official configuration the borders of the metropolitan municipality have been expanded to coincide with those of the province: there are now no independent settlements within the Istanbul province; all villages and rural centres have been made into neighbourhoods within the megalopolis. The prospect of endless growth in this same vein is a recipe for creating a geographical monster covering the entire area between the Marmara and the Black Sea coasts and gnawing into the remaining woodlands in the north of the city.

This development may signify a future that is closer to the European model, with the centre city as the tourist showcase, full of restaurants, cafés, and entertainment venues which share the space with upscale residential neighbourhoods; a primarily middle-class, expensive and marginalizing core, with its poorer neighbourhoods waiting to be gentrified, or officially 'regenerated' under the auspices of TOKİ. Circumscribing the centre in the proximate belt are the old peripheral neighbourhoods, which once were the *gecekondu* areas and are now turned into *banlieus* of middle- and lower-income habitation. In the outer belt, alongside what used to be industrial zones, are large areas dedicated to more genteel lifestyles in gated compounds with swimming pools, university and hospital campuses, but also high rises for the aspiring middle-class and the new housing for the poor built by TOKİ.

Much of this expansionary dynamic into the outer areas is an effect of the optimism that prevailed prior to the crash of 2008, and thus in no small measure a product of speculative frenzy. As in East Asia prior to 1997 and in the United States in 2008, Istanbul has also ended up with an enormous bubble of excess real estate, in the form of office buildings, shopping centres, and middle-class residential development. These new developments sought to avoid the proximity of old *gecekondu* areas and tapped into new areas, with the result that there is now a large surplus of residential, commercial and office space, chasing after a strained clientele. The crisis demonstrates that Istanbul has in fact become part of the global arena in that its real estate bubble was fuelled directly or indirectly by the rapid financialization and credit expansion experienced in the world economy prior to the crash. A glut of office towers, unoccupied apartment blocks, and boarded up shopping malls now dot the landscape, causing a sharp downturn in construction sector employment (historically the biggest source of demand for unskilled labour) whose dimensions are yet to be revealed. The chances are that there will be a long wait until the existing stock finds use through attrition, upgrading and expanding activity. The new apartments in high rises represented a transition to more modern dwellings by the new service-sector employed generations whose parents lived in the ramshackle neighbourhoods of former shantytowns. Expansion of credit allowed them to invest in new residences as well as the modern conveniences of apartment living. This is why the cessation of new construction and land development may signal a larger downturn for the city in terms of absorbing investment and creating employment, portending an unavoidable period of relative stagnation.

The Fragility of Success

Globalization is not only the intensifying of networks penetrating national spaces; it also signifies the eventual transformation of political preferences as the globalizers gain ascendancy (see Robinson, 2001). Political actors commit to the prospective rewards of the capitalist market and relinquish their past commitments to the nation, the civic entity, or the community – in other words, to the social. They merge with new clients whose resources and control over the media produce a new kind of persuasion. As corporate elites integrate into global circuits, politicians become convinced that accommodating the needs of the new economy will become the key to satisfying their constituency. Unequal rewards are naturalized in the rhetoric of the market while polarization, segregation, and exclusion are treated as unfortunate damage. One dimension of this new hegemony is the willingness to institute markets to rule over the allocation of all goods, leading to a deepening of commodification.[4] In the case of Istanbul in the new century, the crucial new market was in land. It can be argued that what has allowed the city's particular kind of development in this decade has been the final commodification of land. In the process leading to this state of affairs, Turkey's long history of hesitant and controlled engagement with the market constituted a formidable impediment. Both the political obduracy of Ankara politicians when faced with the possibility of Istanbul escaping their orbit, and the structural resistance to land being transformed into a simple commodity, were instances of barriers that had to be cleared for the project of neo-liberal globalization to succeed. As Istanbul's performance was the inevitable showcase and the gauge for this success, this is where the political transformation had to prove its capability.

The political transformation led to the repair of the traditional tension between the city and the seat of power in Ankara. Much has been written on the historical origins of this mutual suspicion, which often translated to an uncomfortable co-existence between the bureaucrats and the businessmen, if not to frequent undermining of the capitalist prerogative, in the name of accommodating the popular will. In its most recent incarnation, it was the state elite of Ankara hoping to undermine Recep Tayyip Erdoğan when he became the mayor in 1994. Until the elections in 2002, the uneasy condominium between Erdoğan's Justice and Development Party (AKP) and the state elite in Ankara meant that Istanbul's globalization proceeded in a haphazard manner, often suffering reversal and finding paths that amounted to muddling through. Since then, however, Istanbul's businessmen have found willing partners in politicians in their quest to valorise the city's real estate and to commodify and market its attractions. This was the key to the formation of successful urban coalitions which translated to more rapid growth in the expansion of globalized sectors, employment, and housing.

The institution of a market in land was also the result of a political shift, with a somewhat longer history, that led to the production of commodified space, as well as to the determination by the market of the spatial allocation of income and status groups in the city. Until recently, most empty land around the city was owned by the state. As Istanbul's population increased and migrants flooded into the city, political authorities found it expedient implicitly to permit the occupation of public land for the construction of informal housing. Eventually, however, land became scarce and the politics of populist accommodation started to falter: municipal governments became more interested in consolidating their built spaces as prosperous neighbourhoods out of which they could collect taxes. Local politicians received the message that the best use for the land from here on would be the most revenue generating option – not the one maximizing opportunities for political patronage. After populist inclinations were defeated, land became the object of speculation to an unprecedented scale; it was no longer the object of small corruption but of large-scale developments, determining the direction of spatial expansion of the city. The introduction of foreign funds, easier availability of bank loans, and new legislation regulating mortgage, accelerated the process. Infrastructure projects and highway building by the public sector served to blaze the path. TOKİ participated in this development both by cleaning urban land of remnants of squatter housing and by developing low-income housing in the perimeter of the city. Regeneration cleared space with ambiguous status and opened it to gentrification; expansion to the periphery created new spaces for development. Public land was brought onto the market when TOKİ designated areas for development and allowed contractors to build housing in exchange for a share of the units.

The onset of such privatization implies an untrammelled operation of the market, which in turn promises that spatial segregation will now work in the predicted manner, segregating populations along lines of income and capital. The global has worked its magic: under the coherent urban coalition it gave rise to a land market was instituted; with the commodification of land, Istanbul will now become a true capitalist city. New perspectives will be needed to understand its social fabric and to develop an appropriate politics with which to address its inequities.

Notes

1. Brenner and Theodore (2002), p. 13. Here they are quoting a formulation by Alain Lipietz.

2. There are several articles describing the conflict over Istanbul's advertised identity, including Bartu (1999) and Öncü (2010).

3. *Taraf*, 28 November 2009. http://www.taraf.com.tr/haber/44735.htm

4. As Polanyi (1944, 2001) would say, land is a fictitious commodity whose commodification potentially poses greater problems than ordinary goods that are actually *produced* for sale.

References

Bartu, Ayfer (1999) Who owns the old quarters? Rewriting histories of the global era, in Keyder, Ç. (ed.) *Istanbul: Between the Global and the Local*. Lanham, MD: Rowman and Littlefield.

Bartu, Ayfer and Kırlı-Kolluoğlu, Biray (2008) Emerging spaces of neoliberalism: a gated town and a public housing project in Istanbul. *New Perspectives on Turkey*, **39**(Fall).

Brenner, Neil and Theodore, Nik (2002) Cities and the geographies of 'actually existing neoliberalism', in Brenner, N. and Theodore, N. (eds.) *Spaces of Neoliberalism: Urban Restructuring in North America and Western Europe*. Oxford: Blackwell.

Keyder, Çağlar (1999*a*) The setting, in Keyder, Ç. (ed.) *Istanbul: Between the Global and the Local*. Lanham, MD: Rowman and Littlefield.

Keyder, Çağlar (1999*b*) The housing market from informal to global, in Keyder, Ç. (ed.) *Istanbul: Between the Global and the Local*. Lanham, MD: Rowman and Littlefield.

Logan, John R. and Molotch, Harvey L. (1988) *Urban Fortunes: The Political Economy of Space*. Berkeley, CA: University of California Press.

Öncü, Ayşe (2010) Narrating the past and claiming the present of Istanbul, in Diamandouros, N., Dragonas, T. and Keyder, Ç. (eds.) *Spatial Conceptions of the Nation: Modernizing Geographies in Greece and Turkey*. London: I.B. Tauris.

Polanyi, Karl (1944, 2001) *The Great Transformation: The Political and Economic Origins of Our Time*. Boston, MA: Beacon Press.

Robinson, William (2001) Social theory and globalization: the rise of a transnational state. *Theory and Society*, **30**, pp. 157–200.

Scott, Allen J. (2008) Resurgent metropolis: economy, society and urbanization in an interconnected world. *International Journal of Urban and Regional Research*, **32**(3), pp. 548–564.

Yardımcı, Sibel (2005) *Kentsel Değişim ve Festivalizm: Küreselleşen İstanbul'da Bienal* [Urban Change and Festivalism: the biennial in a globalizing Istanbul]. Istanbul: İletişim.

Chapter 2

The Soul of a City: *Hüzün*, *Keyif*, Longing

Engin F. Işın

This chapter engages Orhan Pamuk's claim that *hüzün* is the soul – if there is such a thing – of Istanbul. Assuming first that there is such a thing but becoming increasingly troubled with *hüzün*, the chapter introduces an alternative, if an opposite, soul, *keyif*, more precisely *şehrin keyfi*, as the soul of Istanbul. Yet, growing increasingly sceptical of itself, the chapter opens towards a discussion on why there should be such a thing as the soul of a city. Taking its cue from the phrase, 'reorienting Istanbul', it begins to argue that claims to know the soul of a city – whether *hüzün* or *keyif* – are discursive constructions that orient Istanbul in both senses: it reorientalizes Istanbul as an object of desire while it Europeanizes it by shaping its direction towards the Occident. The question then becomes how these discursive constructions emerge. Answering that question requires understanding how social groups that constitute contemporary Istanbul use such images as strategies of government. If *hüzün* and *keyif* are effects over which social groups struggle to govern the city according to their taste, habits and disposition, understanding how such effects are produced and are made objects of desire becomes essential to understanding literary products such as *İstanbul: Hatıralar ve Şehir* (2003) or *Istanbul: Memories of a City* (2005).

Istanbul's *Hüzün*

It was Orhan Pamuk who introduced his non-Turkish speaking readers to the Turkish word *hüzün*. In his book *Istanbul: Memories and the City* he suggested that *hüzün*, while having Arabic roots, has a distinctly Turkish meaning that is untranslatable to any other language. Since then he published two further articles abridged from the book (2008, 2007). *Hüzün* does not exactly correspond to the meaning of words such as melancholy, nostalgia, sombreness, sadness or even wistfulness, which comes closest to it (Henschen, 2008). While referring in part to all these words, *hüzün* still maintains a distinct sense by identifying a mood where one withdraws into oneself but without necessarily feeling down. For Pamuk this effect is a kind of longing but it is communal. Pamuk explains the peculiarity of this word through its association with the city of Istanbul. It is *hüzün* as communal longing that becomes a central idea in *Istanbul*. Pamuk thinks that being the capital of a disappeared empire, if not culture, Istanbul is rife with symbols and images of longing. Istanbul, it appears, is a city of longing because of its disappearing past. *Hüzün* is longing for the city.

While I found this idea originally compelling when I read *Istanbul* in Turkish, I was much less convinced of the weight Pamuk placed on it when I read *Istanbul* in English. It is this idea – the city of Istanbul as a city of longing and *hüzün* as longing for the city – that I shall call into question. *Pamuk wants to reorient Istanbul away from an outsider gaze and towards an indigenous or authentic essence* – an insider's gaze. *Yet, in so doing he creates a reverse image of the city with an outsider's gaze firmly cast upon it.* There is another mood that defines Istanbul's soul to which *Istanbul: Memories and the City* is a stranger. That mood is *keyif*, or more precisely, *şehrin keyfi* (enjoyment of the city).

When *İstanbul: Hatıralar ve Şehir* (2003) was first published, I was living in the city. After being away for twenty years (since 1983 I was only briefly in Istanbul in 1999), I was doing archival research on Ottoman munificence and patronage. When I read the book with much anticipation – since I had read Pamuk earlier – I thought it perfectly captured not only memories of my youth and my relationship to the city but also that very moment when I was back after such a long absence. I thought it captured that combination of loss and longing twice: that of my youth when I both longed for the city and yet thought it was lost forever and that of my years away from it when I both longed for and lost it. For those months I was in Istanbul, *İstanbul: Hatıralar ve Şehir* became a kind of companion that Pamuk speaks about in terms of his relationship to other writers: a fierce and continuous dialogue. *Hüzün*, I thought, was the mood that determined my relationship to the city. I left the city with that acceptable longing that was *hüzün*.

When I was back in the city in 2005 to complete my archival research, the English translation *Istanbul: Memories and the City* had been published. When I read it in English I was startled by how its effect on me was so different. Rather than agreeing that *hüzün* was the mood through which I experienced the city, I reacted to the idea negatively, finding *hüzün* rather too inward looking, brooding and lethargic to describe the soul of the city and my relation to it. Moreover, the English translation made me realize that this was more than a memoir: it made several and rather large claims about the city and its history. Why did *Istanbul: Memories and the City* have this effect that *İstanbul: Hatıralar ve Şehir* did not? This has been said so many times, and most compellingly, by Eva Hoffman (1989): translation is interpretation and creates new worlds. What *Istanbul: Memories and the City* made me aware of to an extent that *İstanbul: Hatıralar ve Şehir* did not was that the former was about reorienting Istanbul for outsiders. *Istanbul: Memories and the City* addressed a non-Turkish speaking audience and summoned an outsider's gaze upon Istanbul yet again. While Reşat Ekrem Koçu, Hamdi Tanpınar, Yahya Kemal, and Abdülhak Şinasi Hisar were Pamuk's manifest protagonists in *İstanbul:*

Hatıralar ve Şehir, its latent protagonists were Gérard de Nerval, Théophile Gautier, Charles Baudelaire, Michel de Montaigne and Claude Lévi-Strauss. Reading *Istanbul* in English made me realize how much Pamuk was caught 'under western eyes' worrying about what 'foreigners and strangers' think about Istanbul. I realized that the value of *Istanbul* may be that it reveals this problem: 'My interest in how my city looks to Western eyes is – as for most *Istanbullus* – very troubled; like all other Istanbul writers with one eye always on the West, I sometimes suffer in confusion' (Pamuk, 2005, p. 211). As Ayşe Öncü (1999) has illustrated, however, the notion of authentic *Istanbullu* is itself a cultural and social artefact rather than a stable identity as Orhan Pamuk seems to think.

When I left the city in 2005 I was intrigued by this confession of a confusion and wondered whether Pamuk's explanation was adequate. I was back in Istanbul in 2008 – again to complete the archival research on Ottoman munificence and patronage. But I had another aim too: to engage in a ten-day intense street photography workshop and studio. After several daunting and challenging encounters with the city, its people, its sites, and scenes, it was the hundreds of photographs which I took that revealed to me what was troubling with *hüzün*. I will return to that later. For now I will engage with *hüzün* that *Istanbul: Memories and the City* presents.

Pamuk starts with his childhood to come to grips with *hüzün*. Tracing the origins of the word to its Arabic roots in the Koran, he argues that 'the word is meant to convey a feeling of deep spiritual loss' (*Ibid*., p. 81). But over time the word comes to denote, as Pamuk sees it, two different meanings, each evoking a different tradition. The first *hüzün* arises when 'we have invested too much in worldly pleasures and material gain' (*Ibid*.). He thinks this is distinctly Islamic.

The second *hüzün* specifically arises from Sufi mysticism and 'offers a more positive and compassionate understanding of the word and the place of loss and grief in life' (*Ibid*.). A Sufi suffers *hüzün* because he can never be good enough for God. It is this lack that defines Sufi *hüzün*. (Pamuk does not mention that the specific Sufi tradition which he draws upon is urban Sufism.) He argues that while both senses of *hüzün* have dominated Ottoman and Turkish poetry and music for centuries, it is not enough to describe the *hüzün* he feels and its 'enduring power' as it is inextricably implicated in the city in which he is embedded (*Ibid*., p. 82).

Pamuk then advances the central idea of the book: 'The *hüzün* of Istanbul is not just the mood evoked by its music and poetry, it is a way of looking at life that implicates us all, not only a spiritual state, but a state of mind that is ultimately as life affirming as it is negating' (*Ibid*.). He now thinks that, distinct from melancholy, *hüzün* is communal and that is why it is appropriate to apply it to the soul of a city. But he goes even further than that and suggests that the feeling is unique to Istanbul and it binds its people together (*Ibid*., p. 83). What does Pamuk mean by *hüzün* being communal? 'To feel this *hüzün* is to see the scenes, evoke the memories, in which the city itself becomes the very illustration, the very essence of *hüzün*' (*Ibid*., p. 84). It is communal not because it is expressed in public but it evokes collective memories. Pamuk then goes on to provide a long list of scenes that are presumably the essence of *hüzün* but are not much more than everyday occurrences in any city anywhere: sunset on the Bosphorus, fathers returning home with shopping bags, barbers waiting for customers, children playing football on cobblestone streets, crowds rushing to the ferries, men fishing on Galata Bridge and so on. This is when the idea begins to blur before it even gathers some sense of coherence. Is it just that these scenes make Pamuk feel *hüzün* or does he think that people he observes in these scenes must also be feeling it? If the latter, it is rather far-fetched to attribute an affect on such various scenes. It is then not clear why everyone should feel the same as Pamuk.

Perhaps anticipating this kind of reaction Pamuk immediately suggests that the difference between other cities and Istanbul 'lies in the fact in Istanbul the remains of a glorious past and civilisation are everywhere visible' (*Ibid*. p. 91). Again, one would think that Athens, Rome or Beijing would have more claim to *hüzün* than Istanbul. But Pamuk says 'These are nothing like the remains of great empires to be seen in Western cities, preserved like museums of history and proudly displayed. The people of Istanbul simply carry on with their lives amongst the ruins' (*Ibid*.). So the distinct sense of *hüzün* is that in Istanbul people carry on with their everyday lives amongst ruins of the past while elsewhere that past is glorified and monumentalized.

For Pamuk *hüzün* is unlike a solitary melancholy or Montaigne's *tristesse*: 'Montaigne's own sorrow was as solitary as mourning, eating away at the mind of a man who lives alone with his books. But the *hüzün* of Istanbul is something the entire city feels together and affirms as one' (*Ibid*., p. 95). Rather, *hüzün* is more

Figure 2.1. From Eminönü to Pera and Galata (2003). (*Photo*: Engin F. Işın)

like Lévi-Strauss's *tristesse* in that it 'describes what a Westerner might feel as he surveys those vast, poverty-stricken cities of the tropics ... [b]ut he does not see the city through their eyes' (*Ibid*., p. 92). By contrast, Pamuk claims that *hüzün* 'is not a feeling that belongs to the outside observer' (*Ibid*., p. 93). This is where I feel that the idea is about reorienting Istanbul. Pamuk denies the outsider the right or capacity to feel *hüzün* since it is, he seems to imply, an indigenous mood. The gaze of outsiders does not have access to *hüzün* and when they encounter it they express confusion and bewilderment. This is because '*hüzün* does not just paralyse the inhabitants of Istanbul; it also gives them poetic licence to be paralysed' (*Ibid*.). It is this poetic licence that is indigenous if not the authentic mood of the city. Istanbul bears its *hüzün* with honour. Pamuk relays this with his experience of channel hopping on TV when he sees the hero of a typical black-and-white movie made in the 1950s or 1960s. The hero is always implicated in a sad story but Pamuk argues that something strange happens in the reaction of heroes to their own situations. He feels 'it is almost as if the *hüzün* which infuses the city's sights and streets and famous views has seeped into the hero's heart to break his will' (*Ibid*., p. 95). This idea of the hero with a poetic licence to *hüzün*, it turns out, is quite significant as it symbolizes the authentic actor of the city as the hero.

Pamuk identifies Reşat Ekrem Koçu (one of the four protagonists of Istanbul along with Hamdi Tanpınar, Yahya Kemal, and Abdülhak Şinasi Hisar) as the hero whose *hüzün* drives him to an impossible project: *İstanbul Ansiklopedisi* (*Istanbul Encyclopaedia*). The product of a labour of love, *İstanbul Ansiklopedisi* begins its life as a collection of strange facts by Koçu in 1944 and by serialized publications, and contributions by various other heroic Istanbul-lovers, its first volume sees publication in 1958, reaches letter G and eleventh volume in 1973, and ends incomplete in the twelfth volume. Pamuk sees this heroic act of recording the history of the city in a systematic and yet utterly idiosyncratic way as driven by *hüzün*. He says 'Koçu was one of those *hüzün*-drenched souls who helped create an image of a twentieth-century Istanbul as a half-finished city afflicted with melancholy. *Hüzün* is what defines his life, gives his work its hidden logic, and sets him on the lonely course that can only be his final defeat, but – as with other writers working in a similar vein – he did not see it as central and certainly did not give it much thought' (*Ibid*., p. 141). As Koçu grew older he came to realise with sadness that he would be unable to limit his Encyclopaedia to fifteen planned volumes let alone finish it (*Ibid*., p. 145). Pamuk says that Koçu 'failed in part because Istanbul is so unmanageably varied, so anarchic, so very much stranger than Western cities: its disorder resists classification' (*Ibid*., p. 153). 'Without falling into the strange habit of praising Istanbul's strangeness, we acknowledge that we love Koçu *because* he "failed"' (*Ibid*.). Here then we see the essence of *hüzün*: the hero with a poetic licence to conquer the city is bound to fail but that licence allows him to feel *hüzün* not as defeatism, not as failure but as a kind of poetic conquest.

Figure 2.2 Dusk in Beşiktaş (2005). (*Photo*: Engin F. Işın)

But then again just when the idea is gathering some stability and coherence, Pamuk becomes uncertain. He says 'in the last one hundred and fifty years (1850–2000) I have no doubt that not only has *hüzün* ruled over Istanbul, but it has spread to its surrounding areas' (*Ibid*., p. 210). How to judge such a claim? Clearly, not by its truth value but for its truth effect: if Pamuk does not doubt that *hüzün* 'rules over' Istanbul over such a long period of time then we are invited to accept affectively that it has also 'spread to its surrounding areas'. That's fine but then he shocks at least this reader: 'What I have been trying to explain is that the roots of our *hüzün* are European: the concept was first explored, expressed, and poeticised in French (by Gautier, under the influence of his friend Nerval)' (*Ibid*.). That which seems to be indigenous to the city turns out to be European. It was Europeans who longed for the past of the city and wanted to see it memorialized and monumentalized. It was the European gaze that instilled the mood of longing in the city for the city that was lost. It is because 'Western observers [always] love to identify the things that make Istanbul exotic, non-Western, whereas the Westernizers amongst us register all the same things as obstacles to be erased from the face of the city as fast as possible' (*Ibid*., p. 218). It turns out then that the paradox of *hüzün* is that it is the Orientalist gaze turned into the soul of a city. For Pamuk *hüzün* is the Orientalist will to govern over the city.

It is not clear if Pamuk draws this conclusion from his narrative: he leaves out this specific discussion of the European origins of *hüzün* from the two subsequent extracts he published as articles (Pamuk, 2008). But *hüzün* is now beset by irresolvable dilemmas that also inhabit Pamuk's paradoxical approach to the word, if not Istanbul. While insisting that it has Sufi and Islamic origins we are then presented with European, if not Orientalist, interpretations of it. Pamuk embodies these paradoxes himself by, on the one hand, being drawn to European melancholic literature and, on the other, desperately trying to discover an authentic or indigenous mood of the city. That is why I think *Istanbul: Memories and the City* is more than a memoir. It gives an image of the city that withdraws into itself longing for its future to come through longing for its past that is gone. To put it differently, while Pamuk's longing appears as though it is for the past, it is in fact animated (and made legitimate) by imagining a European city to come. Consequently, Istanbul and *hüzün* are perhaps closely associated and it is this association that orients the city in both senses of the term: it reorientalizes Istanbul as an object of desire while it Europeanizes it by shaping its direction toward the Occident. To put it differently, if bluntly, Pamuk's *hüzün*, while appearing anti-Orientalist, produces Orientalist effects. Is it possible to produce a counter-effect? Can we imagine another mood to which perhaps Pamuk, or self-described *Istanbullus* are a stranger? This city is so vibrant, creative and energetic that arguably *hüzün* is only one mood amongst others that defines its soul.

Istanbul's *Keyif*

Now let us return to the photographs. It was those photographs that revealed to me what was troubling with *hüzün*. After shooting hundreds of photographs I was left with the daunting challenge of making sense of them or at least presenting them in a way that made some sense. Originally, I started with the idea that I would shoot 'working streets'. The term has a double meaning. First, it refers to the countless men, women and children of Istanbul who work on the streets for their livelihood. These range from transvestites to peddlers. The following incomplete list is itself a testimony to the characters of the working streets: *tulumbacı, simitçi, macuncu, turşucu, salepçi, tahinci, şerbetçi, kunduracı, sebzeci, saka, hamal, mestçi, leğenci, çakmakçı, hallaç, zerzevatçı, çömlekçi, değirmenci, kağıtçı, muslukçu, bileyci, kopyacı, çöpçü, dilenci, sihirbaz and hiyleci.* There is an enormous variety of characters working the streets and one or another can always be found there. Second, Istanbul's streets work. That is not in the sense that they function well (some do and some do not) but in the sense that they are at work almost all the time. Streets change their character along with the characters that inhabit them. There is a bewildering ebb and flow to these streets with a different cast of characters in each.

Yet, Istanbul has recently 'decided' that to be European (and 'global') its streets must be cleansed (Potuoğlu-Cook, 2006). Working streets are becoming increasingly sterile thoroughfares where street life and its characters are pushed inside the fast-growing shopping malls. Istanbul's street characters are feeling the pressure and are being displaced to look elsewhere to make their living (Candan and Kolluoğlu, 2008). I wanted to shoot and record the remaining vestiges of these characters before they entirely disappear and before photographing becomes an illegal activity as it practically is in these shopping malls.

That was the idea. What happened was somewhat different. The photographs I shot over ten days did not reveal many of the characters I expected to portray. Looking at these photographs again and again revealed, I do not know how, another idea: *şehrin keyfi* (enjoyment of the city). Time and again my gaze was fixed not on Pamuk's heroic (and tragic), but the everyday and yet resilient *acts* of Istanbul's people (especially those belonging to the social groups that are most deprived and marginalized) seeking to enjoy the city against a background of oppression or sufferings (Secor, 2003). These people are the outsiders and strangers to the city, coming to the city for centuries but especially during the twentieth century (Ayata, 2008). Like many large and historic cities, Istanbul is a city of outsiders and strangers who constitute its essence while at the same time being regarded as 'other' by those who claim 'nativity'. While Pamuk laments the disappearance of a specific Ottoman diversity in the city, he fails to observe, let alone rejoice, in the appearance of another, creative and energetic diversity created by its outsiders and strangers. It is not then only the scenes of pleasure – people having their tea, grilling fresh-caught fish from the sea, simply strolling, playing cards, taking a coffee break, or catching a glimpse of many views of the Bosphorus – but the acts of enjoyment by its strangers taking various risks that define *şehrin keyfi* (Bryant, 2005; Henkel, 2007). Amongst the most impressive

Figure 2.3. Şehrin Keyfi in Sarayburnu (2008). (*Photo*: Engin F. Işın)

of these characters of the city are kids. They seem to enjoy the city most with their inventiveness, mischievousness and defiance. Istanbul's children, at least in neighbourhoods where TV has not yet become the dominant pacifier, have not yet been pushed inside as in New York or London – a point made by a great street photographer Helen Levitt (Loke, 2009). In Istanbul the 'other' children still *enjoy* street life mostly on their own terms.

The *keyif* of the city is of a specific kind: it is not communal, as Pamuk thinks about *hüzün*, but public. Its publicness is not one of ostentatious announcement (the further up you go in the class scale the more prominent that becomes), but enjoyment in the presence of others and against all odds. Strolling in the streets with my camera I am often invited to join in the enjoyment rather than stay outside. I have been offered grilled fish, drinks, cake, tea, coffee and even *çiğ köfte* and *rakı* in such occasions. With children it also means either playing with them or even joining in resolving disputes under the watchful eyes of the street elders. I wondered if *keyif* was the mood that defined the city's soul rather than *hüzün*.

As I mentioned earlier, *hüzün* is caught up with the Orientalist gaze. If Pamuk is intent on rescuing *hüzün* from this gaze it is neither successful nor apparent. But is *keyif* without the Orientalist gaze? Edward Said praises Richard Burton remarking that:

> In no writer on the Orient so much as in Burton do we feel that generalizations about the Oriental – for example, the pages on the notion of *Kayf* for the Arab ... – are the result of knowledge acquired about the Orient by living there, actually seeing it firsthand, truly trying to see Oriental life from the viewpoint of a person immersed in it. Yet what is never far from the surface of Burton's prose is another sense it radiates, a sense of assertion and domination over all the complexities of Oriental life. (Said, 1978, p. 196)

In his *Personal Narrative of a Pilgrimage to Al-Madinah & Meccah* (1893) Burton writes:

> And this is the Arab's *Kayf*. The savouring of animal existence; the passive enjoyment of mere sense; the pleasant languor, the dreamy tranquillity, the airy castle-building, which in Asia stand in lieu of the vigorous, intensive, passionate life of Europe. It is the result of a lively, impressible, excitable nature, and exquisite sensibility of nerve; it argues a facility for voluptuousness unknown to northern regions, where happiness is placed in the exertion of mental and physical powers; where *Ernst ist das Leben*; where niggard earth commands ceaseless sweat of face, and damp chill air demands perpetual excitement, exercise, or change, or adventure, or dissipation, for want of something better. In the East, man wants but rest and shade: upon the banks of a bubbling stream, or under the cool shelter of a perfumed tree, he is perfectly happy, smoking a pipe, or sipping a cup of coffee, or drinking a glass of sherbet, but above all things deranging body and mind as little as possible; the trouble of conversations, the displeasures of memory, and the vanity of thought being the most unpleasant interruptions to his *Kayf*. No wonder that '*Kayf*' is a word untranslatable in our mother-tongue!

This exotic, visceral and sexual sense of *keyif* is intoxicating. One of its meanings is being 'high'. As Burton (1893) says 'In a coarser sense "kayf" is applied to all manner of intoxication. Sonnini is not wrong when he says, "the Arabs give the name of *Kayf* to the voluptuous relaxation, the delicious stupor, produced by the smoking of hemp"'. The French word *jouissance* (enjoyment) especially known for its usage by Jacques Lacan also expresses *keyif* (Evans, 1998). But Lacan associated *jouissance* with sexuality while *keyif* exceeds it. (Lacan might say that nothing exceeds sexuality but we will leave that Ottoman-French disagreement aside for now.) *Keyif* is sensual but not necessarily sexual. What *keyif* shares with *jouissance* is that emotional state or mood as a suspension in the present without the past and future or even despite them or perhaps even against them. Said (1978, p. 103) notes that in fact this is called 'bizarre *jouissance*' by the *Description de l'Egypte* through which 'the Orient becomes a living tableau of queerness'.

So there is as much evidence for *hüzün* as *keyif* for originating in the Oriental gaze. Perhaps just as Pamuk, I was caught in the Orientalist gaze – he as insider, I as outsider-insider. Yet, I do think, and hope, that there is more to *keyif* than its Orientalist interpretation as intoxication and delirium, a bizarre *jouissance*.

I would like to think that *keyif*, or more accurately, *şehrin keyfi*, rather than being a bizarre *jouissance* is a mood of defiance. Unlike *hüzün*, which reaches to the past or future or both, *keyif* is about the present. It is about the present and its affirmation. What it defies is the conditions that are supposed to determine one's fate. With *keyif* one plays with fate. *Şehrin keyfi* is such that the city enables this mood of defiance, makes it acceptable and accepted. In *şehrin keyfi* there are always risks taken, some high, some low. But there is no *keyif* without some risk. It is understood that to abandon time and space, past and future, and to affirm oneself in the present (and in the presence of others) has its costs. It is these costs that Istanbul's people bear and not the honour of the poetic licence to failure. *Şehrin keyfi is the enjoyment of the right to the city or, for short, civic enjoyment.*

Much more than *hüzün*, I think *keyif* is associated with Istanbul. The city, this city, is a city of spaces of *keyif*. These spaces put its people in the mood of *keyif*. Or at least they seduce you to dare to *keyif*. The city, this city, is an injunction to civic enjoyment. But now that I disassociated *keyif* from bizarre *jouissance*, the spaces of *keyif* I have in mind are those spaces of the city that invite its inhabitants to affirm themselves in the present if only for a moment. In a city of intense vitality and energy, this not only means seeking relief from that intensity but also managing it by enjoying it.

There are many spaces of *keyif* but I think one of their shared orientations is either catching a view of the Bosphorus or being on it. It seems for centuries mosques, churches, synagogues, fountains, parks, cafes, and many other public spaces have had this orientation. It seems every architect and builder in the city has been in competition to catch a view of the Bosphorus (though often with disastrous consequences in recent decades). Shirine Hamadeh's *The*

City's Pleasures: Istanbul in the Eighteenth Century (2008) captures this quite well. Throughout the eighteenth century, she argues, the Bosphorus in particular was opened up to enjoyment of all classes through the munificence of almost all classes. Hamadeh argues that the explosion of new buildings in the eighteenth century compares to post-conquest rebuilding in intensity and scale. Yet, she argues, the eighteenth century was marked by an opening of the city to broader classes and social groups. The main source of making this argument is poetry especially *tevarih-i manzume* which literally meant histories in verse. They were composed specifically in celebration of buildings and gardens. Each of these poems included a separate section titled 'monuments of benefactors' dedicated to gardens, palaces, houses, fountains, mosques, madrasas, schools, baths, and bridges (Hamadeh, 2008, p. 15). Hamadeh says 'they constitute the richest form of architectural discourse in the eighteenth century and form the bulk of the poems used in [her] study' (*Ibid.*). It is clear that there is another tradition to draw from than the Divan poetry that Pamuk draws from. We can say that it is this built-enjoyment-poetry that Istanbul inherits and builds upon. Every building, it seems, tries so hard to orient itself to the Bosphorus in order to catch a glimpse of its glorious glisten and glitter. Again, it is not scenes such as sipping coffee or tea with a delicious dessert and catching the view that define *şehrin keyfi* but its defiant characters. It is almost as if it is more important to catch that view just at that moment that spares you the hustle and bustle of the city. It is at that moment that you are of the city but not in it.

To have *keyif* means to lose oneself not only in time but space and to attempt an escape only to discover that we are all thrown together in this universe called the city and then say 'I might as well enjoy it'. *Keyif* is neither nostalgic nor hedonistic. It is affirmative. *Keyif* is that defiant mood of the city not because it is indigenous or authentic but because it is at once universal and particular. It is universal because it belongs to the city. Anyone can experience that mood when one affirms with no past or future but only the present. It is particular because in Istanbul particular elements (sites, scenes, and characters) come together to assemble it. It is this assemblage that makes *keyif* possible not because it is indigenous or inaccessible but because it is for anyone who dares, or rather, must dare. Civic enjoyment is an open act.

Is *keyif* another side of *hüzün* rather then being its opposite? What is the relationship between the poetic licence to communal defeatism that makes itself felt as longing to long and defiance of the moment that makes itself felt as suspension in time and space? Is *hüzün* the acceptance of the impossibility of *keyif*? Is *keyif* the acknowledgment of the inevitability of return to *hüzün*? However we may answer these questions it is impossible to imagine Istanbul without its characters drawn from outside (and beyond) who obey (and create) the injunction to enjoy the city and make it bend to their will regardless of how much the city tries to break it. This was exemplified when a newspaper journalist Mine Kırıkkanat (2005) complained about poor people taking over the city's beaches and parks with their grills, underwear and head scarves. A bitterly dividing debate followed this complaint. The issue was, in my words, whether the poor were claiming their right to the city by practising civic enjoyment or the poor were displaying its vulgar taste. The city government eventually caved in and tried to regulate civic enjoyment by imposing middle-class, 'European' standards upon them though without much success.

The secret of the creative energies of the city lies not only in its European *outsiders* whom Pamuk engages but also in its *strangers*. There is an astonishingly telling moment in *Istanbul* when Pamuk actually identifies *keyif* as an 'Eastern fantasy' though without naming it. This is when he argues that *hüzün* is inaccessible to outsiders and that 'Westerners coming to the city often fail to notice it' (Pamuk, 2005, p. 93). To illustrate the point he mentions that Gérard de Nerval 'spoke of being greatly refreshed by the city's colours, its street life, its violence and its rituals; he even reported hearing women laughing in its cemeteries' (*Ibid.*). Being surprised, if not perturbed, at the sight of *keyif*, Pamuk surmises 'Perhaps it is because he visited Istanbul before the city went into mourning, when the Ottoman Empire was still in its glory, or perhaps it was his need to escape his own melancholy that inspired him to decorate the many pages of *Voyage en orient* with bright fantasies' (*Ibid.*). Is Pamuk's longing for the city expressed in *hüzün* not about Ottoman Istanbul but the contemporary Istanbul that his social group – self-defined *Istanbullu*s – mourns?

References

Ayata, S. (2008) Migrants and changing urban periphery: social relations, cultural diversity and the public space in Istanbul's new neighbourhoods. *International Migration*, **46,** pp. 27–64.

Bryant, R. (2005) The soul danced into the body: nation and improvisation in Istanbul. *American Ethnologist*, **32**, pp. 222–238.

Burton, Richard Francis (1893) *Personal Narrative of a Pilgrimage to Al-Madinah & Meccah*. London: Tylston & Edwards.

Candan, Ayfer Bartu and Kolluoğlu, Biray (2008) Emerging spaces of neoliberalism: a gated town and a public housing project in Istanbul. *New Perspectives on Turkey*, No. 39, pp. 5–46.

Evans, Dylan (1998) From Kantian ethics to mystical experience: an exploration of jouissance, in Nobus, Dany (ed.) *Key Concepts of Lacanian Psychoanalysis.* London: Rebus.

Hamadeh, Shirine (2008) *The City's Pleasures: Istanbul in the Eighteenth Century*. Seattle, WA: University of Washington Press.

Henkel, H. (2007) The location of Islam: inhabiting Istanbul in a Muslim way. *American Ethnologist*, **34**, pp. 57–70.

Henschen, T. (2008) Furcht, Angst und Hüzün die entformalisierung zweier ontologischer Begriffe Heideggers durch Pamuks Begriff kollektiver Wehmut [Fear, anxiety and hüzün: the de-formalisation of two of Heidegger's Ontological Terms through Pamuk's concept of collective wistfulness]. *Studia Phaenomenologica*, **8**, pp. 307–330.

Hoffman, Eva (1989) *Lost in Translation: A Life in a New Language*. London: Heinemann.

Kırıkkanat, Mine G. (2005) Halkımız eğleniyor [People's enjoyment]. *Radikal*, 27 July. Available at: http://www.radikal.com.tr/haber.php?haberno=159792. Acccessed 1 March 2010.

Loke, Margarett (2009) Helen Levitt, who froze New York street life on film, is dead at 95. *New York Times*, 30 March, p. A27. Available at: http://www.nytimes.com/2009/03/30/arts/design/30levitt.html. Accessed 1 March 2010.

Öncü, Ayşe (1999) Istanbulites and others: the cultural cosmology of being middle class in the era of globalism, in Keyder, Çağlar (ed.) *Istanbul: Between the Global and the Local.* Lanham, MD: Rowman and Littlefield.

Pamuk, Orhan (2005) *Istanbul: Memories of a City* (translated by Maureen Freely). London: Faber.

Pamuk, Orhan (2007) Hüzün (translated by Maureen Freely), in Webb, Alex (ed.) *Istanbul: City of a Hundred Names.* New York: Aperture.

Pamuk, Orhan (2008) Hüzün-melancholy-tristesse of Istanbul, in Huyssen, Andreas (ed.) *Other Cities, Other Worlds: Urban Imaginaries in a Globalizing Age.* Durham, NC: Duke University Press.

Potuoğlu-Cook, O. (2006) Beyond the glitter: belly dance and neoliberal gentrification in Istanbul. *Cultural Anthropology*, **21**, pp. 633–660.

Said, Edward S. (1978) *Orientalism*. New York: Vintage.

Secor, A.J. (2003) Belaboring gender: the spatial practice of work and the politics of 'making do' in Istanbul. *Environment and Planning A*, **35**, pp. 2209–2227.

Acknowledgments

I would like to thank editors İpek Türeli and Deniz Göktürk for their astute and and precise if not imaginative comments on an earlier draft of this chapter, which were immensely useful. I am also grateful to them for drawing my attention to the reference to Arabic *kayf* in Said's *Orientalism*. Levent Soysal provided critical comments on a later draft that were crucial to address a number of its shortcomings. Fuat Keyman provided insightful (and incisive) comments on an earlier draft. I am grateful to Ariana Keyman for reading early fragments of this piece with great skill and expression to a group of friends at a dinner table. Barbara Godard has made a crucial intervention for which I am really grateful. Evelyn Ruppert, as ever, cast her watchful eyes over several drafts and incarnations.

Part II. Heritage and Regeneration Debates

Recent proposals for government-sponsored urban renewal projects (*kentsel dönüşüm*, urban transformation) in Istanbul tend to target districts with poor residents and seek to improve them by displacing their inhabitants with more affluent newcomers. Whether in the historic inner-city areas or the peripheries, this process, which Neil Smith calls 'gentrification as global urban strategy', takes place in many parts of the world and is an outcome of the changing role of the government, as it becomes an agent of capital rather than its regulator. Despite this convergence, neoliberal urbanism continues to produce diverse results in different localities. The four chapters in this Part converge on heritage-led regeneration and demonstrate how the specific history of Istanbul and current political claims to presence and power shape the manifestations of neoliberalism on the ground. They also show how those subject to gentrification in a country like Turkey, where the informal sector is relatively strong, are more vulnerable. Nevertheless, urban restructuring projects activate local resistance and social movements whose articulation is globally, and usually EU, mediated.

Özlem Ünsal and Tuna Kuyucu in 'Challenging the Neoliberal Urban Regime: Regeneration and Resistance in Başıbüyük and Tarlabaşı', argue that although economic liberalization began in the 1980s, populist policies with which urban land was typically governed in Istanbul in the second half of the twentieth century, gave way only in the past decade to a fully neoliberal mode. Citing particular legislation as the enabler of this shift, they compare two sites – a slum with predominantly nineteenth-century housing near the cultural centre of Taksim-Beyoğlu area, and a poor *gecekondu* settlement in an industrial zone on the Asian side of the city – and compare how these top-down projects, which disregarded existing residents, instigated grassroots mobilization, with varying degrees of success depending on the particularities of the sites.

The following chapter, 'Contestations over a Living Heritage Site: The Case of Büyük Valide Han', by a group of interdisciplinary researchers, Ayşegül Baykan, Zerrin İren Boynudelik, Belkıs Uluoğlu and Burak Sevingen, explores how living and working built environments have been designated as 'heritage', based on the case of an old city *han* in the Historic Peninsula, historically and currently home to small-scale artisanal production and retailing. They highlight the cultural richness and diversity of *han* residents who are disregarded in current development plans for the area, and advocate the acknowledgement of living heritage rather than simply of built form. The presentation of such heritage sites as venues for culture and tourism has become central to the staging of Istanbul as a world city.

In recent years and under Adalet ve Kalkınma Partisi (AKP) or Justice and Development Party, official heritage preservation and regeneration proposals for the Historic Peninsula have taken a new turn by calling for the restoration of the dilapidated building stock to its predominantly residential character during the Ottoman-era to enhance its touristic potential. This means incorporating these areas into the real estate market or gentrifying them. The renewed official focus on the Ottoman past to imagine the city's future began during the 1980s, but acquired new significance with 'Muslim-oriented' parties winning local elections in the mid-1990s and the national elections with AKP in the early 2000s. In addition to the official focus on the Ottoman past, there are many other groups in the city who claim versions of the Ottoman past in the built environment by excavating heritage as well as the through images of the city they cultivate. Jeremy F. Walton's 'Practices of Neo-Ottomanism: Making Space and Place Virtuous in Istanbul' contrasts two strands of neo-Ottomanism articulated by Sunni organizations, prominent in 'Istanbul's civil Islamic sphere of pious foundations'.

Connecting the issues of gentrification, heritage preservation and neo-Ottomanism discussed in the previous three chapters in this Part, the fourth chapter by İpek Türeli, 'Modelling Citizenship in Turkey's Miniature Park', zooms in on the case of the nation-themed miniature park, Miniaturk, which expands the imaginary of Istanbul, and Turkey, beyond its current borders, and offers visitors participation in a naturalized and inclusive past that includes the Muslim as well as the non-Muslim and antique past – and perhaps the possibility of an imagined unified future. Half of this miniature park, located on the northern bank of the Golden Horn is dedicated to Turkey and the other half to Istanbul. The chapter argues that translating the miniature park typology into a viable substitute for the gigantic expressions of state and global capital, Miniaturk also creates a new national pilgrimage site that concedes the rise of Istanbul as Turkey's gateway to Europe and via Europe to the world.

Chapter 3

Challenging the Neoliberal Urban Regime: Regeneration and Resistance in Başıbüyük and Tarlabaşı

Özlem Ünsal and Tuna Kuyucu

Setting the Scene

In March 2008, about 1,000 people from Başıbüyük, a *gecekondu*[1] neighbourhood in the Maltepe district, protested outside the municipality against the 'transformation project' in their neighbourhood. It was the first time that this politically conservative population participated in such a public demonstration.[2] A group of residents carried signs that read 'Our neighbourhood is our pride', and 'Başıbüyük will be a grave for the TOKİ (Toplu Konut İdaresi)'.[3] Later on, the spokesperson of the Istanbul Neighbourhood Associations Platform,[4] delivered a press statement, criticizing the transformation projects because 'their real aim is displacing poor working classes from their neighbourhoods that they formed with so much sweat and labour, and market their living areas to the rich'. When he finished his speech, the angry crowd surged towards the local AKP (Adalet ve Kalkınma Partisi, Justice Development Party) building, shouting 'Başıbüyük belongs to us and it always will'.

Two months later, in May, Ahmet Misbah Demircan, the Mayor of Beyoğlu; Ertuğrul Günay, the Minister of Tourism and Culture, and the representatives of GAP Construction Company were strolling along İstiklal Avenue with journalists towards Beyoğlu Art Gallery.[5] Exhibited in the gallery were the models of a project, and the mayor was soon to give a press conference on the Tarlabaşı Renewal Project, whose aim is the renewal of 278 historical buildings in the Tarlabaşı neighbourhood. Not a kilometre away was the Tarlabaşı Neighbourhood Association[6] where volunteer lawyers were filing documents to take the project to court. They argued that the project breached rules as specified in Law No. 5366 and violated the rights of owners and tenants in the area.[7] 'These kinds of projects must serve the public and must be based on good intentions, but this certainly is not the case in what they want to do with our lives here', said one lawyer.[8] An old lady who had lived her whole life in Tarlabaşı had just received an eviction letter, a day after she agreed to sell her building to the developer.

What is taking place in Başıbüyük and Tarlabaşı (see figure 3.1) signals a radical shift in the governance of urban land in Istanbul from a *populist* to a *neoliberal* mode that extends capitalist market dynamics to incompletely commodified areas. Large-scale redevelopment projects, which enable radical interventions in the economic and social geography of cities, are the most important mechanisms for dismantling the existing market order, property structure and governance regime and building new ones in their place. They are 'the mechanisms *par excellence* through which globalization becomes urbanized' (Moulaert *et al.*, 2003, p. 3). What is interesting in Istanbul is that despite the globalization of the city's economy since 1983, it is only recently that 'urban renewal/transformation' has become a systematic policy. The Istanbul Metropolitan Municipality, with the full support of the central government, is now implementing 'urban transformation projects' (UTPs) in nine neighbourhoods and is seriously considering starting several others in both *gecekondu* settlements and historic inner-city areas. Istanbul's selection as the 2010 European Capital of Culture, its urgent need, in the eyes of business circles and entrepreneurial politicians, to become a 'competitive' global city, and the grave earthquake threat are often presented by state actors as justifications to implement these mega-projects.

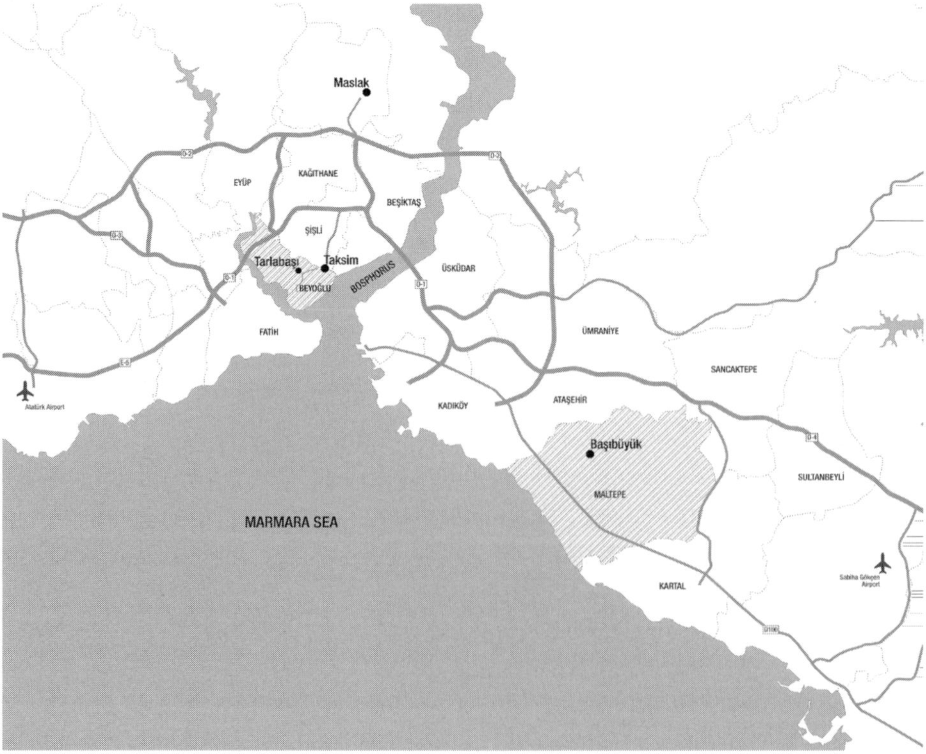

Figure 3.1. Map of Istanbul with Başıbüyük and Tarlabaşı marked. (Drawn by Orhan Kolukısa)

Yet, the socio-economic and physical restructuring of settlements that are inhabited by economically, politically and legally vulnerable groups generate new forms of conflict and contestation between state authorities, private developers and these low-income urbanites, who are becoming mobilized to defend their 'right to the city'. In every 'transformation zone' in Istanbul, local movements have sprung up, with varying power and resources, to challenge the imposition of a new urban regime and economic order on their lives. Even though these highly-localized movements have not yet been able to form a united struggle against the projects, they still have managed to shape their implementation in important ways. They are, in other words, crucial elements in the physical, economic and social remaking of the urban landscape. As such, they deserve a careful and critical analysis.

This chapter analyzes the motivations behind and consequences of the transition to a fully capitalist urban regime in Istanbul by focusing on two UTPs. By redefining property rules and market dynamics in informal housing areas and dilapidated inner-city neighbourhoods, UTPs achieve two major goals: physical and demographic 'upgrading' of particular localities, and the construction of a neoliberal governance regime. The new regime also pushes into the capitalist market a highly vulnerable population, whose livelihood depends on populist redistributive mechanisms. In addition to discussing the roles UTPs play in neoliberalization, our chapter also illustrates the myriad tactics that those who are facing the 'threat' of transformation use to contest the ongoing projects. Because both our cases are still at the implementation stage, it is impossible to show the exact outcomes of these local movements. However, our research reveals that even when the movements fail to stop the projects, they still alter their implementation in significant ways. Thus, we argue that these grassroots resistance movements should be interpreted as powerful *and* novel practices of claiming and redefining places. A focus on the interactions between state authorities, developers and residents captures the contested and contingent nature of urban transformation in Istanbul.

Construction of a Neoliberal Land and Housing Market in Turkey

The relationship between neoliberalization and urban transformation has received remarkable attention since the 1980s (Hackworth, 2007; Moulaert *et al.*, 2003; Brenner and Theodore, 2002; Harvey 1989; Logan and Molotch, 1987). This expanding literature has clearly demonstrated the generation of new forms of urban inequality driven by imperatives of capital in a post-industrial economy (Sassen, 2001; Knox and Taylor, 1995). UTPs, which open to investment potentially profitable spaces (for example, old industrial zones, waterfronts and inner-city slums), function as an integral part of this economy and trigger highly unequal consequences for different groups: while generating fast returns for investors and local governments,[9] they also instigate dynamics of displacement and dispossession for disadvantaged communities (Lehrer and Laidley, 2008). These processes have even more dramatic effects in developing countries where large tracts of urban land, inhabited by legally and socio-economically vulnerable populations and regulated by populist market dynamics, are being redefined through radical neoliberal reforms (Roy, 2004; Sajor, 2003; Thirkell, 1996).

Until 2002, housing provision for low-income groups in Turkey relied on legalizing unauthorized land appropriations and inner-city squatting (Şenyapılı, 2000; Keyder, 2000; Buğra, 1998; Erder, 1996; Tekeli, 1992; Öncü, 1988). 'Building amnesties' and retroactive plan revisions helped to sustain a populist coalition between industrialists in need of cheap labour, political parties seeking loyalty, and lower-class urbanites searching affordable housing. However, the gradual commercialization of these informal housing areas also led to the consolidation of a lucrative, yet hierarchical, market structure, where those with economic and/or social capital enjoy the benefits of populist mechanisms while the majority is left unprotected in quasi-legal markets (Buğra, 1998; Işık and Pınarcıoğlu, 2002).

What ended this populist regime was the emergence of new powerful actors in the form of large developers, real-estate investment trusts and various state agencies, whose interests lie in a fully commodified market in which exchange-rights trump over use-rights. In this context, *gecekondu* zones and inner-city slums become particularly attractive for redevelopment for two reasons: legal ambiguities in their property regimes and their perceived status as centres of crime and decay. Such areas are now seen as 'tumours that have surrounded our cities', as the Turkish Prime Minister described them,[10] which implies that their removal necessitates surgical operations such as large-scale renewal projects.

Despite rapid economic and urban restructuring during the 1980s and 1990s, a *fully* neoliberal system could not be instituted in Turkey until 2001 due to the continuation of populist mechanisms in land/housing markets. Given the political utility of distributing urban rent through populist channels, no political party could terminate such a vital source of political gain. Only as a result of a major economic crisis in 2001 did the populist urban regime come to a definite end (Cizre and Yeldan, 2005). In the aftermath of the economic collapse, the newly elected AKP, with direct financial and administrative support from the IMF, embarked on a major neoliberalization programme aimed at extending capitalist market relations and dynamics to a wider domain of social and economic life. In the specific areas of the governance of cities and urban real-estate markets, the AKP passed a series of legal/institutional reforms after 2002, that have eradicated the 'incompletely commodified' populist regime in these domains and made them much more aligned with the neoliberal system.

One set of reforms concern the prevailing *gecekondu* policy. With the passage of the new Criminal Code in 2004, *gecekondu* construction was made a criminal offence, punishable by five years in prison.[11] As a consequence *gecekondu* demolitions speeded up; between 2004 and 2008, 11,543 units in Istanbul were demolished, a record high for any period.[12] Then, in 2005, the Municipality Law[13] was passed, authorizing district municipalities to implement 'transformation projects' in derelict, obsolescent and unsafe parts of cities. These projects would renew such areas through a 'demolish-rebuild' approach and transfer the 'rightful owners' into mass housing areas, built by the TOKİ. Because of their precarious physical and legal status, *gecekondu* zones became ideal targets of 'transformation'.

In addition to reforming the *gecekondu* regime, the AKP also restructured the TOKİ, making it the key actor in constructing a neoliberal regime. With numerous legal reforms between 2002 and 2008, the TOKİ became the sole agency to regulate the zoning and sale of all state-owned urban land.[14] These reforms authorized the TOKİ to construct 'for-profit' housing on state land either by its own subsidiary firms or through public-private partnerships, in order to raise revenues for low-income housing construction.[15] Furthermore, the TOKİ acquired the power of making planning/zoning revisions and the right to expropriate property in *gecekondu* transformation zones. With these vast regulatory and financial powers, the TOKİ is accomplishing two crucial goals: constructing a formal land/housing market for low-income households *and* privatizing valuable state-owned land.

Another reform concerns the regulation of 'historic and natural protection zones', which have been protected by 'conservation laws'.[16] With the passage of Law No. 5366 in 2005, district municipalities were authorized to implement regeneration projects in 'derelict' and 'obsolescent' areas within protection zones. Similar to other cities, historic 'inner-city' zones in Istanbul experienced rapid deterioration and are mostly inhabited by populations unable to meet their housing needs elsewhere. The new law makes it possible to restore and/or demolish-and-rebuild buildings in accordance with the historic characteristics and development potential of the area. Municipalities are authorized to implement the projects through a partnership with the TOKİ or private developers.

The last set of reforms to institute a capitalist market structure in real-estate have been implemented in the areas of housing finance and foreign direct investments (FDIs) in real-estate. The level of integration of financial sectors with the housing sector have historically been very low in Turkey. Even though the country ranks at the top in Europe in terms of home ownership rates (59.8 per cent in 2000),[17] use of housing credits has been exceptionally low, constituting only 0.75 per cent of GNP in 2001 (Türkiye İstatistik Kurumu, 2000; GYODER, 2007). To reform the system, the AKP in 2007 passed what came to be known as the 'Mortgage Law' which made mortgage-issuing institutions legally operating capital market institutions.[18] As a result, housing credits jumped to 4.7 per cent of GNP in 2007 (GYODER, 2007).[19] In the area of FDIs in real-estate, the AKP passed two laws in 2003 which made it legally possible for non-nationals to purchase real-estate in Turkey, which previously had been prevented.[20] As a result, between 2003 and 2006, foreign direct investment in real-estate increased 43 per cent annually.[21] In fact, after privatizations and mergers, non-residents buying property in Turkey have been the largest source of FDI flowing into Turkey.[22] By 2007, the volume of FDI in real-estate had reached $3 billion.[23]

In short, these reforms have laid the foundations of a fully formalized and commodified urban regime which creates vast opportunities for state agencies, private developers, national and global financiers and other related actors. By eradicating the informal rent-distribution mechanisms, they have fundamentally altered the populist relationship between the legally and economically vulnerable urbanites and the state. UTPs are prime instruments in implementing this new regime, especially in *gecekondu* and inner-city slum areas.

Mapping Başıbüyük and Tarlabaşı

Looking at the socio-economic and physical geography of Başıbüyük and Tarlabaşı reveals their potential for rent-generation. Başıbüyük is a working-class neighbourhood situated on a hill in the Maltepe district, commanding a spectacular view. Maltepe until the mid 1990s remained one of the most important industrial zones of Istanbul's Asian side (Tekeli, 1992). The population of the neighbourhood rose to about 20,000 with migrations from central and north-eastern Turkey.[24] What made it suitable for settlement by migrants was the availability of vacant state-owned land and the proximity to factories. Although the neighbourhood has had infrastructure and services since the 1970s, the inhabitants could not obtain land titles and/or building permits, making all buildings in Başıbüyük illegal.

Two factors have complicated the property structure in Başıbüyük. First, most inhabitants have actually purchased their land from third parties in the informal market with state acceptance and/or involvement, which makes them believe that they should be eligible for formal ownership. Second, about half of the 990 buildings in Başıbüyük have *tapu tahsis* documents, obtained in the 1980s with amnesty acts that successive governments passed. These documents confer to *gecekondu* 'owners' a *de facto* use right, thus providing partial legality.[25] As a result, an ambiguous and confusing property regime emerged in Başıbüyük, where half of the inhabitants are 'occupiers' with no rights whereas the other half are *de facto* owners of their land, on which they built illegal dwellings. In addition, an unknown number of tenants reside in this legally ambiguous area. This complexity of tenure is one of the most important tools used by the municipality in the implementation of a project in Başıbüyük.

With Istanbul's geographic and economic transformations after the 1980s, Başıbüyük became a poor *gecekondu* settlement located in a thriving real-estate zone. With significant reserves of unused land, the Maltepe district has been marked as a potential development area in the Master Plan of Istanbul, making it a prime site for up-scale residential projects.[26] The district is close to Kartal, another de-industrializing zone, soon to be turned into a central business district (with office buildings, luxurious housing, hotels and a marina) designed by the renowned architect Zaha Hadid. The NarCity and KİPTAŞ[27] housing blocks, constructed for upper-middle classes, a private university and major transportation networks are some of the developments surrounding Başıbüyük today.

Tarlabaşı, on the other hand, is like the missing piece of a jigsaw puzzle in Taksim, the cultural hub of Istanbul. Despite the neighbourhood's proximity to the thriving Taksim area and its protection by special regulations since 1993 because of its nineteenth-century housing stock, it has been undergoing a rapid process of socio-economic and physical deterioration. The 'ghettoization' of Tarlabaşı is triggered by two major historical incidents: the deportation of its non-Muslim residents in 1964[28] and the demolition of more than 300 historic buildings for the construction of Tarlabaşı Boulevard during the office of Mayor Bedrettin Dalan in 1986 as part of a broader project to link the newly developing business areas to the main airport via the city centre (Ekinci, 1994; Çeçener, 1995).[29] Cut off from the social and economic flow of Taksim, Tarlabaşı entered a phase of dilapidation, as socio-economic decline also triggered physical decay. The decline of exchange values in the area after the 'Dalan demolitions' speeded up the 'slummification' of the area, which by the mid-1990s became a low-cost living area for internally displaced Kurds,[30] undocumented immigrants and various marginalized groups. The neighbourhood's decline became even more striking as surrounding areas like Galata and Talimhane were gentrified and attracted major investments (Behar and İslam, 2006).[31]

The predominant property structure in Tarlabaşı, despite its slum-like characteristics, is *de jure* ownership. The buildings that used to be owned by non-Muslims until 1964 were transferred to a new landlord class under the auspices of the state after the expulsion of these minorities. Rural migrants mostly benefited from this process either by purchasing the buildings from their official caretakers (*kayyum*) or by extra-legally appropriating them and retroactively becoming legal 'owners'.[32] Following this process of property transfer, a lucrative rental market emerged in the area, where the new owners rented out extra-dwellings, either formally or informally. According to research conducted in the project area, 75 per cent of Tarlabaşı's inhabitants are tenants, 20 per cent are property owners and the remaining 5 per cent are occupiers.[33]

Based on the 1990 and 2000 national surveys, the populations in both neighbourhoods are 'significantly below' the Istanbul average in educational attainment, income and job security (Güvenç, 2005). The majority of Başıbüyük residents only completed primary school and there is also a sizable illiterate population.[34] In Tarlabaşı, the illiterate population, which has greatly increased since 1990,[35] constitutes an even larger percentage. Most of the Başıbüyük inhabitants work in construction and (informal) manufacturing jobs, but there also exists a growing (female) population employed in the low-end service sector. In Tarlabaşı, however, the majority hold low-end service jobs, with minimal social security.[36] Services are over-represented in the occupational structure of Tarlabaşı because of its proximity to Taksim, where the concentration of entertainment centres generates such (low-paid) jobs. A mixture of small privately owned businesses such as grocery stores and marginal jobs such as garbage picking and prostitution exist alongside the service sector.

Implementation of the Projects

The Başıbüyük and Tarlabaşı projects represent the two types of UTP currently implemented in Turkey. Despite their similarities, there are important differences in their goals, implementing actors and processes. The Başıbüyük case is a '*gecekondu* transformation project', enabled by the new Municipality Law and implemented through a partnership between the Istanbul Metropolitan Municipality, the district municipality of Maltepe and the TOKİ. Accepted in February 2006, it aims to demolish *gecekondu*s on 400 *dönüm*s[37] in the neighbourhood and transfer the 'rightful owners' to apartment blocks built by the TOKİ within the neighbourhood. Currently, six of these blocks have been built on 35 hectares, which used to be a park (see figure 3.2).

Each building consists of fifty flats of either 80 or 110 square metres. 'Rightful owners' are given the option to purchase these units with state-subsidized long-term credit. Because their existing homes are illegal, they are only offered a 'demolition value', determined by the Ministry of Public Works and Settlement, for their homes. In most cases, the demolition value constitutes less than a quarter of the value of the new apartments. The *tapu tahsis* holders receive partial compensation for their land.

Figure 3.2. The site in March 2008, before the construction started. (*Source*: Tuna Kuyucu)

The project generated a strong reaction among the population, resulting in the formation of a neighbourhood association, widespread protests and frequent clashes with the police. The municipality, in response, started formal negotiations with the association in May 2008. Bargaining for a fair price for *gecekondus* continued, while the construction of the six blocks advanced (see figures 3.3 and 3.4). The association also initiated a legal battle in October 2008 against the project, arguing that it violates public interest, principles of urban planning and rules of democratic governance. In December, the Administrative Court ruled in favour of the association and ordered a temporary halt to the project. Despite the order, construction continued for three months when building was completed.

The Tarlabaşı UTP, on the other hand, is based on Law No. 5366. Following the designation of 278 buildings in nine blocks as a renewal zone in February 2006, the site was put out to tender by the Beyoğlu Municipality and a construction company which also operates as a private developer, GAP, took the bid in April 2007 (see figures 3.5 and 3.6).

The project, designed by nine 'leading local architects', as the various news releases of the municipality put it, aims to revitalize the area by turning it into a residential, tourist and commercial centre.[38] To protect the historical significance of the area, the project preserves the façades of the buildings, yet significantly transforms their interiors. Property owners are offered either 42 per cent of the surface area of their existing property, or full monetary compensation for their property's current value. The size of apartments to be built range from 35 to

Figure 3.3. Construction site on November 2008. (*Source*: www.toplumunsehircilikhareketi.org)

Figure 3.4. Construction site in November 2008. (*Photo*: Tuna Kuyucu)

75 square metres, which is too small for the majority of Tarlabaşı's households. Although tenants were initially excluded, they have later become eligible to purchase TOKİ units built 35 km away.

The inhabitants responded to the project by establishing a neighbourhood association in March 2007. Because Law No. 5366 rests on principles of 'fair negotiation' and 'participation' of residents in decision-making, official talks between GAP and the neighbourhood association started in February 2008. However, the association withdrew from the talks in July because of inadequate financial compensation and lack of transparency in project implementation. Demonstrating a remarkable efficiency in mobilization, the association now represents the majority of the residents in their bargaining with GAP, who has so far been able to purchase only 20 per cent of the buildings (see figure 3.7). The stand-off between GAP and the association is causing serious delays in project implementation.

Limits of State-Led UTPs and the Question of Resistance

The implementation of the UTPs have a number of drawbacks, raising serious questions about their overall legitimacy and consequences. First, no objective criteria exists in the Municipality Law, or Law No. 5366 to designate areas as 'transformation zones' apart from some vague references to 'dilapidation',

Figures 3.5 and 3.6. Views from Tarlabaşı. (*Photos*: Özlem Ünsalz

Figure 3.7. Mass meeting held by the neighbourhood association at a local coffee-shop in Tarlabaşı (August 2008). (*Photo*: Özlem Ünsal)

'earthquake risk', or 'obsolescence'. Besides, no concrete data regarding what makes the areas obsolescent or dilapidated have been provided in the project brochures, in public speeches, or during the in-depth interviews we conducted with officials. This ambiguity raises doubts about why particular areas are labelled as such. The almost complete exclusion of residents from decision-making processes is another major problem. This primarily results because the relevant laws lack models for participation. Yet, conclusively, this implies that the needs of residents were not considered in the devising of the projects. In fact, interviews with officials revealed that what they view as 'participation' is the inhabitants' acceptance of, or objection to the projects, *after* their official approval. Last, the projects are severely deprived of supportive social and economic programmes, creating a risk of dispossession and geographical relocation of poverty. The fact that the municipalities did not conduct comprehensive surveys regarding the needs of these vulnerable groups *prior* to the projects shows that the UTPs are not designed to improve inhabitants' living conditions.[39]

Due to these major shortcomings, it is likely that the UTPs will deepen poverty and instigate large-scale displacement of existing populations from the neighbourhoods.[40] In fact, this is a definite outcome for tenants, who are mostly excluded from the projects. In Tarlabaşı, tenants have already started moving to other neighbourhoods where they will not face the threat of 'transformation'. We also predict that a substantial percentage of 'owners' will leave their neighbourhoods as real-estate values rise after the projects. In Başıbüyük, for example, many inhabitants fear that they will be unable to afford the new units and the formalized household economy. In several interviews, residents stated that they will sell their new unit and use the cash to build a *gecekondu* somewhere else.

Despite the negative impact of the UTPs on the inhabitants, the resistance movements that emerged in these areas have considerably shaped their implementation. As many analysts have stated, the *particular* ways in which large-scale development projects transform urban areas are determined to a large degree by the existing power constellations, institutional structures and market dynamics prevailing in these areas (Brenner and Theodore, 2002). Localized responses to neoliberal dynamics leave their imprint on the outcome of neoliberal processes in highly contingent ways as analysis of the resistance strategies developed by residents and of bargaining between residents, private developers and municipal authorities clearly reveal.

One of the most interesting and puzzling aspects of the two UTPs discussed here is their potential both to instigate *and* hamper grassroots resistance. In both cases, the strategies that project implementers have used to boost urban rent and redefine property structures quickly transformed the *collective* resistance movements into bargaining processes for *personal* gain. Despite this general similarity, however, the process had different impacts on the movements that emerged in the two neighbourhoods. In Başıbüyük, an established *gecekondu* neighbourhood with strong social networks, the UTP generated a swift and powerful resistance movement, which lost its fervour with an equally surprising speed. A large number of the residents have either already agreed to, or are seriously considering accepting the project terms. In contrast, in Tarlabaşı, a neighbourhood with a relatively lower degree of solidarity, the resistance movement has not experienced such a loss of power. Rather, the association has successfully mobilized almost all residents, including tenants. Furthermore, due to the tactics the association developed, GAP has agreed to revise project details in significant ways, such as agreeing to provide financial support to tenants and providing more space in the project area for small businesses.

These differences can be explained by the different property/tenure structures in the two neighbourhoods and the level of violence used by the police under orders from Istanbul Governorship in project implementation. The strength with which Başıbüyük residents initially acted against the project spurred a high degree of violence by the police. When the residents put up barricades around the construction site, the Governorship deployed around 1,000 riot police so that construction machinery could pass through these barricades. The neighbourhood has been under 'police siege' since then and a 24-hour present police force controls all entries to the neighbourhood. Since the stationing of the police, violent confrontations between the police and inhabitants have occurred, injuring several inhabitants including children.[41] Even though the level of violence decreased after official negotiations with the municipality began, a permanent police force remains in Başıbüyük. The open and frequent use of violence cripples mobilization and generates a highly unequal bargaining structure.

In Tarlabaşı, however, there is no comparable use of state violence, mainly because the UTP does not involve the clearing of an area for construction. Unlike Başıbüyük, overt measures that might spark violence are not taken. Moreover, the public-private partnership in Tarlabaşı also acts as an obstacle to the open use of violence. Since it won the bid, GAP has engaged in an intensive PR campaign to convince both the residents and the public at large that the project is implemented in a democratic fashion.

More important in determining collective mobilization is the prevailing property/tenure structure in these areas. The ambiguity that characterizes Başıbüyük, where half of the buildings have no titles while the other half have *tapu-tahsis* documents, creates a very strong impediment to sustained collective resistance. In the absence of legal guarantees to their property, the residents (especially those with no rights) have a strong incentive to accept the municipality's offers. The municipality effectively uses people's tenure insecurity to persuade them to sign the deals. Furthermore, the existing legal complexity also creates deep divisions between the *tapu-tahsis* holders and the 'occupiers'. Because the former group has some legal security, they are better equipped to refuse municipal offers. Occupiers, in contrast, are more willing to receive an TOKİ unit.

These obstacles to sustained mobilization became very apparent after formal bargaining started between the district municipality and the association. Since then, the appetite for private gain has weakened the 'right-to-housing' based movement, resulting in more residents agreeing to the project. The movement lost power especially after the municipality decided to increase the 'demolition price' for existing dwellings. The bargaining also made the rift between occupiers and *tapu-tahsis* holders very apparent. In the several neighbourhood meetings we attended, the chief point of contention concerned the main beneficiaries of the resistance. The occupiers argue that the resistance strategies only benefit those with *tapu-tahsis* documents. Such rifts, caused by the property structure, make it easier for the municipality to convince more residents.

In Tarlabaşı, a different tenure structure exists, making possible a more effective resistance. Following the expulsion of Greek citizens residing in Istanbul in 1964, which also triggered large-scale immigration of Turkish citizens of Greek ethnicity to Greece, a massive property transfer took place, creating a new landlord class, who became *de jure* owners of the property. Since then, Tarlabaşı became a secure source of income for the new landlords and an affordable area for low-income tenants. The UTP generated a strong reaction from both groups whose economic interests now converged. As a result, a group of owners formed an association in February 2008 against the UTP in the area, which succeeded in collecting more than 200 powers of attorney from owners. Accumulation of these letters still continues, reinforcing the strength of representation of the association. Equally important, many tenants also joined the movement, thus strengthening the association's position *vis à vis* GAP and conferring upon it more legitimacy among the public.

The first significant result of this mobilization was the inclusion of the association in the official talks between the municipality and GAP in March 2008. The legal guarantee that a formal title gives to owners makes them stronger in the bargaining with project implementers, compared to the informal occupants of Başıbüyük. For example, after GAP pledged to make revisions to project terms, the association found these insufficient and left the bargaining table. Following this event, the association effectively prevented individual negotiations between residents and project implementers, a process which weakened the movement in Başıbüyük (see figure 3.8). By blocking the signing of the deals, the association has so far succeeded in suspending the project. This resistance also led GAP to provide financial assistance to tenants purchasing TOKİ units.

Despite the association's success so far in defending the rights of all residents, the coalition between owners and tenants is a fragile one, likely to dissolve if and when owners secure more gains. In an area with such high rates of tenancy and absentee landlordism, it is likely that as the association wins at the bargaining

Figure 3.8. Neighbourhood association posters in Tarlabaşı, September 2008. (The top poster reads: 'Yesterday *we* were the owners and today it will be GAP; this is not a project of renewal and restoration but a project of exile and demolition; the municipality will demolish our homes and GAP will get rich', while the bottom reads: 'Our businesses will be torn down and we will remain jobless; one storey for us, fourteen levels for GAP; no to charity, yes to rights and justice; Tarlabaşı is ours, Manhattan is yours; we have created history and you are destroying it'. (*Photo*: Özlem Ünsal)

table, tenants will be displaced. This (paradoxical) situation is also acknowledged by an influential community organizer; he admits that since it is too late to take the project to court, the most that can be achieved now is to secure maximum gains for the owners who almost certainly will lose their properties. The fact that the association has not yet joined the Chamber of Architects in their lawsuit against the project also shows that the owners want to put as much pressure as possible on the district municipality and GAP in order to maximize gains rather than protecting rights to housing in the area.[42]

To sum up, the form and strength of collective mobilization is a significant variable in manipulating the implementation of projects and determining their consequences. The cases show that effective and sustained resistance movements, especially when joining forces with supra-local organizations, can challenge such interventions and bring about important gains (see also Eckstein, 1990). Yet, as Castells (1983) argues, movements for collective consumption goods in low-income communities tend to be fragile and short-lived, due to important structural impediments to movement building. The resistance movements observed here are particularly threatened by a lack of movement building experience, state violence, and, most importantly, by the prospect for private property ownership that UTPs instigate among inhabitants.

Conclusion

The mounting infrastructural and socio-economic problems caused by Istanbul's unregulated growth makes radical interventions in the built environment via 'renewal' projects a necessity, especially given the major earthquake threat. The current UTPs, however, seem unlikely to accomplish their stated goals – that is eradicating poor quality housing, creating a sustainable urban environment, and improving the lives of the urban poor. Rather, these interventions into urban space and land/housing markets act as tools of 'marketing' certain rent zones to stronger actors. The projects mostly aim at the physical and demographic upgrading of their areas to increase real-estate values, and this violates the current users' right-to-housing.

Our analysis also documents the various tactics of resistance developed by those subject to 'transformation'. These grassroots movements face numerous challenges for sustained mobilization such as state violence, lack of experience, internal divisions and divide-and-rule tactics of project implementers. Despite these obstacles, however, they have achieved some important gains and shaped the projects in critical ways. For example in Başıbüyük, as a result of the efforts of the association's volunteer lawyer and various NGOs, the 5th Administrative Court ordered the UTP to be stopped because 'it violates conceptions of public good, principles of urban planning and rules of democratic governance'.[43] This is a major legal victory that is likely to set a precedent for other projects. In the meantime, in Tarlabaşı, the Metropolitan Municipality initiated a re-assessment of the project as a result of the association's uncompromising attitude. In both cases, without the swift mobilization of the residents and the creative strategies used by the associations, the gains would have been impossible. It is our hope that these struggles pave the ground for the implementation of a more democratic and inclusive 'urban transformation' agenda, which would strive for the social, economic and political empowerment of inhabitants in addition to physical upgrading and rent seeking.

Notes

1. *Gecekondu* is the Turkish word for squatter housing. It literally means 'landed overnight'.

2. The ruling AKP got 72 per cent of the votes in Başıbüyük in the 2004 local elections and 55 per cent in the 2007 general elections.

3. Authors' translation. TOKİ, the Housing Development Administration, is the governmental agency responsible for the provisioning of public housing. It also is a central actor in the implementation of UTPs, as we discuss below. Founded in 1984 to provide affordable housing and to regulate rapid urbanization, the TOKİ played a very important role in Istanbul's urban expansion during the 1980s and 1990s, mostly through providing credit to housing cooperatives.

4. Formed in 2006 with twenty-one neighbourhood associations, the platform aims to form a unified front against demolitions and displacement caused by UTPs.

5. The GAP Construction Company won the tender for the renewal project in Tarlabaşı. It is part of the Çalık Holding Group, known for its close ties to the ruling AKP. One of us (Özlem Ünsal) attended the opening of the exhibition.

6. The Tarlabaşı Neighbourhood Association was formed in March 2008, in response to the Tarlabaşı Renewal Project.

7. Law for the Protection, Renovation and Sustainment of Dilapidating Historic and Cultural Real-Estate.

8. This quote is taken from the speech by Ayhan Kayar, one of the volunteer lawyers, at a mass meeting held by the neighbourhood association to inform the inhabitants at a local coffee shop on 24 May 2008.

9. The existing research on the long-term profitability and sustainability of such mega-projects yields mixed and inconclusive results. Despite this reality, however, they remain highly popular among 'growth coalitions'.

10. *Radikal*, 9 April 2006.

11. Law No. 5237. For further details see: http://www.tbmm.gov.tr/kanunlar/k5237.html.

12. We thank Cem Bico for providing this figure.

13. Law No. 5393. For further details see: http://www.tbmm.gov.tr/kanunlar/k5393.html.

14. Law No. 4966 (2003); Law No. 5162 (2004); Law No. 5582 (2007); Law No. 5793 (2008). TOKİ became the sole agency to regulate all state-owned urban land except that held by the military. Please refer to http://www.tbmm.gov.tr/develop/owa/kanunlar_sd.sorgu_baslangic for more details on the laws addressed.

15. Since 2003, 65,808,239 square metres of land have been transferred to TOKİ, with no cost (*Radikal*, 27 May 2008). Between 2003 and 2008, the TOKİ constructed 340,000 housing units, 50,000 of which are in Istanbul, 317 trade centres and thirty hospitals, in addition to numerous other structures. Available at www.toki.gov.tr.

16. Such as Law No. 2863. For further details see: http://www.kom.gov.tr/dosyalar/Mevzuat/2863.pdf.

17. This, of course, is an inflated number due to high rates of *gecekondu* 'ownership' in Turkey.

18. Law No. 5582: Law for Making Changes to Existing Laws Regarding Housing Finance, see: http://www.tbmm.gov.tr/kanunlar/k5582.html.

19. The deflationary policies of the AKP and the rapidly falling interest rates are also important factors in this increase in the use of housing credits.

20. Law No. 4875 (Foreign Direct Investment Law) and Law No. 4875 (Law to Make Changes in Various other Laws). The former law allowed foreign firms to purchase property in Turkey and the latter one, by reforming the existing Land Title Act (of 1934), made it possible for foreigners to purchase non-movables up to 30 hectares. See: http://www.tbmm.gov.tr/develop/owa/kanunlar_sd.sorgu_baslangic for more details on the laws addressed.

21. Data gathered from GYODER Report (2007) and Colliers International Report (2007).

22. In 2003, the share of real-estate in total foreign investments was 11 per cent, whereas by the end of 2007, it jumped to 21 per cent (*Ibid.*).

23. Even though the Constitutional Court stepped in twice (in 2005 and 2007) to overturn the law that allows foreign individuals to purchase property, the AKP, after each of these interventions, enacted new legislations that legalized these sales.

24. Türkiye İstatistik Kurumu (TÜİK) (2008).

25. According to the Zoning Law, a *tapu tahsis* document guarantees a future *de jure* property right, either to the property that they occupy or to another dwelling built elsewhere. If an informal area receives a formal plan, then *tapu tahsis* documents *may* be turned into formal deeds.

26. Here we are referring to the Masterplan of Istanbul, which has been approved by the mayor of Istanbul as of 15 June 2009. A detailed report on the masterplan is available at http://www.ibb.gov.tr/tr-TR/kurumsal/Birimler/SehirPlanlamaMd/Documents/yonetici_ozeti/CDP_YONETICI_OZETI_15.06.2009.pdf. The 2008 report of Colliers International focuses on the potential of the area for private developers, particularly emphasizing '*gecekondu* transformation projects' and unused factory buildings as the strengths of the region. The report is available at: http://www.colliers.com/Content/Repositories/Base/Markets/Istanbul/English/Market_Report/PDFs/2008_Turkey_Real_Estate_Market_Review.pdf .

27. KİPTAŞ is the Metropolitan Municipality's subsidiary construction firm.

28. In 1964, a large number of Greek nationals and ethnically Greek citizens of Turkey were forced to leave the country, in the midst of a political crisis between Greece and Turkey. See Akar and Demir (1994) for more details.

29. Bedrettin Dalan served as the mayor of Istanbul between 1984 and 1989. He is known for his radical redevelopment plans with respect to transport links, tourism areas and business districts which caused much controversy regarding the protection of the city's historical and natural assets.

30. According to the 2000 census, south-eastern Anatolians, the majority of whom are Kurdish, constituted the largest group in Tarlabaşı.

31. In Talimhane, the value of buildings increased to $3,000,000 with major tourism investments (see *Yeni Şafak*, 13 August 2007).

32. These first 'voluntary' migrants to arrive, who were mostly from central Anatolia and the Black Sea region, are distinct from the internally displaced Kurds from eastern and south-eastern Turkey. Pushed out of agriculture with the mechanization of production, they moved to Istanbul in search of industrial jobs.

33. Kentsel Strateji A.Ş. Report, 2008.

34. In Başıbüyük, only 18 per cent of the population completed a degree higher than primary school and 22 per cent do not even hold a primary school diploma. Nine per cent of the neighbourhood is illiterate, with women making up 80 per cent of this (TURKSTAT, 2000).

35. The inflow of internally displaced Kurds to Tarlabaşı, some of whom do not speak Turkish, is an important reason for the high rates of illiteracy.

36. One survey carried out in Tarlabaşı in 2008 reveals that 64 per cent of the residents lack social security (Kentsel Strateji, 2008). We predict the percentage to be higher than this.

37. Roughly 400,000 square metres. This constitutes about half the neighbourhood.

38. According to the project catalogue, the project comprised of 52 per cent residential, 12 per cent commercial, 17 per cent tourist and 14 per cent office buildings.

39. For a more detailed discussion regarding the shortcomings of the UTPs, see Kuyucu and Ünsal (forthcoming).

40. Bartu-Candan and Kolluoğlu (2008), in their analysis of another UTP in Istanbul, demonstrate the same process.

41. *Birgün*, 28 February 2008; *Radikal*, 2 March 2008; *Radikal*, 20 March 2008;

42. The Chamber of Architects has been one of the leading opponents of the Tarlabaşı Renewal Project since its public announcement. In addition to lending administrative and technical support to the Association, the Chamber has also taken the project to court in July 2008.

43. Case no. 2007/1203.

References

Akar, R. and Demir, H. (1994) *İstanbul'un Son Sürgünleri* [Last Exiles of Istanbul]. Istanbul: İletişim.

Atayurt, U. and Kuyucu, T. (2008) Başıbüyük Direniyor! [Başıbüyük Resists!]. *Express*, May, pp. 36–41.

Bartu-Candan, A. and Kolluoğlu, B. (2008) Emerging spaces of neoliberalism: a gated town and public housing project in Istanbul. *New Perspectives on Turkey*, **39**, pp. 5–46.

Behar, D. and İslam, T. (2006) *İstanbul'da Soylulaştırma: Eski Kentin Yeni Sahipleri* [Gentrification in Istanbul: The New Owners of the Old City]. Istanbul: İstanbul Bilgi Üniversitesi Yayınları.

Brenner, N. and Theodore, N. (2002) Cities and territories of actually existing neoliberalism. *Antipode*, **34**(3), pp. 349–379.

Buğra, A. (1998) The immoral economy of housing in Turkey. *International Journal of Urban and Regional Research*, **22**(2), pp. 282–302.

Castells, M. (1983) *The City and the Grassroots: A Cross-Cultural Theory of Urban Social Movements*. Berkeley, CA: University of California Press.

Çeçener, B. (1995) *İstanbul'un Kültür ve İmar Sorunları* [The Cultural and Development Problems of Istanbul]. Istanbul: Mimarlar Odası İstanbul Büyükkent Şubesi.

Cizre, Ü. and Yeldan, E. (2005) The Turkish encounter with neoliberalism: economics and politics in the 2000/2001 crises. *Review of International Political Economy*, **12**(3), pp. 387–408.

Colliers International (2007) *2007 Turkey Real Estate Review*. Available at: www.colliers.com/Content/.../2007_2_MARKET_REVIEW.pdf. Accessed 2 March 2010.

Eckstein, S. (1990) Poor people versus the state and capital: anatomy of a successful community mobilization for housing in Mexico City. *International Journal of Urban and Regional Research*, **14**(2), pp. 274–296.

Ekinci, O. (1994) *İstanbul'u Sarsan On Yıl* [Ten Years that have Shaken Istanbul]. Istanbul: Anahtar Kitaplar.

Erder, S. (1996) *İstanbul'a Bir Kent Kondu: Ümraniye* [A City has landed on Istanbul: Ümraniye]. Istanbul: İletişim Yayınları.

Güvenç, M. (2005) 1990–2000 İstanbul Haritaları [Maps of Istanbul, 1990–2000]. Unpublished research project, İstanbul Bilgi University.

GYODER (2007) *Gayrimenkul Sektörü ve İstanbul İçin Öngörüler, 2015* [The Real-Estate Sector and Projections for Istanbul, 2015]. Istanbul: Gayrimenkul Yatırım Ortaklıkları Derneği. Available at: http://www.gyoder.org.tr/BrowseServices.aspx?MainCatID=9&SubCatID=11.

Hackworth, J. (2007) *The Neoliberal City: Governance, Ideology, and Development in American Urbanism*. Ithaca, NY: Cornell University Press.

Harvey, D. (1989) From managerialism to entrepreneurialism: the transformation in urban governance in late capitalism. *Geographiska Annaler Series B*, **71**, pp. 3–18.

Işık, O. and Pınarcıoğlu, M. (2002) *Nöbetleşe Yoksulluk: Gecekondulaşma ve Kent Yoksulları-Sultanbeyli*

Örneği [Taking Turns with Poverty: Informal Urbanization and the Urban Poor: The Case of Sultanbeyli]. Istanbul: İletişim Yayınları.

İstanbul Beyoğlu Municipality (2008) *50 Soruda Tarlabaşı Yenileme Projesi* [Tarlabaşı Renewal Project in 50 Questions]. Istanbul: İstanbul Beyoğlu Municipality.

Kentsel Strateji A.Ş. (2008) Tarlabaşı Stratejik Sosyal Plan 2008–2010 [The Strategic Social Plan for Tarlabaşı, 2008–2010]. Unpublished research report.

Keyder, Ç. (2000) Liberalization from above and the future of the informal sector: land, shelter, and informality in the periphery, in Tabak, F. and Crichlow, M. (eds.) *Informalization: Process and Structure*. Baltimore, MD: Johns Hopkins University Press, pp. 119–132.

Keyder, Ç. and Öncü, A. (1994) Globalization of a third world metropolis: Istanbul in the 1980s. *Review*, **17**(3), pp. 383–421.

Knox, Paul L. and Taylor, Peter J. (1995) *World Cities in a World System*. Cambridge: Cambridge University Press.

Kuyucu, Tuna and Ünsal, Özlem (forthcoming 2010) Urban transformation as state-led property transfer: an analysis of two cases of urban renewal in Istanbul. *Urban Studies*.

Lehrer, U. and Laidley, J. (2008) Old mega-projects newly packaged? Waterfront redevelopment in Toronto. *International Journal of Urban and Regional Research*, **32**(4), pp. 786–803.

Logan, J.R. and Molotch, H. (1987) *Urban Fortunes: The Political Economy of Place*. Berkeley, CA: University of California Press.

Moulaert, F., Rodriguez, A. and Swyngedouw, E. (eds.) (2003) *The Globalized City: Economic Restructuring and Social Polarization in European Cities*. Oxford: Oxford University Press.

Öncü, A. (1988) The politics of urban land market in Turkey: 1950–1980. *International Journal of Urban and Regional Research*, **12**(1), pp. 38–64.

Peck, J. and Tickell, A. (2002) Neoliberalizing space. *Antipode*, **34**(3), pp. 380–404.

Roy, A. (2004) The gentlemen's city: urban informality in the Calcutta of new communism, in Roy, A. and AlSayyad, N. (eds.) *Urban Informality: Transnational Perspectives from the Middle East, Latin America and South Asia*. Lanham, MD: Lexington Books, pp. 147–170.

Sajor, E. (2003) Globalization and the urban property boom in Metro Cebu, Philippines. *Development and Change*, **34**(4), pp. 713–741.

Sassen, S. (2001) *The Global City: New York, London, Tokyo*. Princeton, NJ: Princeton University Press.

Şenyapılı, T. (2000) Enformel sektör: durağanlıktan devingenliğe – gecekondulaşmadan apartmanlaşmaya [The informal sector: from stability to dynamism – from gecekondus to apartment blocs], in Akder, A.H. and Güvenç, M. (eds.) *Yoksulluk: Bölgesel Gelişme ve Kırsal Yoksulluk, Kent Yoksulluğu* [Poverty: Regional Development and Rural Poverty, Urban Poverty]. Istanbul: TESEV, pp. 164–165.

Smith, N. (2002) New globalism, new urbanism: gentrification as global urban strategy. *Antipode*, **34**(3), pp. 427–451.

Tekeli, İ. (1992) Development of urban administration and planning in the formation of Istanbul Metropolitan Area, in Tekeli, İ. (ed.) *Development of Istanbul Metropolitan Area and Low Cost Housing*. Istanbul: Turkish Social Science Association.

Thirkell, Allyson J. (1996) Players in Urban informal markets; Who wins? Who loses? A case study of Cebu City. *Environment and Planning*, **8**(2), pp. 71–90.

Türkiye İstatistik Kurumu (TÜİK) (2000) *Bina Sayımı İstatistikleri* [Building Survey Statistics]. Available at: www.tuik.gov.tr.

Türkiye İstatistik Kurumu (TÜİK) (2008) *Adrese Dayalı Nüfus Sayımı Sonuçları* [Residence-based Population Census Results]. Available at: www.tuik.gov.tr.

Chapter 4

Contestations over a Living Heritage Site: The Case of Büyük Valide Han

Ayşegül Baykan, Zerrin İren Boynudelik, Belkıs Uluoğlu and Burak Sevingen

During the first decade of the twentieth century, Georg Simmel characterized the metropolis as an interrelated domain of the built environment, communal lives, technology and political institutions. Contemplating on Simmel's theories, David Frisby (1999, p. 106) argues that spatio-temporal disruptions and transformations following the first decade of the twentieth century seriously hindered our ability to come to terms with the past and the present of the metropolis, resulting in a 'systemic erasure of memory traces'. Istanbul, as other cities around the world, has had its share of 'disruptive' acts of city planning and lost, so to speak, traces of its past.

Contemporary remedy to further erasure of memory comes in the form of policies that designate 'historic' areas of the city and certain built forms as heritage sites and develop and implement plans towards their preservation and protection, under the tutelage of not only local and national governments but also international organizations. These policies and plans certainly have their merits. Nevertheless, these measures, historically informed in their claims and enactment, embody the danger of disregarding recent histories of social and communal structures and of reproducing histories and heritage that reflect the demands of contemporary economic and political interests.[1]

In this chapter, we draw on a large-scale multi-disciplinary study which we carried out in Istanbul's Historic Peninsula, at Büyük Valide Han (Grand Mother Sultan Han and BVH from now on), a monumental building situated in what is referred to as Hanlar Bölgesi (Hans District) in Eminönü and still serving as home to small-scale artisanal production and retailing.[2] We aim to examine the impact which contemporary policies and practices of cultural heritage preservation have on the future of BVH's built form and the life-worlds it embodies.

We begin with a brief history of this Han and excavate its past as a prelude to the present. Then we provide a brief account of policies and practices of urban planning and governance as they apply to the Historic Peninsula in general since they have direct relevance to the current plight and prospects of BVH. Next we offer a purview of the building's contemporary architecture to reveal the ways in which its history is shaped by the lived experiences of those who have made the spaces of BVH meaningful for their work and 'life-worlds' through inventive, and at times 'disruptive', interventions. Finally, we present a sample of ethnographic notes on the occupants of BVH as a way to document the life-worlds which have characterized much of the twentieth century of BVH and are constitutively embedded in the memory of Istanbul at large.

A Short History of BVH: Then and Now

Han is a generic name for physical structures built to house shops of local trade, artisanal production and residential rooms. Similar in their monumental form to caravanserais, they were commonly situated within cities across the Ottoman Empire while the latter were built on long distance trade routes between cities; today they are considered sites of heritage. In the complex societal structure of the Ottoman Empire organized around the *millet* system, spaces of the city reflected the complexity of the social order. Neighbourhoods were culturally homogeneous whereas the *han* provided a public place for the daily social and cultural encounters of religiously and ethnically different groups, and were instrumental in the making of public encounters place-bound, with merchants, journeyman, and artisans of different backgrounds accepting each other's presence with both friendship and distance. Many *han* were built by people of political power, often to provide financial support for the operation of institutions essential to the public good such as mosques for worship, *medreses* for teaching, and *imarets* for welfare; and hence, were centrally located for the functioning of the economy of the city.

One such *han* is BVH, located in the 'Han District' in the Historic Peninsula of Istanbul, a dense area of traditional businesses throughout the history of the Ottoman Empire and later in the Republican era (figure 4.1). The area is surrounded by Çemberlitaş, Eminönü, and Beyazıt and is also known as the Mercan neighbourhood. Today listed as a historic heritage site, BVH was built in mid seventeenth century as a *waqf* or *vakıf* (charitable foundation) to provide resources for the upkeep of the Çinili Mosque in Üsküdar, founded by Kösem Mahpeyker Sultan, the wife of Ahmed I, and the mother of Murad IV and İbrahim I, thus the name 'Büyük (Grand) Valide' meaning the mother of the sultan. With two levels and three courtyards, it is a city*han* (as opposed to caravanserai), with approximately three hundred cells and rooms to house one thousand horses or mules (Benli, 2007, p. 7) (figure 4.2). From the beginning, it housed several manufacturing trades and both long-distance and local commerce.

Throughout its history, it has been a significant space for newcomers and immigrants from across the Ottoman Empire and also from Iran, and after the Republic was founded, it provided home and work for those coming from the peripheral cities of Anatolia, especially due to rapid urbanization in the second half of the twentieth century. Young men from rural regions with links to communal networks established by those who had come earlier often stopped here in search of jobs.

BVH's physical form changed continuously during the Ottoman centuries. The most dramatic alterations, however, took place in the twentieth century, with cells often divided into several units due to increased demand for space, and industry's reliance on the informal economy and small-scale manufacturing carried out in small workshops rather than large-scale factories. Several new

Figure 4.1. Historic Peninsula and the Han District in Istanbul. (*Source*: www.sehirrehberi.ibb.gov.tr/)

Figure 4.2. A general view of BVH, July 2008. (*Photo*: B. Sevingen)

structures which were attached to the building were made of industrially-produced building materials such as concrete and aluminium. In 1951, the newly established *Gayrimenkul Eski Eserler ve Anıtlar Yüksek Kurulu* (Higher Council for Historic Monuments and Antiquities) placed BVH under protection. The monument was designated a historic heritage site by Law No. 2863 in 1982, which was legally ratified in 1983.[3]

Despite legal protection, not much has been done in the way of preservation. Alterations to the original plan have continued and no major restoration has taken place to this day; and the building is fast dilapidating.

Today BVH houses small-scale metal works, manufacturing, textile production, and wholesale textile trading and warehouses. Under discussion by the official bodies at various levels of the government and the city administration are several proposals to restore the physical character close to its original form and to change the functions it embodies by removing the metal works, textile manufacturing, wholesale trading and warehouses. The aim is to replace them with 'traditional crafts', office space, retailing and other businesses which cater for culture industries and tourism.

Given the prospect of enormous change, our research team, consisting of a sociologist, architect, art historian and an anthropologist, undertook a study of BVH to record its social, cultural and physical memory. The study took place from 2006 to 2009, and involved ethnographic documentation and participant observation. In addition, we recorded the use of space by outsiders (artists, tourists and so on), and references to BVH in historic archives, in literature and the press throughout the twentieth century. We investigated the public policy measures and did extensive interviews with prominent actors in the decision-making bodies and academia, as well as mapping the physical space *vis-à-vis* the experiences of the actual occupants.

The findings of this study challenge the planning practices that still advocate restoring historic buildings such as BVH to their original form and ignore the extent to which physical space is actually the accumulated memory of not only the economy and history of the building but also everyday lives of those who continue to inhabit the building and the city. Our findings and conclusions pertain to the prospects of heritage work in Istanbul and the city's transformation at large under urban policies and regimes whereby historic spaces acquire new symbolisms and currency in an urban economy heavily reliant on culture industries and tourism.

Planning Policies and Sites of Heritage

In our overview of the legal framework of historic preservation and urban governance in Istanbul as they pertain to sites such as BVH we follow Newman and Thornley (2005) and focus on urban regimes and policies rather than particular policies or projects.

The establishment of the *Gayrimenkul Eski Eserler ve Anıtlar Yüksek Kurulu* (Higher Council for Historic Monuments and Antiquities) in 1951 occurred when the preservation of material culture and heritage was becoming a worldwide concern. The Council included scholars and prominent figures from the fields of architecture and archaeology and met on a needs basis to inspect and report on damage to historic buildings. This, however, does not mean an absence of concern for preservation and heritage. Even as early as in 1937, records were kept of alterations to historic buildings and monuments, including BVH, by *İstanbul Asar-ı Atika Müzeleri Müdürlüğü* (Istanbul Directorate of the Museum for Antiquities).

The speed of industrialization and the emergence of Istanbul as the primary city of industrial growth from the 1950s onwards resulted in the increased role of the historic district as a business district. To accommodate growth, roads crisscrossed the Historic Peninsula and high-rise apartments were built, resulting in the demolition of innumerable sites that would today be seen as cultural heritage.[4] *Hans* had become an integral part of the change and provided much needed space for small-scale manufacturing and trade. Their existing structures expanded with *ad hoc* additions and divisions to provide space for producers and merchants who could only afford low rents but needed to be close to the city centre. Further, and perhaps more intrusive, urban change and massive spatial transformation followed the introduction of economic liberalization and structural adjustment policies in the aftermath of the military coup in 1980 (Keyder and Öncü, 1993, p. 19).

The post-coup political and economic regime brought about drastic changes in urban politics. According to Keyder and Öncü, in the 1980s three significant developments took place. Firstly, new financial resources were created for metropolitan governments, making them better able to invest in infrastructure and restoration projects. Secondly, the state started subsidizing housing at unprecedented levels. Thirdly, a new, two-tier model of metropolitan governance emerged through a law enacted in 1984, which strengthened the powers of the mayor by expanding his authority to the entire metropolitan area and keeping district authorities responsible only for day-to-day operations (*Ibid*., pp. 23–26). The mayor of the Istanbul Metropolitan Area also took over the jurisdiction rights for administrative action and the resources, which were formerly the responsibilities of the Master Plan Bureau and the Water Supply and Sewerage Authority, organs of the central government (*Ibid*., p. 26).

The Law for the Conservation of Assets of Cultural and Natural Heritage (*Kültür ve Tabiat Varlıklarını Koruma Kanunu*) enacted in 1983, and modified subsequently, created the Boards for the Conservation of Assets of Cultural and Natural Heritage (*Kültür ve Tabiat Varlıklarını Koruma Kurulları*) under the Ministry of Culture and with the right to decide on the listing of heritage sites. However, the law caused controversy as it contained articles which gave the municipalities the right to act on heritage sites and to confiscate property and implement policies for rehabilitation, revitalization and urban renewal, albeit in accord with joint decision-making and approval by the Ministry of Culture and the Boards for the Conservation of Cultural and Natural Heritage (Benli, 2007, pp. 215–216). Nevertheless, especially with UNESCO's declaration of four parts of the Historic Peninsula (Sultanahmet Archeological Park, Süleymaniye and Zeyrek districts and the city walls) as a World Heritage Site in 1985, it became crucial that a policy of conservation was articulated and put into effect *vis-à-vis* the Historic Peninsula.

The spatial transformation of Istanbul during the 1990s under neoliberal policies demanded new management techniques and urban policies.[5] Management of the transformation of the Historic Peninsula in particular became problematic. The area, functioning as a business district, was rapidly losing its residential neighbourhoods and 'degenerating'. Small-scale manufacturing contributed to the pollution and to the unruly development of the existing structures, disfiguring the historic essence of streets, residences and *hans*. However, how to confront such issues did not have ready-made answers. Plans for the region since 1990 have addressed these issues with an eye on both cultural heritage and the interests of local residents. The larger economic interests and higher goals of making Istanbul a 'world city' have, however, worked against the noble goals inscribed into plans. The only substantive opposition came from the representatives of professional organizations such as the Chamber of Architects and Engineers, who protested against altering the city space for good.

In 1990, the Istanbul Metropolitan Municipality approved a 1:5,000 scale 'Istanbul Historic Peninsula Conservation Master Zoning Plan' which, if implemented, anticipated significant changes to BVH and the entire Han District. Gündüz Özdeş, the chief planner, stated the aim of the plan as decentralizing manufacturing, wholesale trading and the warehouses and creating 'a living historic, cultural, tourist, commercial and recreational space, integrated with a local population of restricted dimensions' (Özdeş, 1999). The plan, informally referred to as the Özdeş Plan, included reverting dilapidated housing areas to their 'traditional' residential character and diverting traffic to the outer zone using new bridges and underpasses. The following description is informative of Özdeş's vision for the urban space of the *hans* and their periphery:

> The picturesque character of the small shops adjoining the *hans* would have been preserved, but a number of ramshackle additions that spoil the aesthetic perspective of the historic environment, together with parasitic buildings in the courtyards, would have been removed, and small rest and recreation areas would have been created among buildings providing accommodation and other tourist services. (*Ibid.*, p. 45)

The Chamber of Architects and Engineers took the 1990 Plan to court. After a long trial and many appeals, the plan was discarded by the courts for its violation of the laws on various technicalities. The substantive point of objection involved the possible damage the new traffic routes would cause to archaeological and cultural sites. More significantly, the decision challenged the rights of the Board for the Conservation of Cultural and Natural Heritage to decide what constitutes heritage, as well as zoning, prioritizing and conservation.[6]

In July 1995, the Conservation Board designated the Historic Peninsula an urban and historic heritage site. One segment of the Peninsula was given the status of Urban and Archaeological Site and the space within the interior of the Topkapı Palace walls on the eastern tip of the Peninsula (*Sur-ı Sultani*) was declared Primary Archaeological Site, hence preventing any possibility of development in the area – a primary feature of the earlier Özdeş Plan. In 2002, the Municipality offered a new plan and in 2004 published the results of quantitative and qualitative surveys based on households and (commercial) occupants of the Historic Peninsula.[7] The qualitative survey and expert reports evaluating the plan revealed that neither the inhabitants of the Historic Peninsula nor the experts were content with the potential presence and benefits of the tourism industry. However, there was consensus on the urgent need to rehabilitate derelict neighbourhoods and restore the monumental architectural sites to their former grandeur. Experts asserted in their evaluations the necessity of decentralizing the wholesale commerce, warehouses and other manufacturing practices and easing the heavy traffic in the Han District. There was also an expressed desire for the revival of residential districts.

In essence, the 2002 Plan, not so different from the 1990 Plan, viewed the Historic Peninsula as composed of small segments with particular and dense thematic characters that, taken all together, defined the whole as a mosaic. It claimed that certain functions such as manufacturing and warehousing had negative effects on the mosaic. The plan proposed removal of these functions from the district, along with the transient population which these functions attracted to the district. On the other hand, a new set of thematic functions that would be sustainable and aesthetically and historically harmonious with the mosaic were suggested to fill the spaces vacated by the removal of others (İstanbul Büyükşehir Belediyesi, 2003*a*, p. 76). The plan emphasized the need to make changes that would be in agreement with the 'cultural and historic identity' of the district (*Ibid*., p. 3). Monumental buildings were to be cleared of all additions and renovated using 'original' material and traditional styles and techniques. Commercial districts were to be divided into four different zones, the first being the traditional zone of the Han District, where crafts with 'traditional identity' (twenty-seven such crafts are listed as examples) were to occupy space for retailing (*Ibid*., p. 110) (figure 4.3).

Not unexpectedly, legal battles ensued and crucial questions regarding who is to decide on what constitutes cultural heritage, how the resources will be spent and how the agency of those who live in these places will be realized await answers.

Figure 4.5. The second courtyard of BVH today. (*Photo*: B. Sevingen)

The boundaries that separate the worlds built around these three courtyards are also markedly different from each other. A door passage – a vaulted entrance – separates the first triangular courtyard from the second. The border between the second and the third courtyard, which actually belongs to the next *han*, Sağır Han, is simply a corridor-like passage in the western corner. BVH is in fact a combined entity, comprising of Valide Han and Sağır Han, though today they have separate administrations and well-defined borders for their service zones. Sağır Han is less deserted than Valide Han, presumably because it is much easier to access from the street. Perhaps because of its density and ruin in the past, it is also less dilapidated. Throughout BVH, one comes across an abundance of stairs built at various times and places dividing the cells vertically and horizontally or providing passage to different parts of the building and to the roof, with no apparent rationale for the way they are located (figure 4.6).

At BVH one encounters people of different religious sects, nationalities, ethnic origins, age groups and lines of work who accept each others' presence in respect and cordiality. The contrapuntal presence of these differences can also be felt in the physical character of the space. Different layers exist together in the architectural layout of BVH. The Byzantine tower stands integrated into the rest of the building as if it were a montage. A contemporary *masjid* has replaced the former mosque in the second courtyard. An incredible variety of building materials are used side by side. This overall hybrid and chaotic picture only reveals the intricate complexity of lives and history built in and around BVH. The seemingly disparate elements are so interwoven that it seems impossible to take one off without the whole collapsing.

Figure 4.6. The abundance of stairs across BVH. (*Photo*: B. Sevingen)

Buildings are often conceived as elements frozen in time, little account being taken of the transformative power of everyday practices on the physical environment. To redress this omission, at least in our study of BVH, we introduce its occupants, and to an extent, their everyday practices, as provided below in the summary account of the ethnographic survey we conducted.

Daily Life in BVH

Witnesses to the last half century at BVH recount the presence of a wide variety of occupations, such as the leatherworkers, carpet dyers, traders of cardboard, chest- and scale-makers, and producers of sacks (*çuvalcılar*), as well as various crafts specializing in textiles and metals. Many of these lines of work have become obsolete or had to be relocated to other sites, some due to decreasing demand for their goods, others because of technological developments, competition, absence of new masters or apprentices, or changing city regulations.

Businesses come and go but the management of the *han* and its responsibilities and role are passed down from generation to generation. An important figure in the daily life of BVH's inhabitants is the *odabaşı* (innkeeper), whose responsibilities include the up-keep of the building and maintaining security, in return for monthly fees. The innkeeper holds in his hands various operations of economic importance at BVH. In an informal way, he gets a cut in the profits from transportation activities and shares kinship with the owners of the coffee shops. The current innkeeper, who owns a space himself, defines his relation to BVH as an earned privilege, entitling him to its management. Another centuries-old operation is the transportation of goods, undertaken by the organization of *hamals* (porters). As late as 1980s, the team of porters at BVH is said to have had as many as fifty members, offering easy access to work for those who recently arrived in the city. In the last couple of years, their number has dwindled to eight. Viewed as indispensable and essential in the past, today many merchants consider the self-imposed and rigid rules of the porters to be obstructive and inefficient.

The majority of businesses in the first and second courtyards of BVH are in wholesale trade of textiles, of fabrics or ready-made clothing for men's wear, headwear and uniforms. During their heyday, from the early 1960s until the 1980s, the weaving workshops that dominated the upper floor of BVH often worked in double shifts to complete the orders from merchants from Anatolia. Today, merchants still consider BVH as an important commercial centre for textiles. However, the emergence of new centres of commerce and industry in a number of Anatolian cities has partially deprived the merchants operating from BVH of their most important customers, the Anatolian businessman. Furthermore, city officials are relocating businesses in BVH to newly designated sites of commerce and manufacturing in the peripheral zones of the city, where they face problems of transportation, high rent, and loss of customer base. In addition, Chinese imports contribute to their hardship by reducing profit margins and even leading to bankruptcies. Nevertheless, textile traders consider owning a shop in BVH prestigious as well as crucial to their livelihood since BVH allows them to be near their most prized customer base, namely mid-size businesses without powerful brand names.

Located on the upper floor of BVH there are about 150 rooms, the majority of which are locked up and deserted. The number of active workshops does not exceed twenty-five, mainly occupied by craftsmen specializing in textiles and metalwork. Elders of BVH speak of times when weaving looms on the upper floor of BVH were so many that conversations were impossible because of the clang and clatter. Younger tenants recall how scary it was for them as children to walk around the spinning machines. It has been a while since weaving ceased to be the primary line of work at BVH. When city officials deemed manufacture in historic *hans* against regulations, the majority of the masters quit their jobs or left BVH. Their departure is said to have started in the 1980s and been completed by mid-1990s. As of three years ago, only three weaving masters and two active looms were still present. One of these masters, who left BVH for part-time employment at larger workshops, sold his sixty year old loom to an industrialist intending to use the machine for exhibition at his weaving factory. Another master sold his loom as scrap metal and now works as a waiter at a BVH restaurant. The only active loom is an eighty year old machine, operated by the single remaining weaving master who has an unstable work routine.

Hat-makers have a long history at BVH. The five workshops that still exist operate as sub-contractors and work in conjunction with a number of merchants who bring orders. The craftsmen produce a variety of models, ranging from 'the Anatolian' style men's hats to promotion caps bearing logos of large corporations (figure 4.7). The workshops have four to ten master-workers and apprentices. It is customary for the master-workman to switch between jobs at different ateliers in BVH, when demands of the market are less than stable.

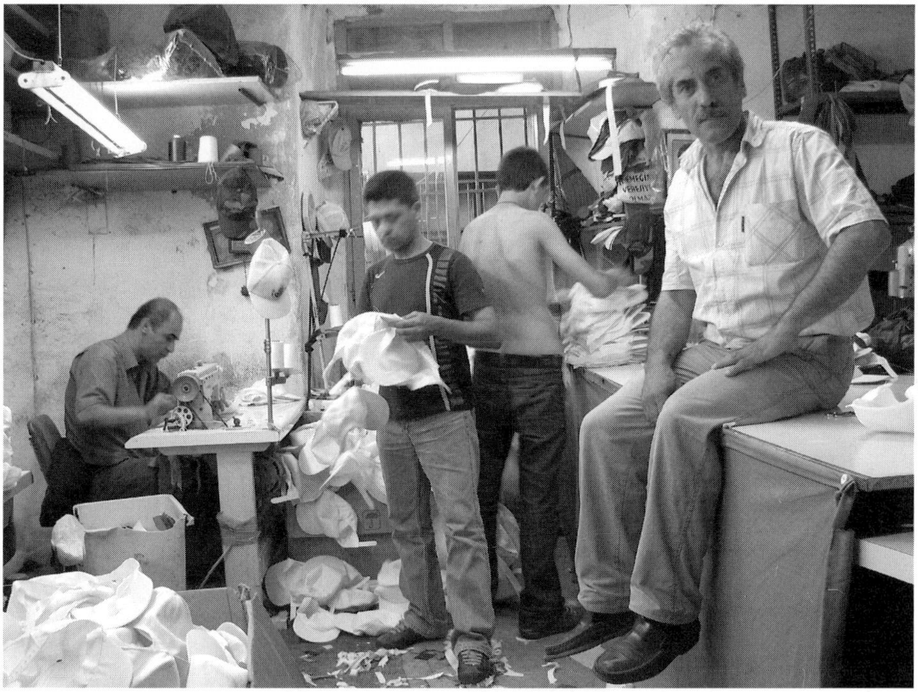

Figure 4.7. Flexible production of caps at BVH (*Photo*: B. Sevingen)

Turners (*tornacı*) and polishers currently constitute the predominant group among the master metalworkers at BVH, their number being close to twenty. Their oldest members, who came to work at BVH as apprentices, are now in their seventies. Some of these craftsmen can be seen turning and polishing products as diverse as ashtrays, flag staffs, minaret-tips and pipes. Some are makers of water pipes, others specialize in accessories for shop displays, and some make lamps and lanterns (figure 4.8). Their workshops are also their showrooms, aiming to appeal to the tourists who visit BVH. Foundry men are a separate category of metalworkers. The three small foundries of this kind employ one to three craftsmen. They cast decorative objects, parts of water pipes and even small machine-parts like iron-presses. The foundry men, who are self-employed, remark that contrary to much larger and technologically advanced factories

Figure 4.8. A maker of souvenir items (*Photo*: B. Sevingen)

Figure 4.9. Space used for recreational purposes coming down from the roof (*Photo*: B. Sevingen)

operating elsewhere, their workshops can survive because of their ability to accept orders of only a few hundred pieces.

In addition to retail shops and workshops as described above, a number of rooms at BVH are occupied for other purposes, such as coffee-shops, where many of the inhabitants stop by during the day for conversation, as office space, depots and even as a studio by a part-time artist who uses the space for recreational purposes (figure 4.9).

Tenants who are old enough to remember BVH from three decades ago recall the busy and exhausting but merry times, when work was plentiful, earnings were more than decent, and relationships were close. Craftsmen who came to BVH as little more than children remember their masters often as tough teachers and bosses and their years of training as one of hardship but as a privileged time of preparation for a professional life. When on a break at the coffee shops, tenants occasionally receive news from old friends and recall anecdotes of success and failure. The occupants of BVH somehow take care of their own, and the most unfortunate still find residence in the makeshift lodgings in its cells. Those who bear witness to the recent history of BVH with their narratives of good old times are significant for a better understanding of identity, pride and spirit of the tenants of BVH but also in understanding BVH as a bounded place where particular life-worlds continue to find a space for work, shelter and community against the pressures of the world outside.

Conclusion

In this chapter we offered a view of BVH from interdependent perspectives of policy-making, architectural change, and the ethnographic presence of diverse trades and people. Our presentation does not exhaust the possible ways of exploring the histories, policies and everyday practices that have shaped the architecture and memory of BVH. Nor does it do justice to the people who have occupied BVH as their workplace and livelihood. It is a glimpse of the centuries-old practices that came to make BVH what it is today, as well as being a summary of the larger research we conducted at the site. In this short narrative, we aimed to underline that during the last two decades, policy debates over urban space management in Istanbul, at BVH in particular, have been closely related to the neoliberal economics of the post-1980 era, the demands of real estate markets and the wish lists of business and governmental bodies which envisage the future of the city in culture industries – unmistakably following the global patterns of employing culture and heritage industries as a primary way to generate resources for the urban economy. It is as if the agenda is to turn the Historic Peninsula, and BVH, into an inert heritage site and reduce public life to 'a kind of vacant lingering in the environs of history' (Crinson and Tyrer, 2005, p. 60). Our study proposes that such an urban renewal practice will reduce the Historic Peninsula and BVH to a common and uncharacteristic copy of a tourist site located somewhere else in the world, devoid of the memories and identities accumulated over history. Therefore, we suggest, that acknowledgement of the heritage of BVH – or acknowledgement of BVH as a heritage site – has to take into account not only the preservation of the (original) form but also the historic manner in which time has taken its course on and in this space, including age-old superfluities and alterations that do not seem attractive to display, as well as the occupants who have become redundant in the new economy. Unless we are able to live with imperfections, we face the danger of crossing into a realm that is 'too gentrified, prettied and transferred into museum and theme [park]' (Wilson, 1997, p. 132).

Notes

1. For further elaboration of this point in the case of Istanbul, see Çınar (2001).

2. The research on which this chapter is based is funded by TÜBİTAK (The Scientific and Technological Council of Turkey). For further information regarding the research and its findings, see http://buyukvalidehan.yildiz.edu.tr.

3. For the decree of the law, see *Kanun No. 2863 Kültür ve Tabiat Varlıklarını Koruma Kanunu* [Law No. 2863 Law to protect cultural and natural heritage], published in *Resmi Gazete* (Official Gazette), 23 July 1983. For further analysis of the outcomes of the law in question, see Örnek Özden (2006, pp. 651–660).

4. For detailed accounts of the transformation, see Kuban (1998) and Tekeli (1998).

5. See Keyder (2005) for an extensive account of the changes.

6. For a detailed discussion of the plan and the legal battle which followed, see İstanbul Büyükşehir Belediyesi (Istanbul Metropolitan Municipality) (2003*a*, pp. 47–53; 2004, pp. 447–449).

7. For coverage of BVH, see especially, İstanbul Büyükşehir Belediyesi (2003*a*, 2003*b*, 2004).

8. For the concept of 'life-world', see Dallmayr (1989, 1991).

9. See Goodwin (1971) who identified the drawing as being by A. Közen.

10. We borrow the use of 'contrapuntal' from Edward Said (1994, p. 59), who employs it to point to the simultaneous awareness of different histories read against each other.

References

Benli, Gülhan (2007) İstanbul Tarihi Yarımada'da Bulunan Han Yapıları ve Avlulu Hanların Koruma Sorunları [Preservation Problems regarding the Han Buildings and Hans with Courtyards in Historic Peninsula of Istanbul]. Unpublished PhD Thesis, Yıldız Technical University.

Çınar, Alev (2001) National history as a contested site: the conquest of Istanbul and Islamist negotiations of the nation. *Comparative Studies in Society and History*, **43**(2), pp. 364–391.

Crinson, Marc and Tyrer, Paul (2005) Clocking off in Ancoats: time and rememberance in the post-industrial city, in Crinson, Marc (ed.) *Urban Memory: History and Amnesia in the Modern City.* London: Routledge.

Dallmayr, Fred (1989) Life-world: Variations on a theme, in White, Stephen K. (ed.) *Life-World and Politics: Between Modernity and Postmodernity*. Notre Dame, IN: University of Notre Dame Press.

Dallmayr, Fred (1991) *Life-World, Modernity and Critique: Paths between Heidegger and the Frankfurt School*. Cambridge: Polity Press.

Frisby, David (1999) Culture, Memory and Metropolitan Modernity, in Korneck, Inge, Illetschko, Georgia and Musner, Lutz (eds.) *The Contemporary Study of Culture*. Wien: Turia und Kant.

Goodwin, Godfrey (1971) *A History of Ottoman Architecture*. Baltimore, MD: Johns Hopkins University Press.

İstanbul Büyükşehir Belediyesi (Istanbul Metropolitan Municipality) (2003*a*) *Tarihi Yarımada Eminönü-Fatih 1/5000 Ölçekli Koruma Amaçlı Nazım İmar Planı Raporu* [Report on the 1/5000 Scale Conservation Master Zoning Plan of Historic Peninsula, Eminönü – Fatih], Vol. 1. Istanbul: Planlama İmar Müdürlüğü.

İstanbul Büyükşehir Belediyesi (2003*b*) *Tarihi Yarımada Eminönü-Fatih 1/5000 Ölçekli Koruma Amaçlı Nazım İmar Planı Raporu* [Report on the 1/5000 Scale Conservation Master Zoning Plan of Historic Peninsula, Eminönü – Fatih], Vol. 2. Istanbul: Planlama İmar Müdürlüğü.

İstanbul Büyükşehir Belediyesi (2004) *Tarihi Yarımada Ticaret Erbabı, Hane halkı ve Müşteriler Araştırması-1 Kaynak Taraması Kantitatif ve Kalitatif (Fatih-Eminönü)* [A Qualitative and Quantitative Survey of Sources: Members of Trade, Households and Consumers of Historical Peninsula, Fatih-Eminönü]. Istanbul: APK Daire Başkanlığı Araştırma Müdürlüğü.

Keyder, Çağlar (2005) Globalization and Social Exclusion in Istanbul. *International Journal of Urban and Regional Research*, **29**(1), pp. 124–134.

Keyder, Çağlar and Öncü, Ayşe (1993) *Istanbul and the Concept of World Cities*. Istanbul: Friedrich Ebert Foundation.

Kuban, Doğan (1998) *İstanbul Yazıları*. Istanbul: Yem Yayınları.

Newman, Peter and Thornley, Andy (2005) *Planning World Cities*. Basingstoke: Palgrave.

Örnek Özden, Elif (2006) Kentsel Sit Alanı İlanı 'Mutlak Korunuyor' Anlamına Geliyor mu? [Does designation as urban heritage site imply absolute conservation?]. *Gazi Üniversitesi Mühendislik Mimarlık Fakültesi Dergisi* [Journal of the Faculty of Engineering, Gazi University], **21**(4), pp. 651–660.

Özdeş, Gündüz (1993) Regarding the new layout of the old Peninsula: urban development plan and its basic strategies. *Istanbul: Selections*, **1**(2), pp. 40–45.

Özdeş, Gündüz (1999) *Türk Çarşıları [Turkish Bazaars]*. Ankara: Tepe Yayınları.

Said, Edward W. (1994) *Culture and Imperialism*. London: Vintage.

Tekeli, İlhan (1998) Turkiye'de Cumhuriyet Döneminde Kentsel Gelişme ve Kent Planlaması [Urban Development and Planning in the Republican period in Turkey], in Sey, Yıldız (ed.) *Tarih Vakfı Bilanço '98: 75 Yılda Değişen Kent ve Mimarlık* [Changing City and Architecture in 75 Years: A reassessment by Tarih Vakfı in 1998]. Istanbul: Tarih Vakfı.

Velioğlu, Selim (1990) Diş Mekân Yaşantısına Bağlı Olarak Mimari Ölçekteki Fiziksel Biçimlenişi Değerlendirmeye Yönelik Bir Model [A Model for the Evaluation of Physical Form in Architectural Scale with Reference to Life in Open Spaces], Unpublished PhD Thesis, Mimar Sinan University, Istanbul.

Wilson, Elizabeth (1997) Looking backward, nostalgia and the city, in Westwood, Sallie and Williams, John (eds.) *Imagining Cities: Scripts, Signs, Memory*. London: Routledge.

Chapter 5

Practices of Neo-Ottomanism: Making Space and Place Virtuous in Istanbul

Jeremy F. Walton

Piety, Publicness, and the Aesthetics of Space and Place

How is public space defined in relation to questions of piety? What types of power and politics underpin these definitions and demarcate the possibilities of publicness in relation to religiosity? In the era of neoliberal urbanity and piety, these questions are global in a strong sense – from Falun Gong prayer sessions in Hong Kong to Hindutva organizers in Mumbai's slums, from suburban American mega-churches to the problematic minarets of Switzerland's mosques, urbanity and religion are increasingly entangled, both conceptually and politically. With this panoramic context in mind, this chapter examines how civil Islamic institutions in contemporary Istanbul can work to promote a novel, pious aesthetic of space and place in relation to secular ideologies of the city and spaces within it. For the purposes of this exploration, 'piety' identifies the nexus of practices, ideals, and sensibilities that constitute the discursive tradition of Islam (Asad, 1986); following Jürgen Habermas (1989) and his interpreters, I invoke the term 'publicness' to describe the abstract modes of sociality and discourse that define modern political subjectivity. My broad conceptual aim is to query how practices and categories of space and place form and inform the relationship between piety and publicness.

Explicit political contests over the nature and culture of urban space in Istanbul have become familiar to Turks of all backgrounds in recent years. The controversy in the mid-1990s over whether to build a mosque in Taksim Square, Istanbul's definitive secular space associated with both the heritage of the early Turkish Republic and the high culture of European modernity, was a prominent opening salvo in this debate (the mosque ultimately came to naught; see Çınar, 2005, pp. 110 *et seq.*). More recently, a group of young men affiliated with an Islamist-nationalist 'lodge' (*ocak*) performed prayer (*namaz*) in Istanbul's pre-eminent tourist destination, the Hagia Sophia Museum, in order to protest about

Pope Benedict XVI's visit to Turkey in November 2006, thereby interrogating both the state's definition of the museum as a site of secular heritage and the lingering association of Hagia Sophia with Christianity (Soncan and Cansev, 2006). Finally, the debate over the legality of the Muslim headscarf (*başörtüsü, türban*) in state-affiliated spaces, and universities in particular, continues to provoke contests over secular and pious practices in public space (see Göle, 1996). On the whole, these obstreperous confrontations over how and where Muslim piety can and should emerge in urban public space are part of a broader, neoliberal reorientation of the relationship between religion and secularism in Turkey in the wake of the economic and political reforms of the 1980s (Öniş, 2004).

In this chapter, I marshal a rather different perspective on the relationship among Islamic piety, secularism and public space in contemporary Turkey. Without dismissing the importance of explicit, politicized contests over urbanity, I focus on pious practices and ideologies of space and place that are more quietist, less confrontational, and, in certain respects, more accommodating to secular definitions of publicness. These practices and ideologies emerge from within Istanbul's civil Islamic sphere of pious foundations; broadly, I characterize them as neo-Ottoman. While other social scientists have noted the salience of neo-Ottoman ideologies to contemporary Sunni actors, organizations and political parties in Turkey (Navaro-Yashin, 2002, p. 96; Çınar, 2005, p. 142; Özyürek, 2006, p. 156), my analysis offers an original take on how practices of and debates over spatial aesthetics within urban civil society institutions specifically articulate pious neo-Ottomanism.[1] This consideration encompasses many scales, from the global horizons of cosmopolitan Islam articulated in the luxury of hotel conference halls to more circumscribed, local criticism and praise of specific Istanbul districts and neighbourhoods, public monuments, places of business and worship, and, at the most microscopic level, particular architectural and decorative features of built structures.

In pursuit of this analysis, I invoke Henri Lefebvre's distinction between space and place (Lefebvre, 1991; see also Zieleniec, 2007, pp. 70 ff. and Cresswell, 2004), broadly understood as contexts in which social relations produce and achieve articulation through spatial relationships, and the abstracted, categorical effects of these relations. For shorthand, space might be glossed as process, place as product; in a somewhat different formulation, space is an act of representation – in Lefebvre's terminology, 'signification' (Lefebvre, 1991, p. 115) – while place is a represented object. With this distinction in mind, I explore how specific spatial practices articulated by pious foundations aspire to produce the city of Istanbul and particular sites within it as definitively neo-Ottoman places. Crucially, this focus on space and place as they relate to questions of piety and secularism offers a reorientation of recent critical literature on secularism and religion generally (Asad, 2003; Mahmood, 2005). Rather than emphasizing the manner in which liberal publicness excludes religious practices (Asad, 2003, p.

183), I draw attention to piety's creative commensurations with secularity as they come to exist in specific spaces and as particular places.

My focus on the relationship between spatiality and piety among Istanbul's urban civil society organizations also troubles the all-too-easy dichotomy of secularist and Islamist as conceived within Turkey. My ethnographic contexts are drawn from foundations affiliated with two of Turkey's prominent modernist Islamic communities (*cemaatler*),[2] the Nur Community and the Gülen Community. Although these two communities overlap, Nur organizations tend to advocate a quietist revival of individual piety inspired by the works of Said Nursi (1873–1960), while institutions associated with Fethullah Gülen (b. 1938) marshal a more cosmopolitan project of pious pedagogy and inter-religious dialogue.[3] Both these communities – and the Gülen Community in particular – remain objects of sharp public controversy in contemporary Turkey – for instance, 'Gülen-ists' (*Fethullahçılar*) along with the governing AKP (Adalet ve Kalkınma Partisi) were allegedly targeted in a secret military coup plot (*Taraf Gazetesi*, 2009), and secularist press organs often comprehend Gülen study-houses (*dershaneler, ışık evleri*) as sites of Islamic 'brain-washing' (*beyin yıkama*) (e.g. Aydoğan, 1999). Nonetheless, the emergence and organization of these groups within Turkish civil society allows them a productive distance from more overtly politicized debates over the relationship among piety, secularism, and urban space and place. Moreover, as we will see, the neo-Ottoman practices and representations that these organizations participate in many of the same modes of publicness as Turkish secularism, even as they also call into question principal secularist presuppositions and emphases.

Say '*Çarşaf!*': Kemalist Fantasies of Urbanity

In recent years, a bevy of social scientists has drawn attention to the fraught and ideologically over-determined nature of space and place within the urban texture of Istanbul. On the whole, this critical inquiry into the ideological topography of Istanbul has focused on secularist alarm, ignorance, and indignation in relation to the newly 'Islamicized' spaces of the city. Yael Navaro-Yashin (2002, p. 63), for instance, relates a telling anecdote from an ethnographic daytrip to the neighbourhood of Fatih, accompanied by a bohemian friend, Fatma who is overwhelmed by the seeming foreignness of the piety in public that characterizes Fatih: '"Are these Turks?" she first asked. "Yes, probably", I said. Other people on the street were asking the same question'. In a similar vein, Jenny White (2002), Esra Özyürek (2006, p. 133), and Alev Çınar (2005, p. 115) each recount the indignation and fear that percolated in secularist circles following the election of the Welfare Party (*Refah Partisi*) to the head of the Istanbul municipal government in 1994. More recently, Cihan Tuğal (2009) has explored how localized pious business networks in Istanbul both orient the city in relation to broader capitalist formations and call these formations into question. This interest in the intersections of political ideology, economy piety, aesthetics and built space rides upon the crest of the broader tide of interdisciplinary analysis of Istanbul's emergence as a neoliberal, global city.[4]

On the whole, recent ethnographic focus on the relationship between urban space and secularism highlights the inseparability of Kemalist politics and Kemalist aesthetics in contemporary Turkey (Bozdoğan and Kasaba, 1997). This integration of the aesthetic and the political is second nature to students of Turkey, and received ample, vivid confirmation throughout my own research. One of the more peculiar moments during my time in Istanbul involved a photo session for an article on my research and impressions of Turkey, which was eventually published in the popular weekly magazine, *Haftalık* (literally, 'Weekly'). My interview with the *Haftalık* reporter herself took place at the Ara Café, a well-known establishment located on a meticulously hip alleyway in the 'cosmopolitan' neighbourhood of Beyoğlu.

While the young woman interviewing me was evidently intrigued by my close relationships with 'fundamentalists' (*dinciler*), we also discussed other aspects of my research and my passion for the music of famous figures of Anatolian rock music from the 1960s and 1970s, such as Cem Karaca and Erkin Koray. Following our repartee, the reporter took several photographs of me sipping tea to accompany the text of the interview. I was therefore surprised when she called me several weeks later to request a second photo session with a staff photographer. Less than an hour later, I squeezed into the backseat of a Fiat hatchback, accompanied by the photographer, his assistant, and the chauffeur, and immediately began to inquire why the *Haftalık* editor had requested new images. The photographer replied mildly: 'We just wanted something more representative of your research'.[5] Weaving through the heavy, late-afternoon traffic, the Fiat sped down the hill from Taksim, past the dilapidated façades of Ottoman-era Greek and Armenian homes of Tarlabaşı and toward the Golden Horn. I soon recognized our probable destination: the neighbourhood of Eyüp, a well-known pilgrimage destination. The photographer confirmed my suspicions: 'It would be wonderful if we could get a photo of you beside a covered woman'. After a lengthy, convoluted ballet in front of the entrance to the Tomb of Eyüp Sultan, the photographer managed to capture his desired image: the tall, foreign researcher smiling awkwardly beside a justifiably suspicious, devout woman wearing a *çarşaf* or black sheet-like over-garment – one of the principal icons of 'fundamentalism' (*aşırı dincilik, şeriatçılık*) that is such an anathema to the secular Turkish gaze (Çınar 2005). While I had already argued with him against such a photograph, noting that my research was neither about so-called fundamentalism nor gender, he pleaded editorial impotence – the specifics of our absurd photo shoot had been dictated from above by a senior *Haftalık* editor, and, besides, as he noted with a matter of fact shrug, 'This sort of thing is what readers expect from an article about religion'[6] (figure 5.1).

As this anecdote rather uncomfortably demonstrates, my own experience of

HABER YAŞAM

Okyanusu aşıp buralara kadar gelen Amerikalı Jeremy Walton "İslami Tarikatlar" konulu doktora tezini hazırlıyor. Türkçe'yi anadili gibi konuşuyor. Fethullahçılar, Nurcular, Süleymancılar, Aleviler ve solcular üzerine araştırma yapan Walton ile Türkiye'yi konuştuk.

Bu Amerikalının tez konusu İslami tarikatlar

FETHULLAHÇILAR, SÜLEYMANCILAR, ALEVİLER... İSTANBUL SOKAKLARINDA DOLAŞAN AMERİKALI JEREMY GENÇ BU KONULARI ARAŞTIRIYOR

Chicago Üniversite-si'nde antropoloji doktorası yapar-ken bir anda Türkiye'de "İslami Tarikatlar" konusunda araştırma-ya yapmaya karar vermiş bir Amerikalı, Jeremy Walton. İlk olarak 2003 yılında Türkiye'ye kı-sa süreliğine gelen Walton, bura-dan o kadar çok hoşlanmış ki, kendi deyimiyle tam bir "Türkiye aşkı" doğmuş içinde. Sonra Ame-rika'ya geri dönmüş ve Şubat 2005'te tekrar Türkiye'ye gelene dek büyük bir hasret çekmiş. Şimdi bir buçuk senedir Türki-ye'de ve çalışmalarını sürdürü-yor. Bu sırada Jeremy, sular sel-ler gibi Türkçe konuşuyor ve "in-şallah" kelimesini cümlelerinden hiç eksik etmiyor. Türkiye'den Amerika'ya döndüğü dönemde radyoda Türkçe pop müzik prog-ramı yapan Jeremy'nin en sevdiği gruplar arasında Hardal, Ma... Işıklar ve Replikas var.

Türkiye'ye ne zaman geldiniz?

İlk olarak 2003'te geldim. ...

34 2006/168 HAFTALIK

Figure 5.1. Photograph of Jeremy F. Walton in front of the Tomb of Eyüp Sultan printed above his interview with *Haftalık*. (*Source*: Harmancı, Destan (2006) Bu Amerikalı'nın tez konusu İslami Tarikatlar. *Haftalık*, 168, 23–29 June, pp. 34–36. Courtesy of *Haftalık*)

Istanbul was marked by the same secular anxieties, fantasies and topographies of urban space that other ethnographers of the city have identified. I too found Istanbul to consist of a warp and woof of neighbourhoods and districts, each ideologically marked by the binary contrast between the secular and the pious. However, I also worry that excessive attention has been devoted to the secular anxieties, fantasies, and repugnance that are both causes and effects of this geography. In contrast, I focus on the various ways in which civil Islamic actors criticize, eulogize, and fantasize the city of Istanbul and many of its particular spaces.

Together, the pious practices of space and place that I examine articulate a coherent chronotope (Bakhtin, 1981) – a specific narrative ideology and practice of integrated space and time – of Istanbul as a devout, neo-Ottoman city. This chronotope is marked by a nostalgic longing for an idealized past in which piety and urban existence were seamlessly, unproblematically yoked together. In certain respects, this modernist nostalgia (Boym, 2001) resembles that of Orhan Pamuk's novels and memoirs (Pamuk, 2003; Engin F. Işin, Chapter 2), although Pamuk's object of longing is the Istanbul of the 1950s rather than that of the Ottoman era. Unlike Pamuk's Istanbul, however, the neo-Ottoman chronotope of the city characteristic of Turkish civil Islam is also profoundly optimistic – as I will show, it imagines and attempts to produce a reintegrated, pious urban space and place that is also enveloped within a broader chronotope of global civil Islam and the universal community of believers, the *ummah* (*ümmet*).

Neo-Ottoman Authenticity and the Spatial Aesthetics of Nostalgic Devotion

The offices of the Istanbul Science and Culture Foundation (*İstanbul İlim ve Kültür Vakfı*, İİKV) are located in a resplendently restored sixteenth-century *medrese*, or Qur'anic school, designed by Mimar Sinan, the Ottoman Empire's pre-eminent celebrity architect, and tucked away among the more prominent tourist attractions and municipal government offices in Sultanahmet, the so-called historical peninsula that is the crown jewel of both Istanbul's and Turkey's historical pride (as well as the object of endless advertising campaigns) (figure 5.2). Over the course of our numerous conversations, Faris Bey, the director of the İİKV, emphasized the advantages of the organization's new home – the Foundation changed locations in early 2007 – at the Rüstempaşa Medrese in explicitly aesthetic terms. Beyond the practical advantages of the new location – centrality and more space – Faris Bey praised the aesthetics and historicity, or better yet, the aesthetic historicity, of the *medrese* as consonant with and

Figure 5.2. Rüstempaşa Medrese. (*Photo*: Jeremy F. Walton)

supportive of the Foundation's aspirations and activities – primarily publishing and organizing support for the writings and moral philosophy of Said Nursi.

The structure of the Rüstempaşa Medrese does of course date from the Ottoman era. However, the neo-Ottoman aesthetics that the members of the İİKV both perceive and practice within the *medrese* do not derive simply from this fact. Rather, the neo-Ottoman aesthetics that I describe in this section produces the *medrese* as a space in Lefebvre's sense – here, spatial qualities come to embody and reorient social relations. In particular, these neo-Ottoman aesthetics help to produce the *medrese* and Foundation as a space that is both public and pious. There is an ambiguity to this achievement. As a public space, the İİKV inevitably participates in broader, secular modalities of publicness. For example, Esra Özyürek (2006) has persuasively argued that nostalgia is one definitive characteristic of the Kemalist public sphere in contemporary Turkey; in the case of the İİKV and other urban pious foundations, nostalgia for the Ottoman replaces 'nostalgia for the modern' (*Ibid.*) as a characteristic principle of public space. We might say that secular and pious publicness share one common historicity, nostalgia, even as the object of this nostalgia differs. In a schematic sense, then, neo-Ottomanism is a negotiation between broader public discourses and pious sensibilities.

My first arrival at the Rüstempaşa Medrese was a festive affair: the Foundation had invited a swath of luminaries and supporters from throughout Istanbul Nur circles to a celebratory meal and inaugural *Risale-i Nur* class at the new location. Coincidentally, I had returned from a visit to Ankara that very day, and proceeded from the bus station to the İİKV without dropping by my apartment beforehand. As any seasoned coach traveller in Turkey well knows, long routes on buses that serve ample tea and Nescafé without offering the convenience of an onboard water closet can create quite a strain on one's bladder; by the time I reached the *medrese*, I was aching to go to the bathroom, and immediately asked Fatih, the young foundation doorman (*kapıcı*) to direct me to the facilities. Fatih paused, and blushed noticeably: 'I don't think you'll like them very much'. I was in no mood to argue over niceties, however, and insisted that he show me the restrooms immediately. As I relieved myself, I understood the cause for Fatih's embarrassment: the toilets at the *medrese* were original, sixteenth-century installations. Unlike 'European' style toilets, with their seats and water basins, these 'Ala Turka' toilets were mere sunken cavities in carved marble slabs, cleverly-sloped to encourage run-off. When I returned to Fatih's office, I assured him that I had had ample experience with 'Ala Turka' toilets, and that he had no cause to worry. He began to beam in subtle pride: 'You know, everything here is original, even the toilets. Mimar Sinan designed the system himself, it is one of the first functional plumbing systems in the world. Even though European toilets, like the one in the old offices, are more convenient, we prefer these original, Ottoman ones'.

Fatih's proud, if highly coloured, contrast between the toilets at the old İİKV offices and those in the Rüstempaşa Medrese was representative of the general opinions of class members and Foundation functionaries concerning their new location. The offices of the Foundation had been located on the second floor of a rather anonymous apartment building in the area of Balmumcu, one of the many flourishing white-collar districts that dot the European side of Istanbul. While class participants were content with the old offices, they were generally lukewarm in their assessments, using adjectives such as 'anonymous', 'sufficient', and 'good enough' (*anonim*, *yeter*). There was an even greater degree of ambivalence, occasionally approaching disdain, for Balmumcu, the Foundation's previous neighbourhood. Balmumcu, a district within the municipality of Beşiktaş, is decidedly bourgeois and corporate – its centre, straddling the broad Barbaros Boulevard, is home to a variety of upscale hotels, financial institutions, high-rise offices, and luxury goods outlets, and the neighbourhood is only a brief bus or taxi ride away from Taksim, Istanbul's premiere nightlife district. Most of my friends at the İİKV were unenthusiastic about the previous offices' location within Istanbul's socio-cultural topography; they regularly criticized other residents of the neighbourhood as 'excessively secular', 'irreligious' and 'immoral' (*fazla laiktir*, *dinsiz*, *ahlaksızlar*).

In contrast to this ambivalence and scepticism over the Foundation's earlier location, nearly everyone associated with the İİKV voiced unanimous enthusiasm over the move to Rüstempaşa Medrese. Class members unequivocally preferred the location of the *medrese*, in the district of Sultanahmet, which remains one of the districts of Istanbul most strongly associated with the Ottoman era (see also Baykan *et al.*, Chapter 4, this volume). The *medrese* itself was the object of even stronger adulation. Foundation members consistently praised its unique, octagonal floor plan, with smaller cubicles and larger conference rooms surrounding a garden equipped with a fountain at its centre, emphasizing the 'peaceful' and 'spiritual' atmosphere of the *medrese* (*huzurlu*, *ruhsal*, *ruhani*, *manevi*). Prior to the İİKV's occupation, the *medrese* had been left derelict for over fifteen years and therefore required extensive renovation. When I queried him concerning the ongoing renovations, Faris Bey emphasized that the Foundation had consulted architectural historians in order to ensure that all alterations would be in keeping with Mimar Sinan's original plan. I encountered this premium on 'originality' and 'authenticity' constantly in conversations about the new Foundation headquarters, in particular in reference to the fountain at the centre of the *medrese* garden and the largest of the conference rooms, where *Risale-i Nur* classes were thenceforth held. The imperative of authenticity, as Charles Taylor (1994) has argued, is a cardinal aesthetic principle of the liberal politics of recognition more generally; in this respect, authenticity, like nostalgia, is a site of commensuration between piety and publicness. Moreover, as Baykan *et al.* (this volume) demonstrate, an emphasis on authenticity orients the politics of built space and questions of heritage throughout Istanbul, and on the 'historical peninsula' especially.

These positive statements concerning the Ottoman authenticity of the Rüstempaşa Medrese should be understood in conjunction with a less explicit aesthetics of absence. In recent years, Esra Özyürek (2006) and Yael Navaro-Yashin (2002) have traced both the proliferation of Kemalist iconography (especially representations of Atatürk) in Turkish public space and the 'privatization' of secularist imagery that was previously associated solely with the state (see also Hart, 1999). Suffice to say that the iconography of Kemalist secularism remains a crucial means of invoking one hegemonic mode of publicness in contemporary Turkey. While this is especially true of such spaces as government buildings, universities, and city squares, it is also noticeable in less institutionalized venues of publicness, spaces where the very distinction between public and private is a matter of debate and conscious practice. It is therefore remarkable that there is no trace whatsoever of Republican history and Kemalist affiliation anywhere in the grounds of the İİKV.

At the risk of conflating two different historical dispensations, one might accurately characterize the decoration of Rüstempaşa Medrese and the offices of the İİKV as Spartan. With the exception of a few framed calligraphic representations of the names of God, written in Arabic and printed or painted on fine sheets of marbled paper, the walls of the *medrese* are entirely bare. The calligraphic prints are crucial to the creation of a nostalgic, pious aesthetic. As my friends at the İİKV often noted, both calligraphy and the practice of marbling paper through a process of water colour painting (*ebru sanatı*) are 'traditional Ottoman arts' (*geleneksel Osmanlı sanatları*). Other than these few works of devotional art, both the interior and exterior of the Foundation are austere, lacking entirely in ostentation. As I noted above, the restoration to the marble *medrese* building have attempted to recreate 'authenticity': concrete reinforcements to the structure's columns and ogees have been made as unobtrusively as possible, the walls have been painted in solid rose, pink and ivory tones, and even the electric light fixtures are disguised as tulip-shaped candle holders. Each of these aesthetic details functions to create an atmosphere that is understood by the Foundation members themselves as at once Ottoman and pious.

While the pious aesthetics of space articulated by the İİKV are particularly rich and illustrative, they are by no means unique. Indeed, I could have marshalled a similar analysis based on any number of other Sunni foundations in Istanbul. Many of the city's Islamic foundations and associations are housed in Ottoman structures, a fact that organization members consistently comment upon positively. On a broader plane, the preponderance of Sunni civil society organizations in Istanbul are located in the city's older 'Ottoman' neighbourhoods, particularly those on the 'Historical Peninsula' and adjacent to the Golden Horn: Sultanahmet, Fatih, Eyüp, and, especially, Süleymaniye.

How, then, do the politics articulated by the aesthetics of pious space relate to secularist-Kemalist ideologies of space in Turkey? As I noted above, a distinct aesthetics of absence characterizes the spaces of civil Islam in Istanbul: the panoptic gaze of Mustafa Kemal, so prominent in most public spaces in Turkey, does not penetrate into the offices of the İİKV or other Islamic foundations. Put another way, the official insignia of the Turkish state, and representations of Atatürk in particular, construct a hegemonic relationship between secularity and public space in Turkey. The aesthetic and political effect of their absence in the spaces of Islamic foundations is therefore extraordinary: contexts such as the İİKV untangle and interrogate the hegemonic relationship between public space and secularity. There is a definite politics here, but it is not one of explicit confrontation or contestation. Although members of the İİKV and other, like-minded foundations do not mince words in their criticisms of the 'secular' spaces of Istanbul, they prefer disengagement from secular space through the construction of alternative pious spaces and, to echo Michael Warner (2002), pious counter-publics (see also Hirschkind, 2006). Crucially, this commitment to an alternative pious aesthetics of space, with its quietist politics, allows organizations such as the İİKV to remain firmly within the parameters of civility and the Turkish public sphere, even as they decouple publicness from the assumptions and imperatives of Kemalist secularism. Institutions like the İİKV achieve this decoupling through signifying practices of *space* – the renovation of the Rüstempaşa Medrese with an emphasis on the virtue of authenticity is a case in point – that produce coherent *places* which pious actors understand and experience as Ottoman. In the following section, I explore contexts and processes that endeavour more expansively to produce the city of Istanbul as a whole as a pious, neo-Ottoman place. I focus on the manner in which pious civil society organizations produce and practice Istanbul as a particular type of place, in Lefebvre's sense.

Sunni Cosmopolitanism and the Recentring of Istanbul, Seat of Neo-Ottoman Glory

The vast conference room of the luxurious Ramada Kaya Hotel, located in the far western expanses of European Istanbul, was abuzz with anticipation following the intermission for the evening call to prayer. Those of us in the audience had already listened to several hours of welcoming speeches by clerics and representatives from Indonesia, Mauritania, Qatar, Palestine, China, Indonesia, and even a Muslim convert from Germany, as well as the requisite host of Turkish dignitaries. But the main event, the keynote speaker, the immeasurably famous Egyptian-Qatari scholar Yusuf al-Qaradawi, was yet to come.

All told, the agenda of the speeches at the Second General Assembly Meeting for the International Union of Muslim Scholars was rather repetitive and uncontroversial. Al-Qaradawi's keynote address underscored the Assembly's major objects of complaint and calls for intervention: appeals for collective action in support of Palestinians, Iraqis, Pakistanis left homeless by the earthquake of November 2005 and Indonesians affected by the Boxing Day tsunami of 2004 and the more recent Jogjakarta quake of May 2006, and condemnations of aggressive

Istanbul inspires pride in most of its residents, regardless of their orientation and attitudes to questions of Islam and secularism, but the particular, pious imaginary of the city I have outlined here extends beyond mere civic esteem. For the devout Muslims of Turkey's Islamic civil sphere, Istanbul is by definition a religious place; for this reason the Ottoman history of the city maintains a privileged, evidential status in pious imaginaries of the city. This piety of place qualifies it as a possible centre for an Islamic cosmopolitanism of global reach, and demands proud demonstration to the sympathetic visitor, especially if he or she is a Muslim.

Importantly, Islamic cosmopolitanism, with Istanbul as its political and pious epicentre, distances itself from the tumult of contemporary Turkish partisan politics. By aspiring to a separation from the imperatives and intrusions of state institutions and ideologies, civil Turkish Islam has forwarded a tentative reconciliation with the political and aesthetic imperatives of Kemalism and Turkish nationalism. The chronotope of Istanbul as a pious, cosmopolitan centre that I have delineated in this section is an instructive case in point. Pious cosmopolitanism represents a novel achievement on the part of civil Turkish Islam precisely because it is neither nationalist nor anti-nationalist. The very notion of a category of belonging that resists this dichotomy expresses a profoundly new horizon of possibility for Turkish Islam. Like the pious aesthetics of space discussed above, Sunni cosmopolitanism in Turkey succeeds in coexisting with Kemalist and étatist modalities of space and place, even as it also troubles their presuppositions. For Sunni cosmopolitan organizations such as the GYV, both specific sites within the city – Mimar Sinan's mosques, the Ottoman palaces – are uniquely signifying: these spaces produce Istanbul as a place that is defined foremost by its Ottoman heritage, and, therefore, easily integrated into a broader map of the *ummah* at large. Equally important, this imagination of Istanbul as a neo-Ottoman place obscures the spaces that represent the city in contrasting ways – especially Kemalist – without confronting these representations overtly.

Achievements and Limitations of the Neo-Ottoman Chronotope of Istanbul

Pious urban citizenship, rooted in a deep nostalgia for the Ottoman era, the aesthetics that this era inspires, and a contemporary Islamic cosmopolitanism, is an aspiration and cherished ideal for a great many Istanbulites. As I have endeavoured to show through a series of ethnographic examples, this ideal achieves coherence and legitimacy within the types of sociality and organization that the institutions of Turkish civil society encourage. In a sense, these institutions, their discourses, and constitutive modes of publicness seek to organize and fix the definition of the city of Istanbul as Ottoman. This pious chronotope of neo-Ottoman, cosmopolitan Istanbul, like any chronotope, is inhabitable, but only up to a point – in particular, as a formation that aspires to publicness, it necessarily tolerates alternative chronotopes and imaginaries of space and place.

Beyond the play of multiple chronotopes, however, the intractable heterogeneity of urban existence also gives pause to the all-encompassing aspiration of neo-Ottomanism as a public ideology of place and space. To adopt a motif from Michel de Certeau (1984, p. 93), Istanbul consists of 'networks of … moving, intersecting writings [that] compose a manifold story that has neither author nor spectator, shaped out of fragments of trajectories and alterations of spaces: in relation to representations, it remains daily and indefinitely other'. No chronotope, neo-Ottoman or otherwise, can contain or predict all these trajectories. In Lefebvre's terms, the practices of space defy seamless translation into homogeneities of place, even as aspirations to homogeneous place organize particular spaces. And as an ethnographer of and in the city, I have difficulty imagining otherwise. For my paths through Istanbul, too, were diverse, disparate, occasionally uncharted. They often begged contradiction and inspired challenge. Above all, they demanded critical respect for the constant struggles that define the city, the shuttling between singularity and multiplicity that is, in the end, a suitable definition of both urban existence and the ethnographic endeavour within it.

Notes

1. Even as I prepared the final draft of this chapter, *The New York Times* published an essay on the political resurgence of Ottomanism in contemporary Turkey (Bilefsky, 2009). Although I am wary of the article's instrumental reading of neo-Ottomanism as a mere response to the EU's resistance to Turkey's membership application, the relationship between neo-Ottomanism and broader questions of Turkey's geopolitical 'place' certainly deserves attention.

2. In opting to use the term community (*cemaat*) rather than 'movement' (*hareket*), I follow my interlocutors' own preferences – as they often pointed out to me, the designation 'movement' implies a problematic political rationality that they claim not to espouse.

3. Nursi was arguably the most important Muslim thinker of late Ottoman and Turkish Republican history; since his death, his collected writings, the *Risale-i Nur* have animated one of the most vibrant and visible 'new Islamic movements' in contemporary Turkey (see Yavuz, 2003). Gülen and his enthusiasts, for their part, are an offshoot of the Nur Community. By and large, they foreground pious probity as cultural capital and the means to a global network of businesses, schools, and non-governmental organizations (see Yavuz, 2003; Turam, 2007). Fethullah Gülen himself has resided in Pennsylvania since the 1990s, a potent indication of the persistent public ambivalence over his influence in Turkey.

4. Among the principal works that have framed and forwarded the analysis of Istanbul as a 'global city' are Öncü, 1999 and Keyder, 1999, 2008.

5. All quotations from conversation are translated from the Turkish by the author.

6. The actual article, titled 'This American's thesis topic: Islamic sects' ('Bu Amerikalı'nın tez konusu İslami Tarikatlar') appeared in the 23–29 June issue of *Haftalık* (Harmancı, 2006).

References

Asad, Talal (1986) The Idea of an Anthropology of Islam. Occasional Paper Series, Center for Contemporary Arab Studies, Georgetown University, Washington DC.

Asad, Talal (2003) *Formations of the Secular. Christianity, Islam, Modernity.* Stanford, CA: Stanford University Press.

Aydoğan, Osman (1999) Gülen'in Askerleri [Gülen's Soldiers]. Available at http://www.radikal.com.tr/1999/06/23/turkiye/01gul.html. Accessed 23 June 2009.

Bakhtin, Mikhail (1981) *The Dialogic Imagination. Four Essays* (translated by Caryl Emerson and Michael Holquist). Austin, TX: University of Texas Press.

Bilefsky, Dan (2009) Frustrated with West, Turks revel in empire lost. *New York Times*, 5 December. Available at: http://www.nytimes.com/2009/12/05/world/europe/05turkey.html?_r=1&emc=eta1. Accessed 15 December 2009.

Boym, Svetlana (2001) *The Future of Nostalgia*. New York: Basic Books.

Bozdoğan, Sibel and Kasaba, Reşat (eds.) (1997) *Rethinking Modernity and National Identity in Turkey*. Seattle, WA: University of Washington Press.

Çınar, Alev (2005) *Modernity, Islam, and Secularism in Turkey. Bodies, Places, and Time*. Minneapolis, MN: University of Minnesota Press.

Cresswell, Tim (2004) *Space: A Short Introduction*. Oxford: Blackwell.

De Certeau, Michel (1984) *The Practice of Everyday Life* (translated by Steven Rendall). Berkeley, CA: University of California Press.

Göle, Nilüfer (1996) *Forbidden Modern. Civilization and Veiling*. Ann Arbor, MI: University of Michigan Press.

Habermas, Jürgen (1989) *The Structural Transformation of the Public Sphere. An Inquiry into a Category of Bourgeois Society* (translated by Thomas Burger). Cambridge, MA: MIT Press.

Harmancı, Destan (2006) Bu Amerikalı'nın tez konusu İslami Tarikatlar [This American's thesis topic: Islamic sects]. *Haftalık*, **168**, 23–29 June, pp. 34–36.

Hart, Kimberly (1999) Images and aftermaths: the use and contextualization of Atatürk imagery in political debates in Turkey. *Political Anthropology*, **22**(1), pp. 66–84.

Hirschkind, Charles (2006) *The Ethical Soundscape: Cassette Sermons and Islamic Counterpublics*. New York: Columbia University Press.

Keyder, Çağlar (ed.) (1999) *Istanbul: Between the Global and the Local*. Lanham, MD: Rowman and Littlefield.

Keyder, Çağlar (2008) A Brief History of Modern Istanbul, in Kasaba, Reşat, (ed.) *Turkey and the Modern World*, Vol. 4. *The Cambridge History of Turkey*. Cambridge: Cambridge University Press.

Lefebvre, Henri (1991) *The Production of Space*. Oxford: Blackwell.

Mahmood, Saba (2005) *Politics of Piety: The Islamic Revival and the Feminist Subject*. Princeton, NJ: Princeton University Press.

Navaro-Yashin, Yael (2002) *Faces of the State: Secularism and Public Life in Turkey*. Princeton, NJ: Princeton University Press.

Öncü, Ayşe (1999) Istanbulites and others: the cultural cosmology of being middle class in the era of globalism, in Keyder, Çağlar (ed.) *Istanbul: Between the Global and the Local*. Lanham, MD: Rowman and Littlefield.

Öniş, Ziya (2004) Turgut Özal and his economic legacy: Turkish neoliberalism in critical perspective. *Journal of Middle Eastern Studies*, **40** (4), pp. 113–134.

Özyürek, Esra (2006) *Nostalgia for the Modern: State Secularism and Everyday Politics in Turkey*. Durham, NC: Duke University Press.

Pamuk, Orhan (2003) *İstanbul. Hatıralar ve Şehir*. Istanbul: Yapı Kredi Yayınları.

Ramadan, Tariq (2009) My compatriots' vote to ban minarets is fuelled by fear. *The Guardian*, 29 November. Available at http://www.guardian.co.uk/commentisfree/belief/2009/nov/29/swiss-vote-ban-minarets-fear. Accessed 15 December 2009.

Roy, Olivier (2004) *Globalized Islam: The Search for a New Ummah*. New York: Columbia University Press.

Soncan, Emre and Cansev, Ali (23 November 2006) Papa'yı Protesto Etmek için Ayasofya'da Namaz Kıldılar [They prayed in Hagia Sophia in protest of the Pope]. *Zaman Gazetesi*, 23 November, p. 1A. Available at: http://zaman.com.tr/haber.do?haberno=459445. Accessed 15 December 2009.

Taraf Gazetesi (2009) AKP ve Gülen'i Bitirme Planı [The plan to end JDP and Gülen]. 12 June, front page.

Taylor, Charles (1994) The politics of recognition, in Gutmann, Amy (ed.) *Multiculturalism. Examining the Politics of Recognition*. Princeton, NJ: Princeton University Press, pp. 25–73.

Tuğal, Cihan (2009) *Passive Revolution: Absorbing the Islamic Challenge to Capitalism*. Stanford, CA: Stanford University Press.

Turam, Berna (2007) *Between Islam and the State: The Politics of Engagement*. Stanford, CA: Stanford University Press.

Warner, Michael (2002) *Publics and Counterpublics*. New York: Zone Books.

White, Jenny (2002) *Islamist Mobilization in Turkey: A Study in Vernacular Politics*. Seattle, WA: University of Washington Press.

Yavuz, Hakan (2003) *Islamic Political Identity in Contemporary Turkey*. Oxford: Oxford University Press.

Zieleniec, Andrzej (2007) *Space and Social Theory*. Los Angeles: Sage.

Acknowledgements

I would like to thank Amahl Bishara, Deniz Göktürk, Kimberly Hart, Ulrika Martensson, Brian Silverstein, Noah Salomon, Levent Soysal, Kabir Tambar and İpek Türeli for their invaluable contributions to this essay

Chapter 6

Modelling Citizenship in Turkey's Miniature Park

İpek Türeli

The Golden Horn (Haliç) suffered over a long period from industrial pollution. Its rehabilitation became an issue as early as the 1960s. In the second half of the 1980s, the government carried out a 'cleaning' operation which also involved major and swift demolition of properties crowding its shoreline. Green parks were created on reclaimed land along the shores but they remained desolate because of the lack of activity. Then the government started encouraging cultural investment around the Golden Horn. Miniaturk, Turkey's first nation-themed park of miniature models, which opened in 2003, is part of the latter phase of this urban regeneration campaign which focuses on culture-led revitalization (figure 6.1). As its name makes plain, Miniaturk presents a 'Turkey in miniature'. Its main outdoor display area features 1/25-scaled models of architectural showpieces chosen for their significance in the city's and Turkey's history. As a site of architectural miniatures, Miniaturk provides an escape from the experience of the everyday. But it must also be understood in dialectic relation to gigantic new sites of global capital as well as to gigantic older sites of nation building. This chapter seeks to interpret why a miniature Turkey appeared in 2003, and why it has been received with enthusiasm across the political spectrum.[1]

Similar miniature theme parks abound around the world and reveal much about the contexts in which they are situated. Private enterprises like Disneyland have become hallmarks of national experience. Other, state-controlled examples may attempt more explicitly to reconfigure the relationship between a nation and its citizens. Ultimately, the tradition of open-air cultural parks can be traced to the international exhibitions which provided a venue for expressions of national identity in the nineteenth century (Bennett, 1995; Çelik, 1992; Kaufmann, 1989).

Nation-themed miniature parks – such as Miniaturk, Madurodam in The Hague, Beautiful Indonesia Miniature Park (Taman Mini Indonesia Indah) in Jakarta, or Splendid China (Jǐnxiù Zhōnghuá) in Shenzhen – show what their producers (increasingly state enterprises) think their country ought to be rather than how it is (Anagnost, 1997). They seek to represent the nation-state, invite

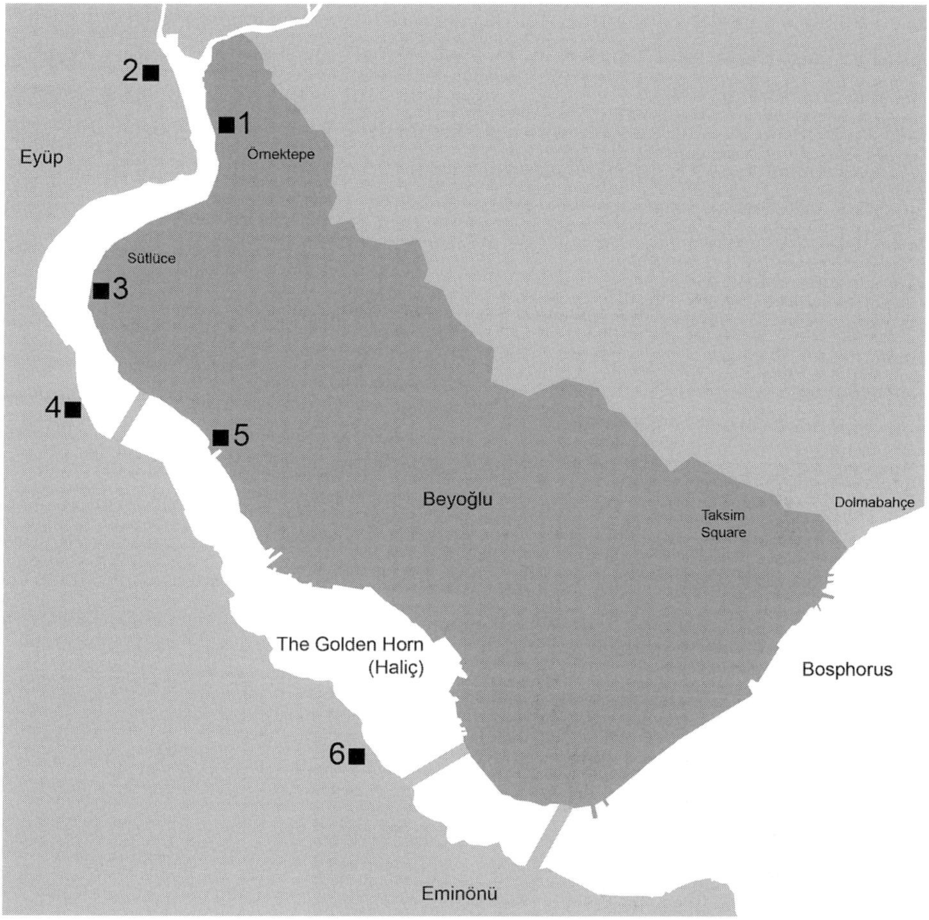

Figure 6.1. Relatively recent institutions of culture located on the banks of the Golden Horn include: Miniaturk in the Örnektepe neighbourhood of the Beyoğlu Municipality (1); Bilgi University's campus on the site of the city's first electric power plant, Silahtarağa (2); Sütlüce (or Haliç) Congress Centre, on the site of the city's former slaughter house (3); Feshane International Fair Congress and Culture Center adopted from the imperial *fez* (hat) factory (4); Rahmi Koç Museum which focuses on transport, industry and communications, adapted from Lengerhane, the Ottoman Navy anchor foundry, and the Hasköy dockyard (5); and Kadir Has University, adapted from the state-run tobacco depot and cigarette factory (6). (*Source*: adapted from aerial photography provided to the general public by Istanbul Municipality, available at: http://sehirrehberi.ibb.gov.tr and from neighbourhood maps provided by Beyoğlu Municipality on its website at: http://www.beyoglu.bel.tr/beyoglu/taniyalim.aspx?SectionId=78)

citizens to believe in the benevolence of the state, an ideological effect similar to that of monumental capitol complexes, especially in the modern capital cities of nascent nation-states. Cities such as Ankara and certain monumental building complexes within them aim to display and legitimize the authority of nation-states. The experiencing of their gigantic scale socializes citizens to internalize the projected nationalist narrative (Vale, 1992). National miniature theme parks, on the other hand, adopt a different palette of spatial techniques of simulation and miniaturization.

The effects of the miniature and the gigantic are dissimilar. One of the earlier modern-day thinkers to dwell on the miniature was Edmund Burke. Writing in the mid-eighteenth century, he associated the sublime with great objects and pain-induced admiration, and the beautiful with small objects of love and pleasure (Burke, 1757, 2008). If monumental projects of nation building were of the order of the sublime, in the current neoliberal era, it is the prestige projects which aim to bring in international investment which cause simultaneous consternation and awe.

To give a striking contemporary example, the publicity campaign for 'Dubai Towers' – twin towers with twisting torsos – proposed by Dubai Properties for Istanbul's Central Business District in 2005, created animated reaction and public debate partially aided by its very own, full-page advertisements of images of twisting everyday objects ranging from computer keyboards to coffee cups (figure 6.2). Designed to promote the 'innovative' form of the proposed towers, these blown-up images of small and painfully twisted objects inadvertently and ironically exacerbated public consternation. As a result of sustained opposition to the Municipality's give-away of state-owned land, and lawsuits against the deal,

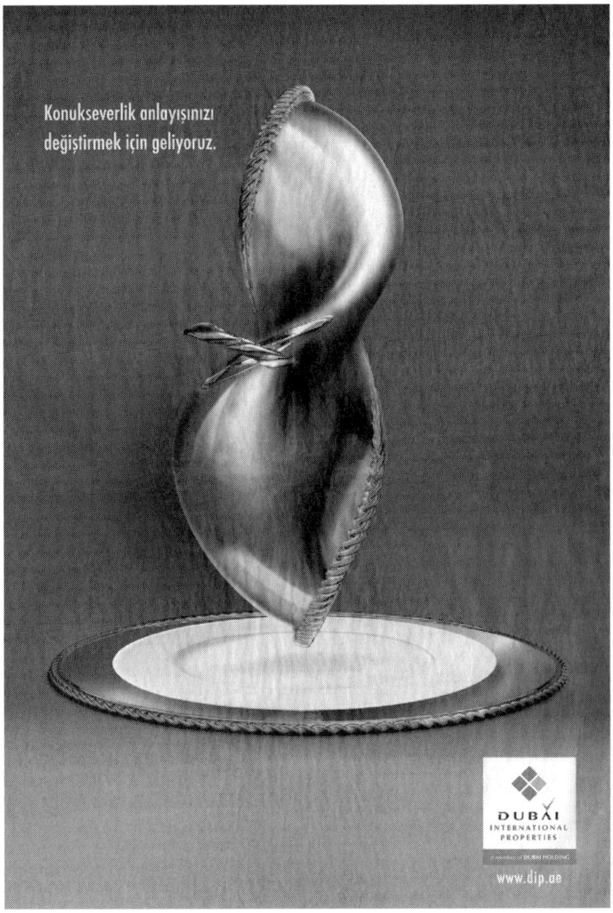

Figure 6.2. Full-page newspaper advertisement by Dubai Properties for its 'Dubai Towers'. *Milliyet*. 24 November 2005. This twisted coffee cup was intended to demonstrate the 'innovative' form of the twin-tower complex proposed by Dubai Properties for Istanbul's Central Business District between Levent and Maslak along the Büyükdere Asphalt. The caption reads 'We are coming to change your understanding of hospitality' [Konukseverlik anlayışınızı değiştirmek için geliyoruz]. The massive advertising campaign backfired.

the investor withdrew from the project. I propose to consider prestige projects of neoliberal urbanism, especially of the Dubai Towers kind, where publicly-owned property is marketed and sold by the government to international capital without any concern for transparency, public participation or debate, and monumental projects of nation building as phases of the 'gigantic', and Miniaturk as their counterpart.

A national miniature themed park is not an 'easy' miniature to interpret. Susan Stewart's discussion of the miniature and the gigantic is helpful in delineating the subjective affects of the two (Stewart, 1984). According to Stewart, the miniature is usually associated with the private collection, while the gigantic with the public. The latter symbolizes authority – i.e., the state, masculinity, and exteriority – and is experienced partially while on the move. The miniature, on the other hand, presents popularity, femininity, and interiority, and is experienced as a transcendental space frozen in time. Miniatures such as miniature railways are appealing partly because their locus is nostalgia for childhood, and (pre-industrial) history (and artisanal labour rather than alienated labour). Such a neat dichotomy, as in gigantic versus miniature, is difficult to apply to the case of miniature theme parks.

Miniaturk and similar nation-themed miniature parks are designed to be experienced while on the move and visitors are immersed in their environments rather than having a privileged and dominant view from above which is typical of the experience of a miniature railway or dollhouse (Errington, 1998). Stewart (1984, p. 64) is, however, helpful in thinking about the miniature theme park because, as she suggests, 'the interiority of the enclosed world tends to reify the interiority of the viewer'. The miniature park appeals to citizens by helping them to imagine their nation in its entirety as an uncontaminated and perfect 'island'.

Since its opening in 2003, Miniaturk's success has been enormous. It has found a place in the city's popular landscape, and Istanbul's guided tours and printed guides now incorporate it next to well-known historical sites. As the press gave it keen and enduring support, the total number of visitors rose to more than two million by the end of its first two years. Following the success of Istanbul, the southern tourist city of Antalya built its own 'Mini-City'.[2] In contrast to the controversy around other global-city projects, however, Miniaturk has not triggered any visible opposition.[3] It was built on publicly-owned land by the Istanbul Municipality's Kültür A.tür A.[4] Even though its construction cost was reportedly acquired from corporate sponsors, it did not appear to represent any particular private interest. Instead, it was presented as a cultural heritage site.

This chapter argues that understanding the place and significance of Miniaturk in the popular historical landscape can shed light on the public reaction to other building projects in Istanbul. The park can also be seen as demonstrating a turning point in Turkish politics, as the 'vernacular politics' (White, 2002) of Islamism moved to the centre, into party politics. Here, Islamism refers to a diversity of outlooks that collectively adhere to the notion that Islam is not private but a public matter and political. Finally, Miniaturk illuminates changing notions of citizenship and national identity in a globalizing city.

The Design of the Park

Miniaturk is located on the banks of the Golden Horn in the Örnektepe neighbourhood of the Beyoğlu district municipality. In addition to being part of a local revitalization campaign that concentrates on the Golden Horn (Bezmez, 2008), it was presented by the officials as participating in a larger process that seeks to establish Istanbul as a 'global brand' (Özdemir, 2003, p. 8). With the rise of tourism worldwide since the late 1970s, places are increasingly turning to theming to compete in a crowded field (Gottdiener, 2001). Whole countries are creating images, icons, and advertising campaigns to turn themselves into 'destination museums' (Kirshenblatt-Gimblett, 1998). Marketing requires easily identifiable signs or images; and nation-themed miniature parks are readily used to market nations. Miniaturk's agenda is to provide a marketable image of Turkey, not only to foreign tourists but also to Turkish citizens.

Miniaturk's site is a strip of infill land between the water and the motorway (İmrahor Avenue) (figure 6.3).[5] In terms of its design, Miniaturk aims to abstract itself from its surroundings. In the words of its architect, Murat Uluğ (2003), the park deliberately seeks to create a 'fairytale-like environment'. A tall fence blocks off all external views of the interior, and the entrance complex, with administrative and commercial functions, a restaurant and a shop, faces a parking lot rather than the street (figure 6.4). From here a carefully controlled entry sequence reinforces the sense of separation. First, a ramp takes visitors to a large raised terrace over a mini-botanical park. And it is only after paying the

Figure 6.3. The location of Miniaturk highlighted in its urban context. (*Source*: Adapted from aerial photography provided to the general public by Istanbul Municipality. Available at http://sehirrehberi.ibb.gov.tr. Accessed 20 December 2009)

Figure 6.4. The entrance complex faces the parking lot. It houses administrative and commercial functions including ticket booths, a restaurant, and a shop. An angular ramp takes visitors up to a raised terrace over a mini-botanical park. (*Photo*: İpek Tureli)

Figure 6.5. After paying the entry fee, visitors are treated to their first overview of the exhibition site from the entry terrace. (*Photo*: Ömer Faruk Şen)

entrance fee and passing through the entry gates that visitors arrive at a vantage point where Miniaturk is revealed to them. From this elevated location, visitors can enjoy a view of the entire park from behind a long balustrade, or take one of two symmetrical ramps down to the ground-level walkways (figure 6.5). The main exhibition area is roughly divided into two by external circular paths within which are meandering paths along which the models are distributed.

Where the ramp to the right, to the side of the motorway, meets the ground/exhibition level, the Mausoleum of Mevlana (built in 1274, Konya) is placed as the first model. This is very symbolic indeed, as the guide explains it was 'chosen to be the monument that greets the visitors in Miniaturk because of the love and tolerance we can hear in the call of Mevlana "Come, come again! Infidel, fire-worshipper, pagan/Whoever you are, how many times you have sinned, come!" This monument bears witness to the multi-cultural nature of Anatolia'.[6] It is difficult to miss the message of multiculturalism in the choice of monuments that are displayed in Miniaturk. The second model is of the Ottoman-era Selimiye Mosque (built between 1568 and 1575, Edirne); the third, of the Republican-era Anıtkabir (1944–1953, Ankara), the Republic's leader Mustafa Kemal Atatürk's mausoleum. The diversity of the buildings modelled signifies the cultural wealth of Turkey. The criteria for selection, visitors are told, are originality and representativeness (figure 6.6).[7]

The literature on theme parks presents these environments as heir to the workings of print in the imagination of people as a nation *à la* Benedict Anderson (Anagnost, 1997). The nation is mapped in these spaces and reproduced through consumption. The designs of nation-themed miniature parks display

Figure 6.6. The models do not all replicate their originals in the same way. In the foreground is the model of Hacı Bektaş Veli Dervish Lodge near the city of Nevşehir. In the immediate background is the model of the Church of Virgin Mary near the city of Izmir. The printed guide says of the latter building that 'only its apse stands today'. Nevertheless, the complete church was modelled in Miniaturk on the basis of references and other resources. This photograph also shows the audio-information panel that is activated by each visitor's smart ticket. (*Photo*: İpek Tureli)

diversity in the way this mapping is materialized. Beautiful Indonesia orders its imaginary nation according to official geographical divisions. Thus, each Indonesian province is represented by a pavilion around a central artificial pond. In the middle of the pond are artificial islands in the shape of the Indonesian Archipelago (see Hitchcock, 1998; Pemberton, 1994; Siegel, 1997, pp. 3–5; Errington, 1998, pp. 194–198). Splendid China arranges its sites roughly to correspond to their relative locations in real life, but without consistency in scale, and the overall outlines of the park somewhat resembles China's official territorial boundaries.[8] The officially declared model for Miniaturk is Madurodam in The Hague. A comparison, however, reveals that the only real similarity between the two is that both display building models at 1/25 scale. Madurodam's major claim is to present a complete built history of The Netherlands. To accomplish this, it takes the form of a city that has grown outward radially from a medieval core. In Miniaturk there is no formal reference to the map of the nation-state, nor was the possibility of growth designed into the site plan. Only seventy-five models were listed in the park's 2003 visitor's guide (thirty-six from Istanbul, thirty-one from Anatolia, and eight from 'abroad') and 105 are listed in its 2010 website.[9] Additions over time have increased the total number and density of the models, and transformed the spatial experience of the exhibition space, making it denser. The overall boundaries and thematic groupings of the park remain.

It is interesting to find in Miniaturk that the exhibition space is organized into two main circular areas, which separate Istanbul from Turkey – the circular area closer to the entrance terrace contains 'Anatolian', and the other, 'Istanbul' sites (figure 6.7). The Istanbul part is organized around a pond that stands in for the Bosphorus. A model bridge, on which visitors can walk, connects across the pond to an elevated restaurant on the roof terrace of a small building which defines the rear edge of the park (figure 6.8). This building is stepped towards the exhibition area to provide outdoor seating, an amphitheatre for special events such as concerts, and inside it are service areas, and a 'Panoramic Museum of Victory' where miniature figures re-enact the War of Independence in 'Anatolia' complemented by a soundtrack. My analysis here is limited to the outdoor areas.

Outside the two circular exhibition areas, of Istanbul and Anatolia, on the bank-side of the park, a third group of models presents a curious selection of buildings dubbed 'Ottoman Geography' ('Abroad' is the title used in the English guide). These models – which include, among others, the Ecyad Castle in Mecca, Damascus Train Station, the Dome of the Rock in Jerusalem, Mostar Bridge in Bosnia-Herzegovina – replicate in miniature buildings outside the current boundaries of Turkey. They were selected, promotional publications state, because they were 'built or renovated during the Ottoman Empire'. This third group clearly seeks to demonstrate the idea that Turkey not only appreciates multiple cultures within the present-day borders of the country but also those once contained within the larger administrative umbrella of the Ottoman Empire.

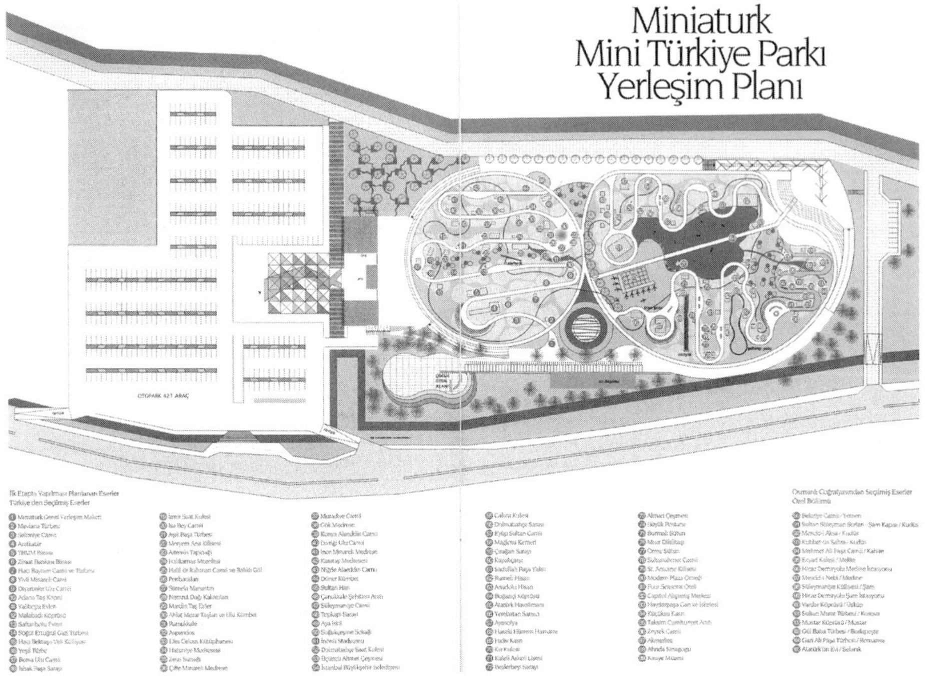

Figure 6.7. Plan published by Miniaturk to solicit models from model-makers. (*Source*: *Miniaturk, Mini Türkiye Parkı Sponsorluk Rehberi*, 2002, by courtesy of Kültür A.Ş.)

Figure 6.8. The only model visitors can touch and walk on is that of the Bosphorus Bridge, which crosses the artificial pond on the site. (*Photo*: İpek Türeli)

Overall, the models do not simulate the conditions of original structures in any consistent way: models that are replicas, restorations, and restitutions of their referents are presented together. Most buildings and sites are presented in ways that rid them off their symbolic content. Little attempt is made to distinguish between the sites based on the communities they serve(d). Some models depict sites that are currently in use, while others are purely historical. Perhaps most startlingly, most of the models are divorced from their original urban contexts. Even buildings located literally next to each other in Istanbul may thus be dislocated and separated in their miniature displays. The most pointed examples of this are from the Dolmabahçe and Sultanahmet areas of Istanbul. In Dolmabahçe, the palace and the clock tower, and in Sultanahmet, the Blue Mosque, Hagia Sophia, Topkapı Palace, and the Cistern share tightly knit urban contexts. However, in Miniaturk these structures are positioned without reference to each other.

Had the main goal simply been to demonstrate multiculturalism, then for instance, Taksim Square – with the Atatürk Cultural Centre (Atatürk Kültür Merkezi, AKM, 1969), Cumhuriyet Anıtı (Republic Monument, 1928) and the Ottoman water distribution (taksim) reservoir (1732) on the square, and the Greek Orthodox Church Ayia Triada (Holy Trinity, 1882) and the French Consulate (formerly Plague Hospital, 1719) nearby – would have been another striking example to demonstrate the complexity and layered nature of the urban fabric in contemporary Istanbul, but in Miniaturk, only the Republic Monument (dubbed Taksim Monument) is reproduced as a singular item, detached from both physical and social contexts. By evading all such contextual references the effect is to 'naturalize' the originals. Another effect that I bring out in my analysis is a tension between the representation of the Republican and Ottoman eras. Nevertheless, it is through tensions and contestations that a consensus can be built.

The Political Context

The decision to build a bounded miniature park where the nation could be viewed in its entirety may be seen partially as a response to anxieties about Turkey's future. As elsewhere, national unity has long been a source of collective paranoia in Turkey. Following World War II, fears for national sovereignty were fuelled by Cold War politics. Since the early 1990s, separatist movements and violent developments in the Balkans and the Middle East have again exacerbated such fears. As a result, the rise of Islamist and Kurdish movements at home have been regarded as threats to national sovereignty, rather than a call for social justice and democracy.

Two principal observations reinforce the notion that Miniaturk may be responding to such anxieties. At the level of image, Miniaturk stage-manages history by painting an ambiguous picture of societal harmony. At the level of production, it attempts to provide a showcase for both the quality and effectiveness of local and the central governments' mutual political vision for the country and the city. The version of history presented at the park is open to different readings according to the visitors' 'viewing positions', influenced among other factors by personal background and the political context. Those involved in creating the park may also have interpreted its purpose dissimilarly. Ultimately, its form has embodied negotiation between a range of politicians, administrators, designers, engineers, builders, consultants and sponsors.

Despite these interpretive ambiguities, Miniaturk has been successfully promoted through advertising and word of mouth. Stories have been written in local magazines and newspapers; articles have appeared in tourist and architectural guides to the city; and images of the park have been posted on billboards (managed by Kültür A.Ş.). With so much coverage and so many actors, it is understandable that certain discrepancies surfaced in terms of credit for its design and realization. This led to multiple opening ceremonies. Miniaturk was opened three times: first, on 23 April 2003; second, by Prime Minister Erdoğan on 2 May 2003; and finally, by the then Mayor Müfit Gürtuna on 29 May 2003, for the commemoration of the 550th anniversary of Istanbul's conquest (Kültür A.Ş., 2003*b*, p. 13). The architect is not usually named in official publications of Miniaturk, but he has published his design drawings in local professional journals of architecture, without reference to the political actors who commissioned the project.

Significantly, the first opening coincided in the official calendar of the Republic with 'National Sovereignty and Children's Day'. By conflating national sovereignty with a celebration of the child, the date not only implies that the preservation of sovereignty should be a concern of children, but casts adult citizens in the role of children in relationship to the state. By comparison, Istanbul's conquest commemoration date, 29 May, is not listed in the official calendar of the Republic (Çınar, 2001, p. 365) but nonetheless is celebrated at the local level with minimum official participation. Instead, the marketing of Miniaturk as a conquest-commemoration project by Mayor Gürtuna must be viewed in relation to public discussions and competing claims on national history.

As Republican history is commemorated via days such as 23 April, 19 May (Atatürk Commemoration, Youth and Sports Day), and 29 October (Republic Day), Muslim civil society groups and Muslim-oriented political parties have demonstrated a need to establish alternative days commemorating important events from Turkey's Ottoman past. The counterpart of this commemoration has been growing nostalgia for the early years of the Republic among citizens who identify with 'Kemalism' (after Mustafa Kemal Atatürk, the founding figure of the Republic, around whom a leadership cult developed) and its doctrine of secularism. Esra Özyürek and Kimberly Hart observe the 'miniaturization', commodification and consumption of Atatürk insignia among a variety of strategies through which the early Republic is remembered and politics is privatized in line with neoliberal symbolism of privatization and market choice (Özyürek, 2005; Hart, 1999). According to Özyürek, Kemalist and Islamist versions of the early Republic in the 1990s competed but provided comparably homogeneous, dominant narratives (Özyürek, 2007).

The backdrop for competing programmes of commemoration involved the reluctance of official nationalist history to recognize the accomplishments of the Ottomans. Especially in its early decades, the Republic attempted to establish itself as a secular, West-oriented nation-state, in historic opposition to the old, Islamic, 'Eastern' Ottoman Empire. And by downplaying conquest-commemoration festivities, it has also sought to avoid upsetting the 'European club' by conjuring memories of a past when the Ottoman Turks were a rival power in Europe.[10] For their part, 'Islamists' in local government have turned to neo-Ottomanist nostalgia precisely because of, and as a reaction to, the secular Turkish establishment. The Ottomans had self-consciously promoted a multicultural society as well as claiming the leadership of Islam. In contrast, the Republican era (1923–1950) was characterized by disrepudiation of the public display of religion. The nascent Turkish state promoted civic nationalism, manufacturing homogeneous national identity out of a very diverse population, but ended up imposing the language – Turkish – and the religion – Sunni Islam – of the majority (Smith, 2005). Accordingly, citizenship emphasized the individual's duties to the nation-state over his or her rights (Kadıoğlu, 2006). However, market reforms in the 1980s, political liberalization in the 1990s, and the EU membership process at the beginning of the 2000s have all contributed to a re-evaluation of accepted norms of citizenship, national identity, and the assumed correspondence between them.

This relationship between the state and the citizen is perhaps nowhere more spatially expressed than in the siting and design of the Anıtkabir. On Atatürk's death, his mausoleum was constructed on an imposing hill crowning the capital city of Ankara. Various features of the building sought to embody and define early Republican ideals about the correct relationship between citizen and nation state (Meeker, 1997; Bozdoğan, 2001). In the years since it was built, the Mausoleum has served as Turkey's official nationalist pilgrimage site. It is where state ceremonies are held, and private and public associations gather there to pay their respects to Atatürk and display their commitment to protect the nation. In the 1990s protests also took place there against such perceived threats to the secular establishment as the headscarf.

One result of the rise of Islamism in the public sphere in the 1990s has been to challenge the state-sanctioned narrative of national identity, and the need to give physical representation to these forces has resulted in efforts to rediscover Istanbul's architectural and urban history as the former Ottoman capital (Jeremy F. Walton, Chapter 5). Alev Çınar argues that the substitute narratives of nationhood produced by Islamist political organizations sought to cast Istanbul as a 'victim' of the Republic – thus the re-enactment of its conquest symbolically served to 'save' it (Çınar, 2001, 2005). She explains that in 1994 the Istanbul Metropolitan Municipality under Refah Partisi (RP, Welfare Party) and an Islamist nongovernmental organization, the National Youth Foundation, jointly organized the conquest celebration as an event which would rival national commemoration days in scope and scale (*Ibid.*).

Among other things, this drew public and academic attention to Islamist claims to public spaces in the city. The victory of RP in the local elections in major cities, and specifically Tayyip Erdoğan's in Istanbul, raised further alarms. And when RP leader Necmettin Erbakan became Turkey's first Islamist Prime Minister in 1996 concerns mounted even further. Some even asked if Turkey was set to turn into a new Iran. In 1997, the army, which sees itself as the guardian of the secular Republic, intervened and ousted the RP (Cizre and Çınar, 2003). The constitutional court then closed down both the RP and its successor, Fazilet Partisi (Virtue Party). These moves by a coalition of the army and other secular elites to block the rise of Islamists swayed electoral support to a reformist faction within the RP. In 2002 this faction, now represented by the Adalet ve Kalkınma Partisi (AKP, Justice and Development Party), won the general elections under the leadership of Istanbul's former mayor, and Turkey's current Prime Minister Erdoğan, in part by promoting an economic programme of 'communitarian-liberal syntheses' (Keyman and İçduygu, 2005, p. 16). In response to questions regarding his new 'reformed' position, Erdoğan defined himself not as an Islamist but as a 'conservative democrat'. Although he used Islamist devices in his daily performance as prime minister, he made it clear that he considered religion a private issue and that AKP would not conflate religion and government, Islam and democracy (Smith, 2005) and would operate within the nation-state's constitutional framework. AKP subsequently used the European Union membership accession process to negotiate and supress potential threats to it from the secularist establishment. It advocated a much milder adherence to religion than its predecessor RP, and apparently recognized a need for societal plurality.

A Ground of Conversation?

In representing cultural 'wealth', Miniaturk seeks to build consensus from contestation. In its version of Turkey, it does not single out the Islamic Ottoman past but overtly privileges Istanbul over Ankara, the capital built to showcase nation-building in the early Republican era. The selection of models in Miniaturk aims to represent major religious communities that cohabited the land – in line with what is today perceived as Ottoman 'cosmopolitanism'. One of the forms of neo-Ottomanist nostalgia that manifests itself is a longing for the multi-congregational and multi-lingual composition of the city with its Christians, Jews and Europeans (Houston, 2001*a*, *b*; Komins, 2002; Mills, 2006, 2007; Özyürek, 2005, 2007; Marcy Brink-Danan, Chapter 16) For Muslim-oriented organizations and individuals, this expresses Muslim hegemony over others. For others, it demonstrates the cultural richness and tolerance of Turkey and its suitability to the new Europe. Miniaturk projects the message of multiculturalism by its selection of Ottoman-era monuments of Muslims and non-Muslims, and pre-Ottoman sites from within the boundaries of Turkey. The incorporation of pre-Ottoman sites from Anatolia, especially those of antiquity, into national patrimony, is very much in line with Republican-era history writing. Yet, the selection constrains those from the Republican era by limiting their number. And the design of individual monuments and their siting negate all societal significance; this is especially true of Republican-era monuments, some of which have immense political significance. The inclusion of monuments from 'Ottoman geography' outside Turkey suggests the park extends the imagination of the nation-state beyond the current borders and claims the Ottoman Empire as historical precedent and cultural heritage, and this reflects the current government's (nostalgic) desire to extend Turkey's sphere of influence to that territory once more. The contradictions in design credits and inaugural dates notwithstanding, the embrace of the park by the Municipality, the visiting public, and the news media all suggest that it does facilitate a ground for conversation.

In a meeting with one of the model-makers and a public relations representative, I was intrigued to discover how the two identified with different versions of national history, even while working together. One of my questions was directed at the model maker.[11] When I asked specifically which his favourite model was, he replied that it was the Mağlova Aqueduct (1554–1562), and gave the following reasons:

> Model-Maker: Not because it is a good model; in fact, it was one of our first… I like it because of the work itself. It shows they had the determination, the belief, and the will to work. I am not saying this because of indoctrination: 'Ottomans, Oh! Ottomans…'. However, every society, like an organism, has a period in which it is alive… For example, if we bring the first and last decades of the Republic together… [Laughs]
>
> Public Relations Representative: Not even bringing the last century together would suffice… Once, I was leading a group of journalists [in Miniaturk]. One asked why there aren't many examples from the Republic. And I pointed to the squatter settlements [overlooking the park] and said, 'There!' He started exclaiming, 'You are indeed Ottomanists, you are this … and you are that…'. His own cameraman reacted. 'Hold on a minute', he interrupted: '[as if] we have them [Republican landmarks worthy of display], and it is Miniaturk that does not display them?'.
>
> Model-Maker: But they [squatter settlements] are not the result of the Republic. They exist because of globalization, because of the conjuncture, because of the Cold War…
>
> Public Relations Representative: They are the result of a homogenizing world. Look at a plaza. Is this building in New York, Paris, in Istanbul, or in Tokyo? One cannot tell. All look the same. But look at a structure from the Middle Ages, and you can tell at a glance where it's from. If one is equipped with some historical knowledge one can even identify the country where it is from. But plazas do not allow this.

In this partial exchange, the model-maker conflates the copy (model aqueduct) with the original (real aqueduct). He does not refer to the original as a functional architectural object or space, but as a symbol of technological superiority, as an artefact of cultural self-confidence. The model-maker also suggests that society is an organism, and that architecture is its reflection. Thus, when an organism reaches maturity, it produces monumental architecture and engineering. The Republic, in such a narrative, becomes a period of delayed replenishment that has been terminated by more powerful processes such as internationalization and globalization that have come from without.

For the public relations representative, however, the organism analogy does not work. Her understanding evolves more along an axis of tradition versus modernity. Once there was a time of heterogeneity, but this was defeated by modernization (which started with the Ottoman Reformations in the nineteenth century, but pursued forcefully under the Republic through the twentieth). She thus externalizes the Republic as an agent of top-down modernization. In this analysis, processes of modernization ultimately dictate cultural homogeneity, and Miniaturk becomes a project of 'resistance' because it reinstates heterogeneity (figure 6.9).

Figure 6.9. Even though the design of the park aims to abstract itself from its surroundings, the surrounding topography impinges in the form of nearby hillside apartment buildings that create an unavoidable backdrop to the models – the source of the public relations representative's disturbance. (*Source*: *Miniaturk: Turkey's Showcase*, 2003, by courtesy of Kültür A.Ş.)

Clearly, the model-maker and the public relations representative exhibit different ideas about Ottoman/Turkish history and how the present built environment has come into being. But they parallel each other in their understanding that architecture is an outcome (rather than, for instance, a cause or an agent) of social processes. Thus, while they do not agree at an ideological level, they have been able to work together to produce representations of architecture that not only display but prove Turkey's exceptionality. In essence, then, Miniaturk offers a consensus ground for two people who do not share the same conception of history, but who have similar anxieties about cultural homogenization and decline. The exchange between them illustrates the park's potential to reveal political differences, just as it facilitates their concealment.

Prime Minister Erdoğan and his senior aides in the Istanbul Municipality, including the current mayor, are eager to shoulder the global-city project, and have been proud to initiate and build prestige projects – referred to as 'plazas' in the previous exchange – and to 'market' Istanbul to international investment. This administration is essentially continuing a project which begun under Anavatan Partisi (Motherland Party) with Turgut Özal as Prime Minister (1983–1989) and Bedrettin Dalan as Istanbul's mayor (1984–1989). The restructuring of the city's economy initiated with the touristification of the city's heritage with a renewed official focus on its Ottoman past, and the opening of numerous five-star hotels on prime sites overlooking the Bosphorus. Through the 1990s and 2000s, the city acquired numerous new hotels, gated communities, high-rise condominiums, shopping malls, entertainment venues, and a new skyline of high-rise buildings, especially in the new Central Business District which now characterizes the city's contemporary face.

Miniaturk has also been used to market the city, but its design and status as a 'heritage' site seem to set it in a different orbit. Thus, when the public relations representative refers to the homogenization of the built environment via 'plazas', she assumes that Miniaturk somehow resists this process. What she chooses not to see is that all projects involving prime sites offered to global capital or leisure environments such as Miniaturk are characteristic architectures of globalization. Both are domestic translations of global types; both entail the rerouting of public sources into private or privatized services; and both are designed to demonstrate Turkey's competitiveness in the global marketplace (figures 6.10 and 6.11).

Miniaturk reflects a desire to imagine a Turkey which displays its cultural wealth and influence with confidence and pride while being clearly bounded and secure. One image that Miniaturk repeatedly uses in promotional publications (possibly to represent its inclusive politics) shows Atatürk's Mausoleum and the Selimiye Mosque together. The coupling brings into mind the controversy around the Mausoleum and the new Kocatepe Mosque in Ankara (modelled on several Imperial Ottoman mosques, including Selimiye) as discussed by anthropologist Michael E. Meeker (1997). In reality, the Mausoleum and the Kocatepe Mosque stand at similar elevations, and crown the two highest hills of the capital city.

Figure 6.10. Miniaturk's 2003 brochure shows, from left to right, the diagrammatic location of the park in the city, an unstaged snapshot of visitors in the park, and a staged shot of an ideal family behind the Byzantine church-turned-mosque-turned-museum of Hagia Sophia under the words, 'The Showcase of Turkey'. (*Source*: By courtesy of Kültür A.Ş.)

Figure 6.11. This photograph was used in several of Miniaturk's publications. The framing foregrounds the Mausoleum, yet establishes the imperial mosque as the centre of attraction. (*Source*: *Miniaturk: Turkey's Showcase*, 2003, by courtesy of Kültür A.Ş.)

They thus represent competing claims to Turkey's national imaginary – the Mausoleum standing for secular modernism, the mosque for a modern Islam. This 'controversy' is complicated today by the fact that army generals who visit the Mausoleum to pay tribute to Atatürk also perform the *namaz* in Kocatepe;

and when AKP leaders visit the Mausoleum, their members with headscarves will not be in sight. Both sides display reluctance but simultaneous consensus.

The composition of the Miniaturk promotional photograph is complex in symbolic terms, and yields two very different readings. According to one, the Mausoleum, the symbol of Republican nationalism, is clearly foregrounded. But the position of the Mausoleum model on flat ground can alternatively be seen as negating its elevated position in relation to the capital city and the nation. The photograph also shows a few spectators giving a passing glance to the Mausoleum while many more crowd around the Kocatepe Mosque. Could the implication be to point out that the Ottoman past is becoming increasingly attractive, in contrast to the Republican one? The angle of view and foreshortening in the image further contributes to this second reading by locating the mosque above the Mausoleum on the printed page. I am not suggesting that the photographer took this image with a definitive discursive claim; but the choice as publicity image suggests the photograph presents the institutional goal of Miniaturk, of building consensus from contestation.

The embodied experience of the park by the park's visitors does not privilege any static point of view. The layout of routes through the park is based on an assumption that visitors will stay for two hours (Kültür A.Ş., 2003*b*, p. 17). During this time, security personnel and visual and audio messages repeatedly remind them to avoid walking on the grass or touching the models. Meanwhile, a specially commissioned musical composition, by Fahir Atakoğlu (known for his patriotic works), is broadcast from disguised speakers, and detailed information on individual landmarks is provided in a booklet, as well as through an audio information system activated by the individual user (and again heard through disguised speakers). The lack of shade, lack of seating, narrow width of paths, and admonishments against touching the models (or even getting close to them) all lead visitors to view the models while in motion. Only on the model of the Bosphorus Bridge are visitors treated somewhat differently. The actual suspension bridge connects Asia and Europe and is traversed only by motor vehicles. But here it serves as a finale of sorts, raising visitors from the ground level to the elevated restaurant terrace, from where they can look back over the park and contemplate the miniature world which has just been presented to them.

Conclusion

The models in Miniaturk are devoid of social life, isolated from any urban context, and laid out in a seemingly arbitrary manner. Instead of trying to depict an overt normative Turkish or Islamic-Turkish character, the park's main rhetorical purpose thus seems to be to indicate the tolerance of Turkey to multiple cultures and ways of life. Miniaturk's final size has been predetermined; it does not allow for extension or change. Finally, Miniaturk differentiates between city and nation. It privileges one specific city, Istanbul, giving it almost equal space in its representation as the rest of the national territory. What remains further peculiar about Miniaturk is that half the original structures modelled in it may still be found in Istanbul. Why would Istanbul be the locus for such a collection of representations when visitors could just as easily visit the originals? This condition suggests that while Istanbul seeks to be a centre of cultural imagination, it is a centre detached from the imaginary of the nation. Likewise, the nation is left with a displaced centre. In trying to re-present an ideal of Istanbul, the park also tries to detach itself from the reality of the city around it.

Visitors understand and judge the park according to their personal reference systems. They may also understand Miniaturk's inclusivity as a form of nostalgia. But, in the end, their appreciation of the park, demonstrated in attendance numbers and positive reviews, reflects a yearning for alternative modes in which to imagine the nation. As the ethos of Republican nationalism fades away, other sites have emerged to challenge Atatürk's Mausoleum as the centre of national symbolism. Miniaturk is clearly one of these. And since it is a miniature, it offers the possibility to grasp the entire nation as if it were an island separated from everyday reality and history.

In their use of scale as a representational strategy, miniature parks paradoxically work in tandem with the promenades of national capitals. In resorting to the

Figure 6.12. 'Historic Opening in Europe!' the title to this double-page advertisement in a Turkish newspaper exclaims to advertise a new housing development, 'Bosphorus City', and its theme of the Bosphorus surrounded by 'mansions'. The popularity of a miniaturized Istanbul has moved on from the representational realm of the exhibition to the world outside of it. *À la* Baudrillard, Miniaturk 'exists to hide' that some of what is outside 'belongs to … the order of simulation'.

miniature and the gigantic, both disrupt the real and the experience of the everyday. The national promenade may present progress-oriented spaces in accordance with an official narrative. In contrast, the national miniature park may level history through the use of seemingly arbitrary layouts. Yet by bringing the lived spaces of the past into the present as precious objects to be looked at, or by removing living spaces from their present contexts, the miniature park may provide an illusionary field on which to imagine a common, nonetheless contested, future. At Miniaturk, citizens join together to consume culture voluntarily, without obligation. Further, Miniaturk is popular precisely because it flattens histories and geographies to bring them together without apparent hierarchy or conflict. Yet, in doing so, it also speaks to an imaginary that seeks to attach to a global world via Istanbul.

As the progressivism and secularism of the early Republican nation-building project is increasingly criticized from within and a new plurality emerges in its place, the national symbolic, the archive of official objects and narratives, is due for renovation through additions that cater for a new national polity. There are many potentially conflicting aspects of this new polity: the recognition of religion; a renewed interest in Istanbul as the potential gateway through which Turkey will join the multicultural European Union; a nostalgia for Ottoman cosmopolitanism; a drive for the bourgeois beautification of the city; and finally, a reconfiguring of the relationship between the state and its citizens in the midst of growing dissent towards the representational quality of the democratic process. Because Miniaturk seeks to fulfil all these criteria in a Turkey striving to reassert itself as one among equals in a globalizing world, it has potentially become a new nationalist pilgrimage site. It is in such contexts that discrepancies between people can be willingly suppressed, and memory, accordingly, stylized (figure 6.12).

Notes

1. An earlier version of this chapter was published in *Traditional Dwellings and Settlements Review* (2006). http://www.ced.berkeley.edu/iaste/tdsr.htm. Accessed 12 March 2010.

2. Mini-City was designed by Emre Arolat Architects for Antalya Metropolitan Municipality and built in 2003–2004.

3. AKP-initiated prestige projects, such as the Haydarpaşa Port, Galataport, and the Dubai Towers all announced in 2005 and in which publicly-owned prime sites in Istanbul were to be privatized, were all met with criticism in the media and have been resisted by civil-society organizations. By contrast, there were few voices raised against public spending on a 'new' site such as Miniaturk, while 'real' heritage nearby awaits reinvestment. See Ekinci, 2004, pp. 53–55.

4. Istanbul Cultural and Artistic Products Company (İstanbul Kültür ve Sanat Ürünleri Ticaret A.Ş. or Kültür A.Ş. in short) is a 'commercial corporation that was established on 2 October 1989, within the body of the Istanbul Metropolitan Municipality to offer cultural, artistic and tourist services'. It is one of the Municipality's numerous corporations. Available at: http://www.kultursanat.org/kurumsal/?main. And also at: http://www.ibb.gov.tr/en-US/Organization/Companies/Pages/KULTURAS.aspx. Accessed 20 December 2009.

5. Aerial photography of Istanbul and a close up of the site is available at http://sehirrehberi.ibb.gov.tr/map.aspx. Accessed 20 December 2009.

6. *Miniaturk Guide*. Istanbul: Kültür A.Ş., p. 3.

7. The official explanation regarding the selection of landmarks represented in the park states: 'Works that found a place in Miniaturk were ones that displayed peculiarities of the era in which they were built, ones that reflected the culture and art of a land that had witnessed thousands of years of heavy invasion, war and destruction, works that had not been destroyed simply because they had been created by those who came before, works that were protected, repaired and enjoyed' (Kültür A.Ş., 2003*a*, p. 27).

8. Promotional publication. *Shenzhen Splendid China miniature scenic spot* (1989) Hong Kong: Chinese Travel Service.

9. See official site of Miniaturk at http://www.miniaturk.com.tr. And http://www.kultursanat.org/kurumsal/?kunyeen&sid=12.

10. The first conquest commemoration festivities were held on its 500th anniversary in 1953. A dedicated association was set up to guide the festivities. The preparations were initiated officially; however, Turkey's President and Prime Minister at the time abstained from attending because, according to the general interpretation today, Turkey had just entered NATO and wished to keep its relations with its Western allies. Since 1953, the commemoration festivities have been repeated on different scales but with minimal state representation.

11. Interview conducted by the author with Miniaturk public relations representative and one of the model makers in Miniaturk. 4 November 2005.

References

Anagnost, Ann (1997) *National Past-Times: Narrative, Representation, and Power in Modern China*. Durham, NC: Duke University Press.

Bennett, Tony (1995) *The Birth of the Museum: History, Theory, Politics*. London: Routledge.

Bezmez, Dikmen (2008) The politics of urban waterfront regeneration: the case of Haliç (the Golden Horn), Istanbul. *International Journal of Urban and Regional Research*, **32**(4), pp. 815–840.

Bozdoğan, Sibel (2001) *Modernism and Nation Building: Turkish Architectural Culture in the Early Republic*. Seattle, WA: University of Washington Press.

Burke, Edmund (1757, 2008) *A Philosophical Enquiry into the Sublime and Beautiful* (edited by James T. Boulton). London: Routledge Classics.

Çelik, Zeynep (1992) *Displaying the Orient: Architecture of Islam at Nineteenth-century World's Fairs*. Berkeley, CA: University of California Press.

Çınar, Alev (2001) National history as a contested site: the conquest of Istanbul and Islamist negotiations of the nation. *Comparative Studies in Society and History*, **43**(2), pp. 364–391.

Çınar, Alev (2005) *Modernity, Islam, and Secularism in Turkey: Bodies, Places, and Time*. Minneapolis. MI: University of Minnesota Press.

Cizre, Ümit and Çınar, Menderes (2003) Turkey 2002: Kemalism, Islamism, and politics in the light of the February 28 process. *The South Atlantic Quarterly*, **102**(2/3), pp. 309–332.

Ekinci, Oktay (2004) Dünya Mirasında 'Miniatürk'! in *İstanbul'un 'İslambol' On Yılı* [Istanbul's 'Islamful' Decade]. Istanbul: Anahtar Yayınları.

Errington, Shelly (1998) *The Death of Authentic Primitive Art and Other Tales of Progress*. Berkeley, CA: University of California Press.

Gottdiener, Mark (2001) *The Theming of America: Dreams, Media Fantasies, and Themed Environments*. Boulder, CO: Westview Press.

Hart, Kimberly (1999) Images and aftermaths: the use and contextualization of Atatürk imagery in political debates in Turkey. *Political and Legal Anthropology*, **22**(1), pp. 66– 84.

Hitchcock, Michael (1998) Tourism, Taman Mini, and national identity. *Indonesia and the Malay World*, **26**(75), pp. 124–135.

Houston, Christopher (2001*a*) *Islam, Kurds and the Turkish Nation State*. Oxford: Berg.

Houston, Christopher (2001*b*) The brewing of Islamist modernity: tea gardens and public space in Istanbul. *Theory, Culture and Society*, **18** (6), pp. 77–97.

Kadıoğlu, Ayşe (2006) Citizenship and individuation in Turkey: the triumph of will over reason. *Cemoti, Cahiers d'études sur la méditerranée orientale et le monde turco-iranien*. Available online at: cemoti.revues.org.

Kaufman, Edward N. (1989) The architectural museum from world's fair to restoration village. *Assemblage*, **9**, pp. 20–39.

Keyman, E. Fuat and İçduygu, Ahmet (2005) Citizenship, identity, and the question of democracy in Turkey, in Keyman, E.F. and İçduygu, A. (eds.) *Citizenship in a Global World: European Questions and Turkish Experiences*. London: Routledge.

Kirshenblatt-Gimblett, Barbara (1998) *Destination Culture: Tourism, Museums, and Heritage*. Berkeley, CA: University of California Press.

Komins, Benton Jay (2002) Depopulated cosmopolitanism: the cultures of integration, concealment, and evacuation in Istanbul. *Comparative Literature Studies*, **39**(4), pp. 360–385.

Kültür A.Ş. (2003*a*) The selection of models, in *The Showcase of Turkey, Miniaturk, The Story of How it Came to Be*. Istanbul: Kültür A.Ş.

Kültür A.Ş. (2003*b*) *Faaliyet Raporu 2002* [Activity Report 2002]. Istanbul: Kültür A.Ş.

Meeker, Michael (1997) Once there was, once there wasn't: national monuments and interpersonal exchange, in Bozdoğan, S. and Kasaba, R. (eds.) *Rethinking Modernity and National Identity in Turkey*. Seattle, WA: University of Washington Press, pp. 157–191.

Mills, Amy (2006) Boundaries of the nation in the space of the urban: landscape and social memory in Istanbul. *Cultural Geographies*, **13**(3), pp. 367–394.

Mills, Amy (2007) Gender and mahalle (neighborhood) space in Istanbul. *Gender, Place & Culture*, **14**(3), pp. 335–354.

Özdemir, Cengiz (2003) *Faaliyet Raporu 2002* [Activity Report 2002]. Istanbul: Kültür A.Ş.

Özyürek, Esra (2005) *Nostalgia for the Modern: State Secularism and Everyday Politics in Turkey*. Durham, NC: Duke University Press.

Özyürek, Esra (2007) Public memory as political battleground: Islamist subversions of republican nostalgia, in Özyürek, E. (ed.) *The Politics of Public Memory in Turkey*. Syracuse, NC: Syracuse University Press, pp. 114–137.

Pemberton, John (1994) Recollections from 'Beautiful Indonesia': somewhere beyond the post-modern. *Public Culture*, **6**(2), pp. 241–62.

Siegel, James T. (1997) *Fetish, Recognition, Revolution*. Princeton, NJ: Princeton University Press, pp. 3–5.

Smith, Thomas W. (2005) Civic nationalism and ethnocultural justice in Turkey. *Human Rights Quarterly*, **27**, pp. 436–470.

Stewart, Susan (1984) *On Longing: Narratives of the Miniature, the Gigantic, the Souvenir, the Collection*. Baltimore, MD: John Hopkins University Press.

Türeli, İpek (2006) Modeling citizenship in Turkey's miniature park. *Traditional Dwellings and Settlements Review*, **17**, pp. 55–69.

Uluğ, Murat (2003) Miniaturk. *Yapı*, **262**, pp. 71–75.

Vale, Lawrence J. (1992) *Architecture, Power, and National Identity*. New Haven, CT: Yale University Press.

White, Jenny B. (2002) *Islamist Mobilization in Turkey: A Study in Vernacular Politics*. Seattle, WA: University of Washington Press.

Part III. The Mediatized City

Cinema never represents cities as given; it plays an active part in shaping the urban imaginary and fabric. The authors of the four chapters in this Part offer in depth analyses and contextualization regarding the production and circulation of moving city images. Their case studies range from the beginnings of cinema in Istanbul to recent box-office hits, festival successes and travel documentary. By considering old and new media and focusing on questions of spectatorship then and now, the chapters introduce a historical dimension to the discussion of cosmopolitan Istanbul, and extend Nezar AlSayyad's argument for a 'cinematic urbanism', which recognizes that 'reel and real', physical environment and virtual images are 'mutually constitutive' to Istanbul.

In 'The Spectator in the Making: Modernity and Cinema in Istanbul, 1896–1928', Nezih Erdoğan stresses the cosmopolitan origins and international trade connections of cinema, pointing out that polyglot non-Muslim citizens with connections to other European cities continued to shape Istanbul's cultural landscape well into the Republican era. The striking multilingualism (French, Armenian, Greek, and Ottoman Turkish) located in documents relating to cinematic exhibition practices attests to this diverse fabric of the city. The digital archive project that Erdoğan conducts, encompassing press clippings, is devoted to making such documents on the early history of cinema accessible – hence, visible. The archive thus serves as a medium allowing contemporary viewers/readers access to 'a [city] space expanding onto other modern histories and cities'.

While cinema entered the Ottoman Empire in the late nineteenth century, and the first films were produced in the 1910s, domestic cinema evolved into an institution only in the post-1950 era. Production expanded from a few films per year before World War II to around several hundred in the 1960s. Film production and locations concentrated in Istanbul. From the mid 1970s the industry went into decline. By the 1990s, film production in Istanbul, and thus in Turkey, had scaled back down to the tens. Then, European, and some American, funds, particularly through the European Council's co-production and distribution fund Eurimages, helped enhance domestic production and furthermore provided films produced with this fund with international visibility. The new generation of directors no longer needed to observe the norms of a fast-paced industry and were thus able to diversify production in terms of style and genre.

In 'Istanbul through Migrants' Eyes', İpek Türeli compares two famous migration-to-Istanbul films in the light of discourses on the city: Halit Refiğ's *Birds of Exile* (1964) and Nuri Bilge Ceylan's *Distant* (2003). Set apart by four decades and considerably different from each other in form and setting, both films tell the story of migrants in a wintry Istanbul and seek to project the city through their eyes, nevertheless from an urbanite perspective. In their engagement with migration, both films problematize belonging to the city and reflect on Turkey's status *vis-à-vis* the world with characters having conflicts about staying and going abroad.

Deniz Bayrakdar and Elif Akçalı in 'Istanbul Convertible: A *Magic Carpet Ride* through Genres', propose to interpret the new Istanbul films within the context of European integration and circulation of value in the globalized city. They make their case through a reading of Yılmaz Erdoğan's *Magic Carpet Ride* (2005), a fast paced cross-section film that employs an abundance of helicam shots and aerial perspectives in establishing unlikely connections across the city, linking an upper-middle class family of academics with the mafia through the circulation of money and cars. *Magic Carpet Ride* provides a sophisticated commentary on discourses of gentrification and commodification that inform debates in urban planning, architectural design and public policy in the process of re-branding Istanbul as a global city or a European Capital of Culture.

In 'Projecting Polyphony: Moving Images, Travelling Sounds', Deniz Göktürk situates Turkish German director Fatih Akın's *Crossing the Bridge* (2005), a documentary on Istanbul's music scene, in the context of debates on the EU accession process and Turkey's 'bridging' role. Göktürk reads the film as a utopian projection of pluralist polyphony. The film's engagement with Roma performers, in particular, opens up questions about mediations of ethnicity and authenticity in world music and cinema. Göktürk relates this vision of cosmopolitanism back to the early days of cinema when travelling operators set 'locations on the move' and 'brought the world to the world'. Spanning the arch to the present, Göktürk ends with new configurations of spectatorship and participation in digital media and art exhibitions that transcend conventions of documentary cinema, using Wong How-Cheong's collaborative video production with children in Sulukule as an example. The role of mediators in intertwining inside and outside perspectives on the city is central to her argument. Göktürk explores the production of locality within globalization, in tune with European perceptions.

Chapter 7

The Spectator in the Making: Modernity and Cinema in Istanbul, 1896–1928

Nezih Erdoğan

Modernity and Cinema: A History of Disruptions?

Judging from the diversity and multitude of publications and events organized in and about it, Istanbul has become many things throughout the ages, including the bridge between East and West and between tradition and modernity; the capital of the Islamic World; and finally, the cultural capital of Europe in 2010. It has gained eventually the status of an icon, in the sense that it has become an epistemological tool, opening up space to discuss a wide range of issues, among which modernity is a prominent one.

Istanbul, unlike Paris of the nineteenth century, was not considered the emblematic city of modernity. However, it staged a kind of modernity blended with a peculiar kind of Westernization. In this context, I suggest setting the beginning of the history of modern Istanbul as 1896, the year the first movie screening took place, which goes counter to Çağlar Keyder's premise that 'the history of modern Istanbul, like the history of modern Turkey, begins with the end of the First World War and the demise of the Ottoman Empire' (Keyder, 2008, p. 504). I have to emphasize that I am not particularly interested in setting the dates right, nor am I unaware of the context within which Keyder is making his point. In this chapter, following Leo Charney and Vanessa Schwartz (1995, p. 1), who observe that 'cinema, as it developed in the late nineteenth century, became the fullest expression and combination of modernity's attributes', I will seek to trace specific conditions of Istanbul's modernity through cinema. Although I am fully aware of the fact that, both as an apparatus and as a socio-cultural institution, cinema has a standardizing and uniforming power, I believe we still need to consider the specific socio-cultural context within which the spectator is constituted and the modern city reconfigured.

This chapter relies heavily on archival materials, mainly newspaper advertisements and pieces of news related in some way to cinema, extracted from the periodicals published in Istanbul between 1894 and 1928. The year 1894 is the date of the first cinema-related news that we were able to locate. It reports of a stereoscopic slide-show that took place in the presence of Ottoman Sultan Abdulhamid II (Anon., 1894*a*). As 1928 is the year the Arabic alphabet was abandoned and the Roman alphabet was adopted, the archival materials examined here are not readily accessible to the contemporary reader who cannot read Ottoman-Turkish in Arabic script. Ironically, this 'violent separation', to use Mahmut Mutman's expression in this context (Mutman, undated), while being considered a radical move towards modernity on the one hand, has suppressed the story of cinema, another big sign of modernity on the other. Many advertisements I have come across are in multiple languages: Ottoman Turkish, Armenian, Greek, Hebrew, and French (figure 7.1).[1] This may suggest that these newspapers and magazines addressed a polyglot readership, fluent in multiple languages.

Enter Cinema

Cinema received a warm welcome in Istanbul. Neither the state nor society showed any significant resistance. As early as September 1896, the Chamber of Science, in response to an official query coming from the government, stated that

Figure 7.1. The newspaper advertisement of Cine Melek. (*Source*: Gökmen, 1991, p. 246)

'from a scientific perspective cinema is for the good of humanity' (Özuyar, 2007, p. 16). Although Sultan Abdulhamid was notorious for his conservatism, he 'gave permission to film and to take photographs of his weekly ceremony *Selamlık*, one of the traditional symbols of his power – the only activity for which he left the palace in order to assist to the Friday prayers in the mosque nearby his residency' (Özen, 2008*a*, p. 145).

Cinema was introduced largely by non-Muslim cosmopolites (*cihani*).[2] According to the records of *Annuaire oriental du commerce de l'industrie, de l'administration et de la magistrature 1910*, non-Muslim communities, including Italians, Jews, French and Russians, appear to have been involved in a range of professions, from selling and marketing film stock to acting and singing on stage, managing music halls, and tuning musical instruments.[3] A great majority of the non-Muslim population had previously established commercial connections in major European cities; therefore, it comes as no surprise that it was non-Muslims who introduced cinema as an innovation coming from Western Europe. Sultan Abdulhamid's daughter Ayşe Osmanoğlu remembers Bertrand, a French illusionist at the palace, who used to go to France regularly to collect items of interest. Once he brought from France a cinematographic apparatus and screened short films for the residents of the palace (Scognamillo, 1987, p. 11). In February 1896, Theodore Vafiadis, a Greek photographer from Istanbul, who had taken an interest in the *cinematographe*, wrote to the Lumière brothers and ordered one, but they had to decline, because the only *cinematographe* at their disposal was the one with which they shot and showed their films (Evren, 1995, p. 19).[4] The first public screenings were realized by a Frenchman, Henri Delavallée, a painter, who had been living in Istanbul since 1893 (Özen, 2008*b*, p. 47). He used Edison's apparatus to show Lumière films in Salle Sponeck, then a well-known beerhouse (figure 7.2). The advertisement promised, in French:

> live photograph/life-sized animated projection/marvellous and striking spectacle/that seized the whole of Paris/for the first time in Istanbul/presentations day and night/5½, 6½, 8½ and 9½/Sunday and Friday matinee. (Scognamillo 1991, p. 12; Evren 1995, p. 30)

The commercial value of cinema was immediately understood. A newspaper article of 1901 relates the invention to entrepreneurship and tells a story of success:

> How does one become rich? Lumière, the photographer. We have so far seen and learned the biographies of celebrities of various professions and arts; how they started in reduced circumstances as apprentices and then how they gradually earned millions by stepping into the world of work. Today, we are going to introduce Lumière, the photographer, who belongs to this community. Just like the others, Lumière is one of those great actors who plays a leading role on the stage of work and who has crowned his labours and efforts with success. (Anon., 1901)

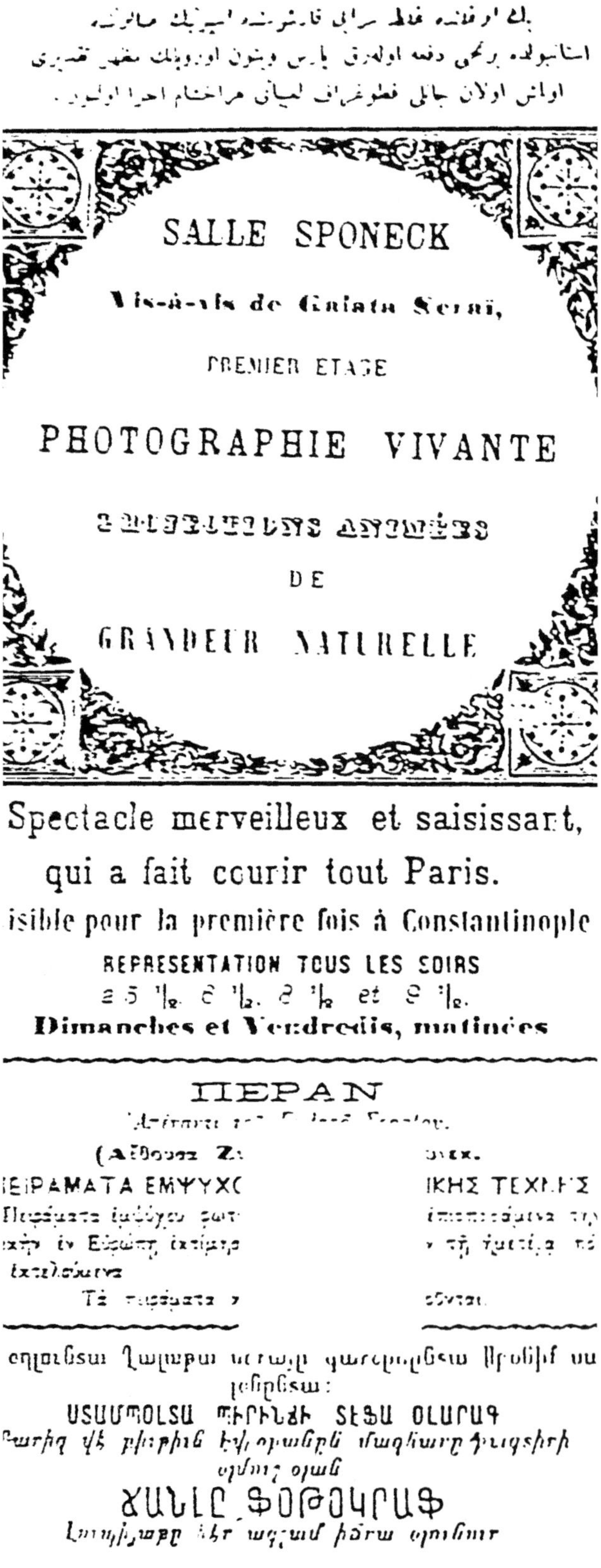

Figure 7.2. Salle Sponeck. (*Source*: Evren, 1995, p. 30)

In addition to this, correspondences of an Armenian film importer, Armenag Utidjian, show that in 1914 his company was doing business with film companies in Berlin, Paris, and London.[5] But it was Sigmund Weinberg, a Romanian Jew from Poland, who stood at the centre of the film and photography business. He sold *appareil photo* in Pera as the representative of Pathé, and in 1908 he opened the first movie theatre, Cinema Theatre Pathé Frères – formerly Anfi Theatre in Tepebaşı, near Pera. Its opening was hailed in the 'Spectacles et Concerts' column of the Istanbul daily *Le Moniteur Oriental* as '*Cinéma-theâtre de la Maison Pathé Frères, de Paris, la maison sans rivale* [the unrivalled movie house]' (Anon., 1908). The opening night's programme exemplifies how shows were punctuated with music; *ouverture-orchestre*, and then natural scenes from China, *La jolie dactilographe*, comic scene, *valse-orchestre*, a series of short films and *pot purri-orchestre*, again films, and finally the closing performance: *marche finale orchestre*.

Not unlike the preceding audio-visual apparatuses, cinema offered erotic pleasures from the outset. In 1910, *Le Moniteur Oriental* announced a special 'blue night'[6] show, *Le cinématographe suggestif*, to which women would not be admitted (Anon., 1910*d*). The Istanbul audience was already familiar with nudity and obscene jokes, common characteristics of the popular entertainment such as the Turkish shadowplay – *Karagöz*. In its early years in Istanbul, cinema was very rarely criticized for its vulgar face. An exception might be a newspaper article which complained that the Odeon Theatre in Beyoğlu (Pera), as the final attraction in the screenings, showed obscene pictures in order to attract more viewers:

> It is not only the female viewers who go to the Odeon and then have to walk out before the last number starts, it is also the young students, children who are not going to the school yet, young men who do not know the good from bad, paterfamilias. These are the significant elements of social morality, so much so that they cannot be sacrificed to a fistful of piastres… We sincerely advise paterfamilias not to send their children to such exhibitions of corruption until the police force [*zaptieh*] does its job. (Anon., 1908)

Women were not allowed to sit together with men. They could either attend women-only screenings or use the seats in the designated zones. Cinema, however, promised erotic possibilities for men as well as women. Movie-going was an excuse for women to leave their homes and stroll in the city, creating a whole new public space consisting of interiors (movie theatres) and exteriors (streets) where they met men. Hakan Kaynar (2009, p. 198) gives a vivid account of how men approached women in their carriages on their way to the cinema. Mahmut Mutman has produced a brilliant analysis of Recaizade Mahmut Ekrem's novel *Araba Sevdası* (The Carriage Affair) (written in 1883, published in 1896), putting the carriage in the centre of his discussions of modernity and defining its owner, Bihruz, the protagonist, as a 'mimic man', who could perhaps be read as an allegory of the film spectator (Mutman, nd).

The early 1900s witnessed a rapid increase in the number of movie theatres.[7] Mustafa Gökmen shows how cinema penetrated throughout the city: from Tepebaşı to Pera, to Galata, and then to Şişli, Taşkasap (the first movie theatre on the Asian side), Beşiktaş, Şehzadebaşı, Sirkeci, Erenköy, Sultanahmet, Üsküdar, Kumkapı, Doğancılar, Balat, Kuzguncuk, and finally the isles nearby: Kınalı Ada and Büyük Ada (Gökmen, 1991, pp. 19–33). As *Annuaire oriental* tells us, in 1922 there were about eighteen movie theatres and fourteen shops selling 'films and apparatuses for cinema'.

When a *flâneur* or *flâneuse* came to the Grand Rue de Pera, today İstiklal Caddesi, for a weekend promenade, he or she would pass by the Emin Coffeehouse and a *bakkal* (the small market at gate no. 1), Sadenis & Co. Beerhouse (no. 3), have an ice cream at Karayanni Brothers (no. 5), buy cigarettes at Nicolaou & Nicolaidis (no. 15), window shop at Levides jewellery (no. 45), buy postcards at Leonidas Perovolidos (no. 61) and then cross the road and enter Cinema Cosmograph, a theatre with 900 seats (no. 50), to watch a film (figure 7.3). In 1928, one year before Walter Benjamin started writing his *Arcades Project*, Ahmet Haşim, a well-known modern Turkish poet, introduced an urban character: *piyade*, the stroller (or, if you like, the *flâneur/flâneuse*).[8] He seems to have been influenced less by Baudelaire than Paul Valéry, who protested against the threat automobiles posed to this newly emerging species. Haşim (1991, p. 86) described the aimless stroller as the 'ornament' of the city streets: 'Since every stroller of the pavements comes

Figure 7.3. The map shows movie theatres (shaded areas) in 1922 which concentrate around Le Grand Rue de Pera. Adapted by Bahar Giray from *Alman Mavileri* (*Source*: Dağdelen, 2006).

out of his or her home in order to see another stroller, at every hour of the day, the street is the bed of a colourful river of humans scattering smells of face powder and bergamot'. Seeing and being seen – Haşim clearly recognized the pull of the visual; the voyeurism and exhibitionism of everyday life converged quickly with that of cinema.

Early screenings took place in buildings that were not specifically designed for cinema. Until electricity arrived in 1911, gaslight was used. Since the films were inflammable, some incidents of fire occurred. In 1922, a survey conducted under the direction of Clarance Richard John, *Constantinople To-Day: A Study in Oriental Social Life*, reported that along with cinemas, coffeehouses, beerhouses, saloons, and pool and billiard rooms were places for popular commercial amusements. By May 1921 there were thirty-two permanent and twelve temporary movie theatres:

> Undoubtedly an increasing number of people frequent cinema shows. The films shown are French, Italian, German and American. On the whole they are cheap and sensational. Very occasionally there appears a good film. The comedies are invariably coarse. There is no censorship of films except political and military. Some of the films are very suggestive and would not be allowed in America or England. (John, 1922, p. 265)

The report concludes the cinema section by stressing the need for the enforcement of existing laws concerning fire control, a board of censors to eliminate immoral scenes from films, and the construction of better built theatres (*Ibid*.).[9]

Although the early film screenings offered a specific cinematic experience, cinema was introduced as yet another form of entertainment among others, or – if you like – a segment in a series of attractions. An advertisement, for example, announces that 'Şevki Efendi Company is going to present *Şükran-ı Nimet* (Thanksgiving), a five-act comedy-drama, then *kanto* (cabaret song), cinematographe, Aşki Efendi's *fasıl* band (*incesaz*)' (Anon., 1910*c*). Drama, music, film, then music again: the menu of a night of entertainment in Istanbul in the 1900s. Again between two musical numbers, a film screening is inserted. No title given, no reference to the content made; *sinematograf* alone is good enough for the attraction. The audience's expectations positioned film somewhere amongst storytelling and music, both of which cinema more or less lacked at the time. Only fourteen years later, another advertisement referred to a film serial, *Red Glove/Kırmızı Eldiven*, as a 'cine-novel whose early episodes are rich in detail' (Anon., 1924). These two advertisements are significant in the sense of how quickly the shift in emphasis from performance as attraction to performance as storytelling took place. As storytelling plunged deeper into the concept of cinema, music had to renegotiate its place, now defining its terms more with reference to film's narrative.

One explanation for cinema's popularity and penetrative forces might be that the traditional entertainment forms, particularly *Karagöz*, bear a striking resemblance to the cinematographic apparatus: the screen, with its aspect ratio, lighting, and sound, must have prepared the audience for the cinema.[10] However, instead of tracing continuities over the traditional/modern axis, I would prefer to put emphasis on the modern subject. The weekend promenade of the Istanbul *flâneur/flâneuse*, which I have tried to reconstruct above, perfectly illustrates how modernity and cinema are articulated in urban life. Stories of the first screenings – accounts of the primal scene – tend to display similar patterns: the shock of a train moving towards the audience, the astonished audience members fleeing their seats. But did cinema develop in the same way in every city? And is there only one way of inventing modern life?

From the 'Muslim Hour' to *Matinee-Soire*: The New Temporalities of the City

By Istanbul audience, I am not referring to a homogeneous or monolithic entity. The make-up of that audience depends on a set of divisions, all of which express attributes of modernity and urban life. An examination of the evidence from newspaper advertisements reveals divisions along lines of age, gender, class, and spatio-temporality. Take, for instance, an advertisement from 1910:

> Sound Cinematographe. In Balat, adjacent to İskelebaşı, this Tuesday, at 7 pm for women only. As of today, every Tuesday and Friday, screenings for women only. Saturday and Sunday nights for men only. (Anon., 1910*a*)

The promise of an experience describes itself in terms of space and time (Balat, İskelebaşı, 7 pm, Tuesday, Friday, etc.) and divides the audience into men, women and children.

A city is a place where diverse modes of temporality can co-exist. *Ezan* (call to prayer), emanating from the minarets, or the chimes of church bells used to punctuate and structure the day. When cinema came to Istanbul, it brought with it a new programme, imposing its own temporality on the city. This was a threat to what Ahmet Haşim would call the 'Muslim hour'. Resonating with the distinction Richard Sennett (1994, pp. 200–206) draws between 'Christian time' and 'economic time'[11] in the High Middle Ages (1000–1300 CE), Haşim draws a sharp contrast between the 'Muslim hour' and modern time: 'The most hidden and influential invasion that renewed Istanbul and surprised its inhabitants was the entrance of foreign clocks/hours to our lives. By "clock/hour", I do not mean the apparatus that measures time, but time itself' (Haşim, 1921).[12] While the foreign hour imposed a twenty-four-hour day, the Muslim hour suggested a twelve-hour one, which was brief, light and easy to live with, and which started and ended with (day)light.

> The bygone hours were hours when our fathers died, our mothers married, we were born, caravans moved and the armies entered enemies' cities. These were tolerant and indifferent

friends who set free the life around us. The arriving foreigners undid our lives, re-ordered it according to an unknown principle and made it unrecognizable for our souls. Just like an earthquake, this new 'measure' shook the landscapes of time surrounding us, adding the night to the day, brought about a new 'day', which is less happy, lengthier, more tiresome, blurry. This was not the happy old hour of the Muslim, but the bitter and endless day of the great civilizations, with all their drunkards, homeless, thieves, and murderers, and whose innumerable slaves were to work underground as much as possible. (*Ibid.*)

Thus, the 'new' time, that is the time of cinema – with its points such as *matinee–soire*, as the French expression was adopted – emerged.

Cinema Le Tual Doryan. Right across the mausoleum of Sultan Mahmud, as a charity for the Source of Knowledge School, in the garden of the said school, every night, from 3 pm to 6:30 pm big show of cinematography. First class 5 piastres, second class 3 piastres, third class 2, for children 2. (Anon., 1910*c*)

Did Haşim realize that the time he mourned was replaced by the time of cinema that he praised? From 3 pm to 6:30 pm. What is the time of the citizens? The time of the city? The advertisement announces ticket prices, classifies the audience, isolates the 'child' as a separate category, and executes a further division. The film programme does not simply inform us about screening hours; it also programmes the spectator, imposing upon him or her the structure of what we would call modernity.

One last point I want to make about time is that cinema enables its spectators to experience a universal temporality, a sense of 'connectedness', an experience which is perfectly enabled by the Internet today. I want to explain this by drawing on a recent film *Vizontele* (directors: Yılmaz Erdoğan and Ömer Faruk Sorak, 2001) which tells a story of modernization: television coming to a small town in a less developed region of Turkey in the mid-1970s. In a dramatic scene, Nazmi, the mayor, gathers the inhabitants of the town and preaches the virtues of having a television: 'The national newspapers arrive two days later than they get read in big cities,' he says, 'When a piece of news gets us excited, it has already been forgotten in the big city. We want to live the same moment with the rest of the country.' I would like to suggest that cinema, in its early years in Istanbul, offered to its audiences a similar sense of contemporaneity and synchronicity. It aimed to fulfil the spectator's desire to be one with the rest of the European world.

Altered States: The Spectator's *Hayret* and Identifying with European Audiences

The train moved, silently of course. Oh Lord! It is coming straight to us. There is agitation in the dungeon-like room. I think those who feared that the train would burst out of the

> screen and run over them left their seats. I was scared too, but curiosity nailed me to my seat. Thankfully, the train quickly passed and went away. (Evren, 1995, pp. 28–31)

Ercüment Ekrem Talu, novelist and journalist, was present at the famous Sponeck screening in 1896. The film was most likely *L'arrivée d'un train en gare de La Ciotat* (1895). Talu must have just turned ten when he attended the screening of this film (his elder brother was with him; when they switched off the lights, 'my hands searched for my brother's hands in the dark'). Therefore, his narrative from the 1940s, repeating the story of the audience taking the image of a train for real, is more or less a memory-work. Referring to *The Arrival of the Train at the Station* as the 'primal scene' of spectatorship, Tom Gunning analyzes a number of stories, focusing on the encounter of the spectators with this new medium. He draws our attention from the images that are taken for real to the apparatus's power of transforming images, a magical attraction that produces the effect of motion: 'The audience's sense of shock comes less from a naive belief that they are threatened by an actual locomotive than from an unbelievable visual transformation occurring before their eyes, parallel to the greatest wonders of the magic theatre' (Gunning, 2002, p. 82).

Gunning's concept of 'astonishment' provides insight into the spectator of early cinema. The power of astonishment stems from that 'unbelievable visual transformation occurring before the eyes' of the viewer. But how long does astonishment last? What is it that remains after the initial shock is over? At this point, following Cemal Kafadar (2006, p. 18), who argues that we cannot afford not to study the Ottoman way of 'seeing' by way of *hayret* (marvel), I would like to introduce *hayret*, which denotes an initiatory experience, first by slightly decontextualizing it and then offering it in a more secular form. To put it very roughly, this is to define what the Istanbul viewer experiences once he or she overcomes the astonishment and begins to marvel at the power of the image. To marvel at a performance – here it is rather the performance of the cinematographic apparatus – is a very common state in which the viewer is believed to be subjected to a process combining mastery of technique (virtuosity) and religious experience.

In an eighteenth-century miniature,[13] the viewer is represented by human figures looking on, without being directly involved in the event taking place. *Hayret* appears as a motif which is commonly represented by human figures who, with their index fingers on their lips, apparently marvel at the happening which is centred in the image. In the case of cinema, we have a more secular form of *hayret*, because the religious aspect is now replaced by 'science and technology'. The newspapers of the late nineteenth and early twentieth century appear to have been very curious about advancements in technology. The headline in a newspaper from 1894 reads, 'The latest inventions of Edison'. Referring to the verisimilitude produced by Edison's phonokinetoscope, the news describes its effect on the viewer as 'his/her finger on the mouth out of amazement' (Anon.,

1894*b*). In 1898, the newspaper *İkdam* announced two amazing [*şayan-ı hayret*] discoveries by a poor Austrian schoolteacher: 'Everyone marvelled when the inventor Szczepanik announced his apparatus telelectroscope' (Anon., 1898). The discourse of the news handles the initiation of the pre-modern subject into the modern by using the language of religious initiation, highly charged with the tension between continuities and discontinuities.

Conclusion: Towards a National and Westernized Spectator

Istanbul has always been connected to other cities. We know that the Ottoman governments, having developed skills in intelligence, were able to acquire knowledge intelligence of every kind. The news of the fall of Granada in Spain in 1492, for example, reached Istanbul immediately. By the nineteenth century, however, the nature of this connection had been transformed into something else. The attention was now directed more towards the European capital cities. Paris in particular served as a reference point in fashion, style and popular entertainment.

Cinema conveyed images from abroad – scenes from the major capitals and landscapes of Europe – and thus fed its audiences' ever-growing desire for the West. For the local audience, the significance of these screenings was twofold. The images offered the local audience the West as an object of desire. However, more importantly, film projection was an experience in itself. Unlike its European and American counterparts, the audience of Istanbul found a point of identification that went beyond mere identification with the camera and the characters onscreen. Recall the Sponeck advertisement above: 'marvellous and striking spectacle/that seized the whole of Paris/for the first time in Istanbul'. The Istanbul audience identified with the audiences of other cities. Spectators might have thought: 'Here, I am watching a film, and looking at images that have already been looked at by European audiences, sitting in a movie theatre somewhere in Europe'. As such, the film audience came to feel part of a larger community, belonging to a space expanding onto other modern cities.

Cinema, which started out as an innovation brought to the country by cosmopolites (e.g. Henri Delavallée and Sigmund Weinberg), soon became the site of Westernization and nationalism blending into each other. Film historian Ali Özuyar (2007, p. 137) draws attention to a regulation issued as early as 1903 dictating that the personnel working at the movie theatres 'wear fez and those who are in contact with the people speak in Turkish'. On 17 February 1914, *Le Moniteur Oriental* issued reports of Izmir students' protests against films being screened without Turkish subtitles (Anon., 1914; Özuyar, 2007, pp. 91–95). Almost a decade later, nationalist fervour resulted in major change as Çağlar Keyder describes:

> When the nationalists, following the expulsion of the Greek Army from Anatolia and the departure of the last British forces, entered Istanbul in October 1922, they put an end to the

> cultural debate. From then on, nationalism was to provide the accepted narrative, attempting to reconcile cultural nativism with Westernizing zeal, while keeping Islam out of the picture. In the Turkish version, and especially under the influence of inter-war authoritarian regimes, nationalism took on a strongly statist and relatively ethnic colouring, which dictated that the Europe-oriented remnants of the Ottoman Christian bourgeoisie had to be excised from the body national in order for the nation-state to begin anew on a healthy basis. (Keyder, 2008, p. 507)

In 1929, six years after the establishment of the Turkish Republic and one year after the Roman alphabet was adopted, university students (*Dar-ül-fünun Hukuk Fakültesi Talebe Cemiyeti*) launched a campaign, supported by both the government and the press, voicing these nationalist aspirations in the form of a slogan: 'Citizen, speak Turkish!' (Bali, 2000). The campaign caused, in addition to fear, unrest and street battles, a tendency to use more Turkish names for cinemas (Gökmen 1991, pp. 13–56).

In this chapter, I have tried to tell a version of the story of 'cinema and the invention of modernity', this time the setting being Istanbul. What makes this story different from its Western European versions is that it ends with a peculiar blend of Westernization and nationalism. The reconciliation of these two lines of change implies the constitution of a new spectator, new modes of spectatorship – altogether a new chapter in the history of cinema.

Notes

1. Although the city was taken over by the Ottoman forces in 1453, Kemal Karpat's study of Ottoman population records reveals that a kind of demographic balance had always been maintained until the Republic was founded in 1923. In 1844, there were 102,532 Muslims and 111,532 non-Muslims (all male), and about fifty years later, in 1896 – the year the first movie screening took place – numbers appear to have quadrupled: 520,194 Muslims, 488,655 non-Muslims (Karpat, 1978, p. 254). In addition, Zafer Toprak (2008, p. 70) claims that in the second half of the nineteenth century, forty languages were spoken in Istanbul, and four in Pera, today Beyoğlu: Ottoman Turkish, French, Greek, and Armenian.

2. My use of the term 'cosmopolite' (*cihani*) defines a certain kind of 'individual whose country is the world' and should not be used to suggest a 'cosmopolitan Istanbul'. In this regard see Eldem's (1993, p.13) discussion of the related terms.

2. I am using the electronic database version extracted from the original source. I would like to express my gratitude to Murat Güvenç, who generously provided me with the files for the years 1910 and 1922.

4. Mustafa Özen does not seem to have been convinced by the Lumière brothers' excuse; he argues that they were already anticipating the commercial success of cinema and therefore they wanted to be the sole distributors of the apparatus (Özen, 2009, p. 184). Özen completed a PhD dissertation on cinema in the Ottoman Empire between 1896–1914 (2007).

5. My warmest thanks to Giovanni Scognamillo for making Utidjian's papers available to me.

6. A 'blue night' is actually a black night, *soirée noir*, where male viewers pick their programmes in sealed envelopes at the box office!

7. After Pathé, one cinema opened in 1911, two in 1912, four in 1913, eighteen in 1914, twenty in

1915, three in 1916, four in 1917, two in 1919, twenty-one in 1920, seven in 1921, six in 1922, and so on.

8. I will, however, stick with 'stroller', in order to avoid suggesting a direct link with Benjamin's *flâneur*.

9. This may be why data provided by insurance companies may not always be reliable, as they tend to neglect districts less likely to be found worthy of insuring because they have a high risk of fire. Movie theatres in such zones may have been left out of their research.

10. The audience was also already familiar with some of the pre-cinematic apparatuses such as *Le Grand Diaroma* (1843), *Cosmorama* (1855), *Lanterne Magique* (1882), and *Diapanorama* (1885) (Scognamillo, 1991, p.11).

11. 'This Christian time knew no idea of individual autonomy such as the corporation defined it. Imitation of Christ rather than autonomy should rule one's actions; the imitation should be strict, because time had little in common with clock time. By the High Middle Ages, the length of a confession had given to what the modern philosopher Henri Bergson calls *durée*, a "being in time" when confessor and sinner emotionally connect. Whether it lasts a second or an hour is of no consequence; the only thing that matters is that it occurs' (Sennett, 1994, p. 206). I thank Levent Soysal for bringing this to my attention.

12. In Turkish, *saat* means both clock and hour.

13. A miniature from the eighteenth century illustrates a miracle of Muhammad (And, 2007, p. 131). In order to prove to his sceptical fellow citizens that he is the messenger of God, Muhammad splits the moon into two with a snap of his fingers. In the miniature, we see the human-faced moon in two halves, Muhammed – his face veiled, with aura in the form of flames over his head, his right arm held towards the moon – and the men who are witnessing his miracle, with their index fingers (called 'finger of *hayret*') over their mouths. The man on the right-hand side is looking at Muhammad and the man on the left-hand side, who is facing Muhammad, is looking above his hand. Neither appears to be looking at the moon, but this is perhaps because the non-perspective system to which the miniature conforms does not allow or necessitate their gaze to rest on the moon. I would argue that they marvel as much at Muhammad's power to divide the moon as at the astonishing image of the divided moon.

References

And, Metin (2007) *Minyatürlerle Osmanlı-İslam Mitologyası* [The Islamic-Ottoman Mythology through Miniatures]. Istanbul: Yapı Kredi Yayınları.

Anon. (1894*a*) Alaturka bir ziyafet [A feast à la turca]. *İkdam*, 7 July.

Anon. (1894*b*) Edison'un son ihtirâ'ları [buluşları] [Latest inventions of Edison]. *İkdam*, 10 December.

Anon. (1898) Bir ihtirâ-yi cedîde [Yeni bir buluş] [A new invention]. *İkdam*, 4 March.

Anon. (1901) Nasıl zengin olunuyor? Fotoğrafçı Lumier [How does one become rich? Lumière the photographer]. *İkdam*, 12 July.

Anon. (1908) Sinematoğrafta Ahlaksızlık [Immortality in cinematography]. *Tanin*, 9 November.

Anon. (1910*a*) Sedâlı [Sesli] Sinematoğraf [Sound cinematography]. *İkdam*.

Anon. (1910*b*) *İkdam*, 31 July.

Anon. (1910*c*) *İkdam*, 18 September.

Anon. (1910*d*) Mavi suare [Blue night]. *Le Moniteur Oriental*, 4 February.

Anon. (1914) İzmir gösterileri [The Izmir demonstrations]. *Le Moniteur Oriental*, 17 February.

Anon. (1924) *Cumhuriyet*, 4 September.

Bali, Rıfat N. (2000) Vatandaş Türkçe Konuş!' veya bir ulus-devletin kuruluşunda dil birliğinin gerçekleşmesi çabaları [Speak Turkish, you citizen! Or attempts at the unity of language in the constitution of a nation-state]. Available at http://www.rifatbali.com/stories/dokumanlar/turkce_konusma_birgun.pdf. Accessed 13 January 2010.

Charney, Leo and Schwartz, Vanessa (1995) *Cinema and the Invention of Modern Life*. Berkeley, CA: University of California Press.

Dağdelen, İrfan (2006) *Alman Mavileri 1913–1914 C: I İstanbul Haritaları* [German Blue 1913–1914. vol. 1. Maps of Istanbul]. Istanbul: İstanbul Büyükşehir Belediyesi Yayınları.

Eldem, Edhem (1993) Batılılaşma, Modernleşme ve Kozmopolitizm: 19. Yüzyıl Sonu ve 20. Yüzyıl Başında İstanbul [Westernization, Modernization and Cosmpolitanism: Istanbul at the end of the 19th and beginning of the 20th century], in *Osman Hamdi Bey ve Dönemi* [Osman Hamdi and his Age]. Istanbul: Tarih Vakfı Yurt Yayınları, pp. 12–26.

Evren, Burçak (1995) *Sigmund Weinberg: Türkiye'ye Sinemayı Getiren Adam* [Sigmund Weinberg who brought Cinema to Turkey]. Istanbul: Milliyet.

Gökmen, Mustafa (1991) *Eski İstanbul Sinemaları* [Old Cinemas of Istanbul]. Istanbul: İstanbul Kitaplığı Yayınları.

Gunning, Tom (2002) Vienna Avant-Garde and Early Cinema. Available at http://www.sixpackfilm.com/archive/veranstaltung/festivals/earlycinema/symposion/symposion_gunning.html. Accessed 10 September 2009.

Gunning, Tom (2004) An aesthetic of astonishment: early film and the (in)credulous spectator, in Simpson, Philip, Utterson, Andrew and Shepherdson, K.J. (eds.) *Film Theory: Critical Concepts in Media and Cultural Studies*. London: Routledge, pp. 78–95.

Haşim, Ahmet (1921) Müslüman Saati [The Muslim hour]. *Dergah*, **1**(3). Available at http://www.40ikindi.com/ikincidonem/kitap/icerik/33.htm. Accessed 20 October 2009.

Haşim, Ahmet (1991, 1928) Sinema [Cinema], in Enginün, İnci and Kerman, Zeynep (eds.) *İkdam; Bütün Eserleri – Bize Göre/İkdam'daki Diğer Yazıları* [İkdam; Complete Works – According to Us/ Other Writings in İkdam]. Istanbul: Dergah.

John, Clarance Richard (1922) *Constantinople To-Day: A Study in Oriental Social Life*. New York: Macmillan.

Kafadar, Cemal (2006) Conference panel, in Bayrakdar, Deniz (ed.) *Türk Film Araştırmalarında Yeni Yönelimler V – Tarih* [New Directions in Turkish Film Studies V – Tarih]. Istanbul: Bağlam Yayınları.

Karpat, Kemal H. (1978) Ottoman population records and the census of 1881/82–1893. *International Journal of Middle East Studies*, **9**(3), pp. 237–274.

Kaynar, Hakan (2009) Al Gözüm Seyreyle Dünyayı: Istanbul ve Sinema [The world of my eye: Istanbul and cinema]. *Kebikeç*, no. 27, pp. 191–220.

Keyder, Çağlar (2008) A brief history of modern Istanbul, in *The Cambridge History of Turkey,* vol. 4, *Turkey in the Modern World*. Cambridge: Cambridge University Press, pp. 504–523.

Mutman, Mahmut (nd) *The Carriage Affair*, or, The Birth of a National Hero. Unpublished essay.

Özen, Mustafa (2007) De opkomst van het moderne medium cinema in de Ottomaanse hoofdstad Istanbul, 1896–1914 [The Rise of the Film in Istanbul, 1896–1914]. Universiteit Utrecht.

Özen, Mustafa (2008*a*) Visual representation and propoganda: early years and postcards in the Ottoman Empire, 1895–1914. *Early Popular Visual Culture*, **6**(2), pp. 145 – 157.

Özen, Mustafa (2008*b*) Travelling cinema in Istanbul, in Loiperdinger, Martin (ed.) *Travelling Cinema in Europe: Sources and Perspectives*. Frankfurt am Main: Stroemfeld/Roter Schriften. pp. 47–53.

Özen, Mustafa (2009) 'Hareketli Resimler' Istanbul'da ['Motion pictures' in Istanbul]. *Kebikeç*, no. 27, pp. 183–189.

Özuyar, Ali (2007) *Devlet-i Aliyye'de Sinema* [Cinema in the [Ottoman] Empire]. Istanbul: De Ki.

Scognamillo, Giovanni (1991) *Cadde-i Kebir'de Sinema* [Cinema on le Grand Rue de Pera]. Istanbul: Metis.

Scognamillo, Giovanni (1987) *Türk Sinema Tarihi* [Turkish Film History]. Istanbul: Metis.

Sennett, Richard (1994) *Flesh and Stone*. London: Faber and Faber.

Toprak, Zafer (2008) Beyoğlu'nda Batı Tarzı 'Yeni Hayat' ve Tüketim Örüntüsü [Western style 'new life' on Beyoğlu and patterns of consumption], in *Beyoğlu'nun Dünü, Bugünü, Yarını* [Beyoğlu: Yesterday, Today and Tomorrow]. Istanbul: İKSV, pp. 68–73.

Acknowledgements

This chapter is part of a research project funded by Istanbul Bilgi University. I would like to express my gratitude to Faik Gür, Şule Tezcan, Özlem Çekmece and Ayhan Han, members of the research team. I would also like to thank Özge Özyılmaz, Murat Güvenç and Giovanni Scognamillo for their continuing support; and Bahar Giray and Ateş Uslu for their valuable assistance. Uslu's assistance was provided by Özyeğin University. My thanks also go to Şirin Tekinay for making this possible. Finally, I want to express my gratitude to Deniz Göktürk, Levent Soysal, İpek Türeli and Nicholas Walter Baer without whose efforts and suggestions the completion of the chapter in this form would have been impossible.

Chapter 8

Istanbul through Migrants' Eyes

İpek Türeli

Cinema is a vital yet relatively untapped source that can be used to study Istanbul from many angles. While it entered Istanbul earlier (Nezih Erdoğan, Chapter 7), domestic cinema became a vibrant institution within the social and economic life of the city only from the 1950s onwards. Transformations in the city, including rural-to-urban mass migrations, housing problems, and class encounters, have all proved to be rich issues for films to draw on. The theme of migration is especially prominent across popular genres such as melodrama and comedy. If we were to make a list of 'migration films', Halit Refiğ's Gurbet Kuşları (Birds of Exile, or Birds of Nostalgia, 1964) and Nuri Bilge Ceylan's Uzak (Distant, 2003) would be among the earliest and latest most well-known examples of internal (rural-to-urban) migration films.[1] Migration was formative in the rapid growth of Istanbul in the second half of the twentieth century so much so that contemporary Istanbul can be considered a 'city of migrants' with most of its adult population born elsewhere in Turkey. Although no longer the driving force of the city's population growth, migration remains central to cultural imagination. How do these two films reflect the changes that occurred in the city in the forty years that separate them? Situating the films within the history of Turkish cinema, this chapter analyses the framing of the city, in particular of key buildings and vistas. Furthermore, it examines how ideas of the rural/provincial and urban are woven into articulations of national identity and citizenship.

Birds of Exile works within mainstream depictions of the figures of the migrant and the urbanite, the country and the city, aiming to reveal Turkey's social reality as the director sees it. A family of older parents, three grown up sons (Kemal, Selim and Murat) and a daughter (Fatma) arrives in Istanbul in search of fame and riches, but they lose everything and the daughter who has drifted into prostitution kills herself; they return home. *Distant*, on the other hand, opts to destabilize a clear duality between the migrant and the urbanite, and rather, points to the complexity of social and psychological relations under the impact of economic globalization. Unable to find work at home, a young man (Yusuf) arrives in Istanbul to find a job. He stays with an older relative (Mahmud) and finds both his relative and the city unwelcoming.

Both *Birds of Exile* and *Distant* feature 'entry to Istanbul' scenes that set the stage for the main characters' relationship with the city as outsiders, and both close with scenes that show their characters' displacement. That they are out-of-place as migrants is necessarily the directors' interpretation and in line with dominant discourses within Turkey on rural-to-urban migration, which tend to overlook the fact that the Turkish Republic is shaped very much by migration. In the early decades of the Republic, this was mainly forced migration determined by state policies to homogenize the population. In the post World War II period, demographic growth and economic developments led to another type of migration, from the countryside to urban centres. Until the 1950s Turkey had remained a rural society with only 20 per cent of its population living in cities; by the 2000 census, however, 80 per cent was urbanized due to internal migration (Kirişçi, 2008). The squatter settlements that the migrants built (called *gecekondu,* literally 'constructed over night') in cities became important mobilization sites for political parties and since the 1980s, as some of these squatter settlements developed into fully-fledged municipalities, they came to be associated with Islam-identified parties. This economically-driven internal migration had a profound effect on the shape and culture of cities and gave way to a verbose discourse about 'integrating migrants' within Turkey and to anxieties of the

Figure 8.1. *Distant* film poster. (*Source*: www.nbcfilm.com. *Courtesy of* NBC Film)

'provincialization' of the city in Istanbul (Erman, 1998, 2001; Şenyapılı, 2006), paralleling debates in migrant-labour receiving countries in Europe about foreign migrants. Just as the social-realist cinema of migration in Germany which has paternalistically dealt with the problem of the integration of foreigners, Turkish cinema has frequently taken on the issue of migration showing that migrants are culturally different from urbanites (Göktürk, 2002*a*). Turkish migration-to-Istanbul films have targeted a national audience consisting of both the urbanites and migrants and thus tended to straddle a fine line between selling images of Istanbul and participating in a civilizing process, simultaneously training their characters in urban behaviour and showing how they are able to resist the temptations of the city.

A Cinema Made in Istanbul

Domestic cinema was a late bloomer in the cultural and economic life of the city. Production expanded from a few films per year before World War II to several hundred in the 1960s to scale back down to the tens in the 1990s. In the immediate post-War period, Turkey aligned itself with the Western block; and America replaced Europe as the paradigm of modernity. Along with popular magazines and imported consumables, Hollywood promoted American lifestyles and star culture. By the 1960s, however, domestic products and a nascent domestic star culture were able to translate and rival that of Hollywood. During the heyday (1960–1975) of Turkish cinema referred to as Yeşilçam,[2] production consisted predominantly of melodramas and comedies; and its audience of families (Abisel, 2005; p. 200). The medium became the major public entertainment that reached mass audiences in urban centres with half of the national audience concentrated in Istanbul (Coş, 1969*a*, *b*).[3] Box office figures gradually picked up until the mid-1970s when a series of factors including the nation-wide spread of TV undermined its sway (Scognamillo, 1998).

Starting in the 1990s but especially in the 2000s domestic production was on its feet. The deregulation of state-controlled TV and radio, partly driven by the neoliberal agenda of privatization and partly encouraged by the EU's pressure to reform and democratize (Çatalbaş, 2000), led to the rise in private channels and created work for directors who could then use their earnings to fund their films (Simpson, 2006). By 2006 domestic films accounted for more than 50 per cent of box office takings. European funding schemes such as Eurimages encouraged multinational co-productions, helped to improve production values and supported distribution, thus rendering Turkish films more visible on an international stage (Göktürk, 2002*a*, *b*; Dönmez-Colin, 2008, pp. 216–218).

Yeşilçam and its audience practices may be defunct but by no means dead. The 1990s witnessed the return of Yeşilçam films and a matching proliferation of Yeşilçam-inspired TV dramas on private channels, Yeşilçam-inspired block-busters in cinemas as well as the emergence of revalorizing studies on Yeşilçam films growing out of communication faculties at private universities that have been opened in ever-increasing numbers since the 1990s.[4] In a Turkey strained by neoliberal economic restructuring and troubled by the rise of sectarian identities, this popular and scholarly interest arose from a re-evaluation of the way identity issues have been dealt with in Yeşilçam films. Within local film studies, Yeşilçam films have recently been interpreted as 'narratives of resistance' (Erdoğan, 1998, 2006) as well as 'our imaginary homeland' (Bayrakdar, 2004). Perhaps the most evocative description of the memory work old films do in the present comes from the novelist Orhan Pamuk who writes in his memoir:

> In the 1950s and 1960s, like everyone, I loved watching the 'film crews' all over the city – the minibuses with the logos of film companies on their sides; two huge generator-powered lights; the prompters, who preferred to be known as *souffleurs* and who had to shout mightily over the generator's roar at those moments when the heavily made-up actresses and romantic male leads forgot their lines; the workers who jostled the children and curious on-lookers off the set. Forty years on, the Turkish film industry is no longer … they still show those old black-and-white films on television, and when I see the streets, the old gardens, the Bosphorus views, and the broken-down mansions and apartments in black and white, I sometimes forget I am watching a film; stupefied by melancholy, I sometimes feel as if I am watching my own past. (Pamuk, 2005, pp. 32–33)

Figure 8.2. *Birds of Exile* film poster. (*Source*: *Artist Sinema Dergisi*, 1964).

Films, especially old films recycled on TV channels, perform as devices of 'prosthetic memory' (Landsberg, 2004). In addition, the relevance of these films for the present comes from their presentations of alternative perspectives or voices that other kinds of documents and hegemonic representations may not readily reveal. In fact, it is this aspect of films around which the 'cinematic city' literature has evolved since the mid-1990s in the US and UK. Much of this literature (Clarke, 1997; Penz and Thomas, 1997; Shiel and Fitzmaurice, 2001; AlSayyad, 2006; and others) attempts to bring to light the 'counter discourse' offered by films (Fitzmaurice, 2001).

Owing partly to their directors' takes on the city, and partly to their low budgets, *Birds of Exile* and *Distant* both use location shootings and amateur actors. In both, but especially in *Distant*, many details are autobiographical, incorporating elements of personal, class, and national experiences. Refiğ's (1934–2009) principal artistic and ideological inspiration self admittedly comes from literature (Türk, 2001, p. 14). Narrative structure dominates his film which tries to reach out to a wide national audience. The siblings take turns to voice their partial versions of the story. In contrast, forty years later, Ceylan (1959–) approaches cinema as an extension and development of his prior pursuit in photography, practising an extremely low-budget cinema free of corporate bonds and commitments that would be screened internationally at film festivals to a specialized audience. Action or scripted dialogues are stripped to bare essentials. Subplots (of the dying fish, the trapped mouse, and the overhauled ship) are used to express the characters' inner worlds. My analysis of the two films in the following sections follow their directors' emphasis, respectively on plot and cinematography.

Open Vistas in *Birds of Exile*

As portrayed in *Birds of Exile*, in 1960s Istanbul, social differences are marked by where the characters live, but public spaces enable encounters across social groups. Refiğ's migrant characters initially display prowess participating in the consumption of the city. They join the urbanites in *flânerie* as they stroll in the boulevards and parks, and socialize in cafes, patisseries, film theatres, and night-clubs. The city provides a stage for leisurely exploration. They can easily change their looks, but not their neighbourhood. Where the migrating family settles, in the historic peninsula with cobbled narrow roads and timber houses in decaying neighbourhoods, is clearly a space for the urban poor. The areas where Kemal's fiancé Ayla and Fatma's boyfriend live, in the northern, 'European' part with wide asphalt streets and modern concrete buildings, are the domain of the affluent middle classes. And with its churches as well as disreputable establishments, Beyoğlu/Pera is for non-Muslim minorities. Finally, there is another, 'other' Istanbul, of the squatter settlements, where rootless peasants take refuge.

When the family arrives by train at the Haydarpaşa Train Station, the camera

Figure 8.3. *Birds of Exile*, composite image of video frame captures. The family settles in a historic but impoverished neighbourhood in the historic peninsula with decaying wooden houses and cobbled, crooked streets.

Figure 8.4. *Birds of Exile*, video frame capture. The view from the old house to the north where the siblings desire to live.

Figure 8.5. *Birds of Exile*, video frame capture. Neighbourhood in the northern part of the city with modern concrete apartment buildings and asphalt wide roads.

keeps to their eye level. The characters fill the frame. The camera captures the siblings' distracted attentions, mesmerized by the view. The opening scenes establish the differences between the family and the figure of Haybeci (tramp) as they all get on the ferry to cross from Haydarpaşa to the historic peninsula. Haybeci tries to evade paying for both the train and the ferry. He further reasons with ticket collectors that he is a citizen of this country and is entitled to a free ride. The family looks down upon Haybeci with disapproval and pity. Yet they share similar dreams of wealth and prestige, and utter similar proclamations of symbolic conquest; they come to 'become shahs to Istanbul'. Both the family and Haybeci look in awe at Istanbul's silhouette, which the latter characterizes, stereotypically, as an immoral woman, as a whore (*kahpe*), predictive of the emotional-erotic cause of the failure of the family to succeed in the city – the association of the city with a seductress is a trope well established in cinema.

In the final sequence shot at the same location, the parents and one of the sons, Murat, defiantly make their way back but a new family arrives dubbed by the same utterances of symbolic conquest. The narrative comes full circle. The camera switches to a high angle that represses the newcomers within the frame, implying their insignificance in this big city. The choice of the Haydarpaşa Train Station as location is significant. As a type, the train station is a symbol of modernity where time is regulated and classes come together. Specifically, however, this train station is the terminus of railways from Anatolia. It was built (1906–1908, by German architects Otto Rittner and Helmuth Cuno) as a link in the Berlin-to-Baghdad railway. It represented an important stage in the incorporation of the Ottoman Empire into the economic system of Western powers. Through its location, projecting into the sea from Kadıköy across the Bosphorus, towards the historic peninsula, the station framed the city from a distance and became the site of departure in many literary and cinematic works. By choosing to introduce his characters through this historically loaded location, the director points to the *resemblance* of their experience to other real and imaginary migrations.

The most didactic discussion on migration takes place in the memorable sequence where the medical students, Kemal and Ayla, discuss their future in the Maçka Park. They decide to get married, but Ayla wants to go to America. The recently constructed Istanbul Hilton Hotel is selectively placed between the characters in the frame. The use of the Hilton Hotel (1951–1955, by the American firm SOM in collaboration with Turkish architect Sedad Eldem), designed to frame the city from its location in Maçka Park (Wharton, 2001), features an important commentary on Turkey's political alliances at the time. By taking Ayla to the *gecekondu* (squatter) settlements, Kemal convinces her not to go to the US for graduate study, but rather to stay in Istanbul and 'mend [their] own home'. The home Kemal shows Ayla is not Anatolia but the *gecekondu*s in Istanbul.

The *gecekondu* neighbourhood looks like a post-disaster settlement, laid out in

Figure 8.6. *Birds of Exile*, video frame captures. Haybeci (tramp) rises from rags to riches. Although he arrives in Istanbul without even a train ticket, by the end of the film, he becomes a *gecekondu* neighbourhood entrepreneur taking the train back to Kayseri to shepherd in more migrants.

Figure 8.7. *Birds of Exile*, video frame capture. Kemal and Ayla decide to get married. The director places the Hilton Hotel selectively between them to show their disagreement on their future and going to America.

Figure 8.8. *Birds of Exile*, video frame capture. By taking Ayla to the *gecekondu* (squatter) settlements next, Kemal convinces her to rather stay in Istanbul.

order, unpeopled, and unplanted. Over the shoulder shots frame the characters looking over to the neighbourhood from a distance, but neither the characters nor the camera enter this space of deprivation. Istanbul with its complex geographies and histories contrasts deeply with the *gecekondu* without reference, identity or history. The squatter settlements function as the antithesis of the city. The future lies in Kemal and Ayla, the educated intellectuals. They want to unite for a shared future, but America, iconically represented by the hotel, stands between their conflicting ideals of conformity and service, reflecting Refiğ's 'anti-migration and anti-Western theories' (Dönmez-Colin, 2008, p. 59).

One apparent effect of urban modernity is the encounter between different classes in public spaces and institutions. In one exemplary sequence, Ayla criticizes Kemal for tipping a beggar on the street. She conflates being poor with being a migrant and argues migrants come to the city to take advantage of the urbanites – speaking out her class prejudices without individual assessment or questioning. She will need to be 'enlightened' about migration first by her father and then by her fiancé. Along with Ayla, the audience is also taught that – following the nationalist myth of a migratory nation – rural-to-urban migration is part of Turkish national identity since, according to nationalist lore, Turks migrated to Anatolia from central Asia and are a migratory nation. Yet, Refiğ's approach to the migrant is not sympathetic. He uses, he says, the 'story of a family, which has migrated from a small provincial town to Istanbul in order to benefit from the opportunities of the city whose soil and stone they regard as golden, without contributing anything from themselves' (Refiğ, 1971, 2003) to garner public opinion against what he sees as 'pillage'. Both in the film and in such statements the director later made, the dependency of the formal sector to the informal one, the vitality of cheap labour provided by rural-to-urban migrants to the more affluent urbanites, go unmentioned. Personal reasons are cited for migration and the supposed failure of migrants in the city as opposed to class relations and social inequalities.

The second effect of rapid urbanization and growth is the proliferation of consumption practices in the city. Before straying off the path, the daughter, Fatma goes to the cinema, to patisseries, and to a dance party. In order to pursue these newly acquired habits, she systematically lies to her family as to where she has been. There are, of course, agents through which the city's threats are mediated and they are usually women without morals. It is the single, working woman next-door, who opens Fatma's eyes. For her brother Murat, it is the woman from the *pavyon* (bar). Murat falls in love with an independent and self-willed woman, Naciye, thinking she is an Istanbulite, but finds out that she works as a prostitute, and is in fact herself a migrant from his own hometown. She prefers her immoral life in Istanbul to an impoverished one in the provinces as Murat's wife. Another figure of the immoral woman is that of the competitor Greek mechanic's wife, with whom Selim has an affair. She seduces him to follow her on the streets of the city's 'European' quarters against the backdrop of a church. Selim soon discovers her attentions were driven by a plot to sabotage the family's car repair workshop. The city which is introduced as a whore seems to turn its women into whores.

The choice of the locations and characters is reflective of concurrent antagonisms towards the Christian-Greek (*Rum*) community and a critical one that needs dwelling on although this affair is but a subplot in the film. Following the exchange of populations between Greece and Turkey in the 1920s, the population of Istanbulite *Rums* dwindled considerably. The enmity between the two countries led to officially-orchestrated mob violence directed primarily at Istanbul's *Rum* community on 6–7 September 1955. In 1964, the year *Birds of Exile* was released, the *Rum* in the city without Turkish citizenship, about 12,000 of them, were deported from Turkey with two day's notice, as retaliation against Greece, because of the plight of Turks in Cyprus. The *Rum* population of Istanbul dwindled from about 100,000 in 1960 to 7,000 in 1978 and 1,500 in 2004 (Kirişçi, 2008, p. 183). *Birds of Exile* seems to support the 'Turkification' of the city by depicting the *Rum* woman as a saboteur.

In *Birds of Exile*, not only do characters migrate or discuss migration but also the newspapers report on the many facets of the phenomenon, all within the diegesis. The sudden sharp increase in the population is accompanied by a rapid rise in housing development. Wooden houses in the historic peninsula are abandoned to a transitory lower middle class while the upper classes opt to live in the newer modern concrete apartments in the northern part of the city. While the film demonstrates all these transformations, it also reflects the anxieties urban modernity engendered through new class encounters and consumption practices. Not only wooden houses but also the cultural integrity of Westernized, urbanite classes appear to be crumbling. In this way, *Birds of Exile* is permeated by a nostalgia in the guise of socialism sometimes and nationalism at others.

Bounded Horizons of *Distant*

In *Distant*, anxiety manifests itself via bounded horizons. The film is predominantly shot indoors and lit by diegetic lighting. The framing of characters within window or doorframes emphasizes the physical limits of spaces and the feeling of entrapment. The camera opts for a distant framing and deep focus. It rarely assumes the characters' point of view, avoiding viewer-character identification. When the characters are outdoors, the camera focuses on them in medium shots. Spatial characteristics are given through city sounds – wind, waves, horns, dog barks. Although Ceylan's camera avoids the aestheticizing pans of the city's scenery, when it does step outdoors, it aestheticizes the mundane everyday, reproducing the city as photographs.

Yusuf arrives in Mahmud's street on foot. This is not a glamorous establishing shot of Istanbul seen from the train station but a bleak view onto a commonplace unpeopled streetscape defined on the sides by contiguous apartment buildings

Figure 8.9. *Distant*, film still. Mahmud and his lover in the living room. (*Source*: www.nbcfilm.com. *Courtesy of* NBC Film)

Figure 8.10. *Distant*, DVD frame capture. Yusuf arrives in Mahmud's street. (*Courtesy of* NBC Film)

– hallmark of Istanbul's urbanization in the post-War period. He walks up the narrow street on which his cousin Mahmud lives. The medium long shot on the narrow and deserted street displays him dwarfed by the surrounding buildings. Yusuf has to stand around on the street the whole day to wait for his host. While clearly he is seeking to establish human contact during this time, he remains unsuccessful to the point of irritating residents. He does not know what to do or how to behave in public space. This figure, of the country bumpkin, is a staple character and loaded figure in many popular comedies (played most memorably by Kemal Sunal and İlyas Salman in Turkish cinema). There is also a great deal of dark humour in the staging of interactions in *Distant*.

Figure 8.11. *Distant*, DVD frame capture. Yusuf cannot see into the distance because of snow that unusually covers the city. (*Courtesy of* NBC Film)

Figure 8.12. *Distant*, DVD frame capture. Overhauled ship. (*Courtesy of* NBC Film)

The Istanbul Yusuf arrives in is abandoned except by a few tourists. It is cold and unwelcoming. Upon his arrival, the city is covered under a blanket of snow – an exceptional sight for Istanbul. Snow helps visualize the anonymity of the city. It also symbolizes Yusuf's distance from the ways of the city and according to the director, his need for a warm place and dependence on the home.[5] Yusuf trails a touristic walk through the city – from Beyoğlu to Sultanahmet to the southern banks of the historic peninsula. He cannot find a way to engage with the city and remains an onlooker/spectator. The city under snow does not allow him to see through onto the horizon and metaphorically into the future. He wants to get a job on a ship that would take him out of Istanbul and Turkey into the world and yearns to meet a woman. Yusuf's repeatedly failed attempts in both employment and love contribute to the feeling of spatial containment. The docks are abandoned. The only ship he encounters is overturned, a ruin of sorts. His imagination of Istanbul as a gateway to the world soon bleaches into whiteness.

In *Distant*'s Istanbul, in what is now a city of migrants, the differences are subtle; they are not read necessarily through contrasts in the looks of the characters or the geography they inhabit, but through the movement and performance of people in space. Paralleling the subtleness of differences, colours are desaturated. Ceylan allows the viewer to make finer distinctions of intensity. In this city, there is no one who can be identified as a rural type from his looks; there are no squatter settlements.

It is through a short business trip made to an idyllic Anatolian landscape that the urban-rural dichotomy is spatially established. The rural landscape is bright and beautiful in its simplicity and expanse. Mahmud is tempted to photograph, but cannot bother to step out of the confines of his car to do so, continuing habits he developed in the city. Inside his flat, Mahmud either stares out of his window or at his TV and computer monitors. His study is well stocked in books or music CDs, but he does not read or listen to music; he watches only news, fashion TV, porn, or Tarkovsky videos. His conversations with old friends reveal he entertained becoming a Tarkovsky-like film director in his youth, but has yielded to commercial work and abandoned his ideals over time – an example of the 'new type of urban intellectual … created in the 1980s with the social and economic changes and the hegemony of the neo-liberalist world view' (Dönmez-Colin, 2008, p. 200, following Algan, 2004; Akbulut, 2005, p. 28). In this sense it is important that Mahmud is an advertising photographer and working in creative industries, adding a self-reflexive dimension to the role of the filmmaker in the city.

Figure 8.13. *Distant*, DVD frame capture. Mahmud watching TV. (*Courtesy of* NBC Film)

Mahmud's city is constrained to interiors. When he does go out, he drives. In relation to this self-containment in interiors, Ceylan explains, '... in the city, if you have the opportunities and if you earn enough money you begin to be reserved. Firstly, you don't like to want something from others and in return you begin not to give anything to others. So you start to live in your own apartment like a prison' (Wood, 2004). Mahmud has adapted to the ways of city life very well but also suffers from some of its neuroses, namely agoraphobia.

Agoraphobia is not only the fear of open spaces as the word's etymology suggests, but also of the crowds, of being alone in public spaces, of entering new spaces, as well as of leaving the homely and the familiar. It was originally labelled as a neurosis in the late nineteenth century in response to certain behaviours the big city engendered (Phillips, 1993; Vidler, 2000). Like its counterpart *flânerie*, agoraphobia is a distinct reaction to urban modernity. The agoraphobic's recourse to the home is an attempt to protect his/her private sphere threatened by new market forces. Immobility becomes an antidote for anxieties stemming from accelerated transactions in the market place. It makes the sufferer housebound. Mahmud resorts to agoraphobic immobility as a way of protection from social interaction. The use of *mise-en-scène* and framing further the feeling of socio-spatial anxiety.

At the beginning of *Birds of Exile*, when the family reaches the rented flat on the second floor of a crumbling wooden house on a sloppy street in the historic peninsula, the siblings feel outside (modern) Istanbul. They desire to be in the picture that their window frames, in the northern part of town, in one of the modern, concrete apartment buildings there. Forty years later, it is a flat in one of those modern buildings that Ceylan occupies. The large windows of his study

Figure 8.14. *Distant*, DVD frame capture. Mahmud's window frames a nondescript view of concrete apartment buildings with the carcass of the Park Hotel in the distance. (*Courtesy of* NBC Film)

Figure 8.15. *Distant*, DVD frame capture. Mahmud by the water bank. Images of Mahmud by the water with the city in the distant background have been used on the covers of several recent books on cinema including Türkoğlu *et al.* (2004) and Suner (2010). (*Source*: www.nbcfilm.com. *Courtesy of* NBC Film)

frame a nondescript view of rooftops. Mahmud does not contemplate the view with admiration, but on the contrary, with apprehension. On the water bank, he stares at a distant historic peninsula reduced to a postcard silhouette.

Mahmud's solitude by the water bank against a background of Istanbul in silhouette mediates a particular idea of the city – isolated and in ruins. In different film posters and film scenes that feature versions of this Istanbul silhouette, three pre-Republican period landmarks dominate the skyline of the city in the distance: the Galata Tower, Hagia Sophia, and Süleymaniye Mosque. In the poster, these monuments are arranged in geographically and topographically impossible configurations. A strange quality emanates from these images of a distant silhouette of the city in the background, and a lonely man in the wintry foreground. These images mirror not only the individual's (Mahmud's) isolation but also the bounded-ness of the city's horizons. Where Mahmud stands is some 'provincial' water bank.

Of particular relevance here is Orhan Pamuk's discussion of Istanbul in his memoir. He argues a collective form of nostalgia for the lost empire that marks Istanbul. Its landscape of ruins is the bearer of this nostalgia rather than the subject which contemplates it. Rather than empire nostalgia, I would like to dwell on the notion of 'provinciality' that Pamuk raises and which has a resonance with *Distant*: 'After the Ottoman Empire collapsed, the world almost forgot that Istanbul existed' (Pamuk, 2005, p. 6). The remains of a glorious and bygone past that *Distant*'s silhouette feature and the distance of the character to the city, together, correspond to the personal loss of artistic ideals that seem to have brought him to the city in the first instance. Province, here, is not simply a geography but a sense of belonging – and simultaneously of exclusion.

A Provincialized City?

Despite lacking references or flashbacks to the past, both *Birds of Exile* and *Distant* are marked by anxieties of change and imbued by nostalgia. Comparison of the geographies depicted by the two films suggests that, from Refiğ to Ceylan, Istanbul seems to have evolved from a modern city that lends itself to *flânerie* and class encounters in public spaces to an interiorized experience of bounded horizons and agoraphobic immobility. Ceylan's Istanbul is a globalizing city increasingly experienced as a dual city of image producers on one hand, and unskilled labourers on the other. The apartment is a protection from the city's pace, but the sharing of this secluded space enables the city to penetrate the interior, thereby annihilating its privacy and security. Throughout the films the camera iconically frames certain architectural pieces, vistas and buildings. In *Birds of Exile*, the choice and framing of Haydarpaşa Train Station and the Hilton Hotel, in *Distant*, the city's silhouette speak to the characters' predicaments.

Between the release dates of *Birds of Exile* and *Distant*, Istanbul expanded tenfold in population. The different takes on migration in these two films closely follow but also critique the larger realm of public discourse on the city. *Birds of Exile* parodies the assimilationist view of 1960s sociological studies. Earlier in the film, there seems to be a clear spatial and social separation between urbanites and migrants. This distinction paradoxically promotes disguise and misunderstanding. Provincial characters can easily change their looks and pass as urbanites. The siblings disguise their identities in their romantic relationships with the urbanites. When truth is revealed, they have to face the consequences. Haybeci on the other hand, never sheds his accent or bizarre looks but is able to climb up the economic and social ladder from a day labourer to a car park attendant, an antiques dealer and finally a squatter settlement entrepreneur – Refiğ documents a well established path to riches here. With all his opportunism and disinterest in imitating/adopting urbanite culture, he represents the much-feared image of the rural migrant. Yet, it is Haybeci that Refiğ allows to stay in the city, without assimilation but excluded. Unlike other educated and cultured characters, he recognizes mass migration as economic opportunity, and perhaps more significantly, as the future of the city. Haybeci and the earlier mentioned figure of Naciye, who prefers to work as a prostitute, create ideological contrast with that of the didactic intellectual, and enable alternative readings.

Refiğ and his generation felt, in the aftermath of Turkey's first *coup d'état* (1960), that the Turkish project of modernity had to be revised and took the initiative to give it a direction in serious journals such as *Yön* as well as using popular mass media. In *Birds of Exile*, Refiğ calls intellectuals into service. He shows the city for the nation and the squatter settlements as its problems (rather than the countryside). By substituting the city for the nation, he defines the citizen as an educated urbanite and suggests his/her commitments are not to the countryside but to the city. He does so, significantly, by excluding the Christian minority. The educated ones (medical students Kemal and Ayla) are to remain while the unqualified (Selim) are to go into further exile, to Western Europe, as guest labourers, or back home (Murat and the parents). Refiğ privileges morality in the criteria of a communitarian model of urban citizenship. His urbanite characters are judged in relation to the work they do for the city and hence the nation. In contrast to Refiğ's anti-migration view, forty years later, Ceylan registers a city of migrants, asserting, '... most of the people living in Istanbul are originally from the country' (Wood, 2004). Ceylan's urbanite character manages to offend his country cousin enough to make him leave the flat but the film remains indeterminate about their future trajectories. In fact, the final scene by the water bank, in which Mahmud smokes the cheap cigarette Yusuf left behind and which he had scorned before, points to an opening.

In the 2000s, the criteria of citizenship in a globalizing Istanbul are dramatically transformed. Much of global city literature informs us that in the present era, capital and information can move across the world between global cities more readily than between cities and their provincial hinterlands. The global city engenders a new professional-managerial group/class to control and manage this capital. It produces and consumes services while also reproducing new groups of professionals. The other side of the professional-managerial class is that of unskilled labourers. A new political economy emerges between these two groups and the city emerges as a political arena where they compete (Işın and Wood, 1999, pp. 91–122).

In cities like New York or London, ethnic and racial minorities, youth and immigrants tend to constitute the latter group; and one may talk about a 'dual city' (Sassen, 2001). Members of each group may be working in the same building but may not encounter each other socially. In a globalizing city like Istanbul, because of the endurance of the cultural legacy of internal migration, conflicts between different groups may be played out in the private realm. The advertising photographer (Mahmud) who serves, and is part of, the creative industries and the professional-managerial class, and the unskilled labourer (Yusuf) can turn out to be country cousins. The basis of their cohabitation is kinship solidarity, characteristic of rural-to-urban migration in Istanbul. The two characters are forced to share the city and the flat but this leads to a sense of boundedness and consequent withdrawal.

Distant's engagement with migration is also a reflection of the endurance of migration as the prevalent trope through which the city and belonging to the city are conceptualized. Ceylan believes rural-to-urban migration is not marginal but the prevalent experience. In relationship to *Distant*, he explains, 'the subject matter is quite typical for Turkey. It happens to everybody' (Ceylan and Wood, 2004). The extent of stay in the city allows one character to condescend the other. The migrant is no longer marginal; and in what is now a city of migrants, the criteria for urbanite status depends on the inflow of newly arriving peasants. Ceylan does not romanticize or disapprove provinciality as a cultural characteristic. Rather, he takes issue with the alienation and depersonalization the city engenders in the urbanite. Ceylan's film unpeels the myth of the urbanite and expresses a futility of intellectual emancipation promoted by Refiğ as well as a critique of its absolute reverse in the form of total detachment. While Mahmud (Ceylan's *alter ego*) tries to differentiate himself from his country cousin by his skill and lifestyle, he comes to realize provinciality – that is spatial confinement – is a state they share together. While Mahmud abandons reflexive engagement with the world around him through photography, Ceylan pursues film making as a form of critical practice.

The films register a shift in perception. Benefiting from opportunities in the city or in the countryside is equally contingent on access to global flows of capital. The 'right to the city' is no longer a given or earned liberty (by means of a work ethic as Refiğ proposes). Istanbul may be a trap, a 'mousetrap', in a 'Turkey [that] is from the small town of the world' (Ceylan in interview with Öğünç, 2003). Istanbul under snow represents Turkey's position *vis-à-vis* the world. It might be an unfit habitat for the migrant but, as a symbol in national imagination, it stands for the entrapment of Turkey in a globalizing world economy; perhaps its overlapping cluster of anxieties of entering the European Union, as well as not entering; of losing the small town innocence of a culturally and economically bounded nation-state.

Notes

1. *Birds of Exile* and *Distant* were selected among the best five films of the last forty years by Turkey's prominent Antalya Film Festival in 2003 on the festival's fortieth anniversary. *Birds of Exile* received the Antalya Film Festival's first Golden Orange in 1964 and *Distant* received the 2002 Golden Orange. *Birds of Exile* was part of the most 'ambitious' retrospective of Turkish cinema held at the Pompidou Centre in Paris in 1996. It was shown in 2000 at the annual London Turkish Film Festival, and in 2003, at the Annual New York Turkish Film Festival. After *Distant* received the Grand Prix at the 2003 Cannes Film Festival, its director Nuri Bilge Ceylan became an international name; the film has been shown in international festivals, in retrospectives of Turkish cinema *and* retrospectives of Ceylan's work abroad, and in art house cinemas around the world.

2. 'Yeşilçam' is named after the street on which film businesses concentrated. It translates literally as 'green pine', perhaps also an allusion to Hollywood.

3. It may be useful to cite some figures to get a sense of the increasing prominence of cinema. The number of cinema tickets sold in Istanbul per person was only 10.3 in 1940, 11.8 in 1950 but increased to 16.5 in 1960 and 23.6 in 1968. This meant that by 1968 on average everyone in the city went to the cinema twice a month. The number of film theatres increased accordingly. By 1968, Istanbul had 150 indoor and 260 outdoor film theatres (Coş, 1969*a*, *b*).

4. State universities' Communication, Radio, TV and Cinema departments have been critically studying Turkish cinema films since the early 1980s. The first annual conference on Turkish film research (*Türk Film AraTürk Film Araırmalarında Yeni Yönelimler [New Directions in Turkish Cinema]* was held in Istanbul in 1999 at a private university. There has been a deliberate interest within Turkish film studies circles since then to analyze early films, and a small, but increasing effort, to make connections between cinema and the city (proceedings, Bayrakdar, 2001–2008). Some of the more local publications that take on cinema and the city include, but are not limited to Öztürk (2002) and Türkoğlu *et al.* (2004). Makal (1987) and Güçhan (1992) are earlier works that concentrate on representations of migration to the city.

5. Nuri Bilge Ceylan Interview, Uzak DVD Commentary.

References

Reviews of *Distant*. Official website of Nuri Bilge Ceylan. http://www.nbcfilm.com/uzak/. Accessed 11 April 2010.

Abisel, Nilgün (2005) *Türk Sineması Üzerine Yazılar* [Writings on Turkish Cinema]. Ankara: Phoenix.

Akbulut, Hasan (2005) *Nuri Bilge Ceylan Sinemasını Okumak: Anlatı, Zaman, Mekan* [Reading Nuri Bilge Ceylan's Cienma: Narration, Time and Space]. Istanbul: Bağlam Yayınevi.

Algan, Necla (2004) İç Basından alıntılar: 'Bir kez daha *Uzak*: Türk Sineması'nda tematik açıdan bir kırılma noktası' [Extracts from domestic press: 'Another time *Distant*: A Thematic Breaking Point in Turkish Cinema']. In *Nuri Bilge Ceylan: Uzak* [Nuri Bilge Ceylan: Distant]. Istanbul: Norgunk. Original edition, *Yeni Sinema* Winter 2003–2004. Available online at http://www.nbcfilm.com/uzak/press_yenisinemanecla.php. Accessed 12 April 2010.

AlSayyad, Nezar (2006) *Cinematic Urbanism: A History of the Modern from Reel to Real*. London: Routledge.

Bayrakdar, Deniz (2004) Türk Sineması; Hayali Vatanımız? [Turkish cinema: our imaginary homeland?], Bayrakdar, D. (ed.) *Türk Film Araştırmalarında Yeni Yaklaşımlar IV* [New Approaches in Turkish Film Studies IV]. Istanbul: Bağlam Yayınları.

Çatalbaş, Dilruba (2000) Broadcasting deregulation in Turkey: uniformity within diversity, Curran, J. (ed.) *Media Organisations in Society*. London: Arnold.

Clarke, David B. (1997) *The Cinematic City*. London: Routledge.

Coş, Nezih (1969*a*) Türkiye'de Sinemaların dağılışı [The distribution of cinemas in Turkey]. *As Akademik Sinema*, No. 2 , pp. 19–20.

Coş, Nezih (1969*b*) İstanbul'un Sinemaları... [Istanbul's cinemas...]. *As Akademik Sinema*, No. 4, pp. 11–20.

Dönmez-Colin, Gönül (2008) *Turkish Cinema: Identity, Distance and Belonging*. London: Reaktion.

Erdoğan, Nezih (2006) Narratives of Resistance: National identity and ambivalence in the Turkish melodrama between 1965 and 1975, in Eleftheriotis, D. and Needham, G. (eds.) *Asian Cinemas: a reader and guide*, Honolulu: University of Hawaii Press (original edition: (1998) *Screen*, **39**(3), pp. 259–271).

Erman, Tahire (1998) Becoming 'urban' or remaining 'rural': the views of Turkish rural-to-urban migrants on the 'integration' question. *International Journal of Middle East Studies*, **30**, pp. 541–561.

Erman, Tahire (2001) The politics of squatter (gecekondu) studies in Turkey: the changing representations of rural migrants in the academic discourse. *Urban Studies*, **38**(7), pp. 983–1002.

Fitzmaurice, Tony (2001) Film and urban societies in a global context, in Shiel, M. and Fitzmaurice, T. (eds.) *Cinema and the City: Film and Urban Societies in a Global Context*. Oxford: Blackwell.

Göktürk, Deniz (2001) Turkish delight – German fright: migrant identities in transnational cinema, in Ross, K., Derman, D. and Dakovic, N. (eds.) *Mediated Identities*. Istanbul: Bilgi University Press.

Göktürk, Deniz (2002*a*) Beyond paternalism: Turkish German traffic in cinema, in Bergfelder, T., Carter, E. and Göktürk, D. (eds.) *The German Cinema Book*. London: BFI.

Göktürk, Deniz (2002*b*) Anyone at home? Itinerant identities in European cinema of the 1990s. *Framework. Special Issue on Middle Eastern Media Arts*, **43**(2), pp. 201–212.

Güçhan, Gülseren (1992) *Toplumsal değişme ve Türk sineması: Kente göç eden insanın Türk sinemasında değişen profile* [Social Change and Turkish Cinema: The Changing Profile of the Migrant to the City in Turkish Cinema]. Ankara: İmge Kitabevi.

Işın, Engin, and Wood, Patricia K. (1999) Cosmopolitan citizenship: contested sovereignties, in Işın, E. and Wood, P.K. (eds.) *Citizenship and Identity*. Thousand Oaks, CA: Sage.

Kirişçi, Kemal (2008) Migration and Turkey: the dynamics of state, society and politics, Kasaba, R. (ed.) *Turkey in the Modern World*. Cambridge: Cambridge University Press.

Landsberg, Alison (2004) *Prosthetic Memory: The Transformation of American Remembrance in the Age of Mass Culture*. New York: Columbia University Press.

Makal, Oğuz (1987) *Sinemada Yedinci Adam: Türk sinemasında iç ve dış göç olayı* [Seventh Man in Cinema: Internal and External Migration in Turkish Cinema]. Izmir: Köprü, Mars Matbaası.

Öğünç, Pınar (2003) Nuri Bilge Ceylan Interview: From *Kasaba* to *Uzak*. *Turkish Time*, No.16.

Öztürk, Mehmet (2002) *Sine-Masal Kentler: Sinematografik Bir Üretim Alanı Olarak Kent Üzerine Bir*

İnceleme [Cine-tale cities: An investigation on the city as a field of sinemotographic production]. Istanbul: Om.

Pamuk, Orhan (2005) *Istanbul: Memories of a City* (translated by M. Freely). London: Faber and Faber.

Penz, François, and Thomas, Maureen (1997) *Cinema & Architecture: Méliès, Mallet-Stevens, Multimedia*. London: British Film Institute.

Phillips, Adam (1993) *On Kissing, Tickling, and Being Bored: Psychoanalytic Essays on the Unexamined Life*. Cambridge, MA: Harvard University Press.

Refiğ, Halit (1971) *Ulusal sinema kavgası* [National Cinema Stuggle]. Istanbul: Hareket Yayınları.

Refiğ, Halit and Karadoğan, Ali (2003) *Halit Refiğ* (edited by A. Karadoğan). Istanbul: YKY (originally published in *Sinema*, 2 February 1965).

Sassen, Saskia (2001) *The Global City: New York, London, Tokyo*. 2nd ed. Princeton, NJ: Princeton University Press.

Scognamillo, Giovanni (1998) *Türk Sinema Tarihi, 1896–1997* [History of Turkish Cinema, 1896–1997]. Istanbul: Kabalcı.

Şenyapılı, Tansı (2006) Gecekondu Olgusuna Dönemsel Yaklaşımlar [Period Approaches to the Phenomenon of *Gecekondu*], Eraydın, A. (ed.) *Değişen Mekan: Mekansal Süreçlere İlişkin Tartışma ve Araştırmalara Toplu Bir Bakış, 1923–2003*. Ankara: Dost Kitabevi Yayınları.

Shiel, Mark, and Fitzmaurice, Tony (2001) *Cinema and the City: Film and Urban Societies in a Global Context*. Oxford: Blackwell.

Simpson, Catherine (2006) Turkish Cinema's Resurgence: The 'Deep Nation' Unravels. *Senses of Cinema*, No 39. Available at http://www.sensesofcinema.com. Accessed 12 April 2010.

Suner, Asuman (2010) *New Turkish Cinema: Belonging, Identity and Memory*. London: I.B. Tauris.

Türk, İbrahim (2001) *Halit Refiğ: Düşlerden düşüncelere söyleşiler* [Halit Refiğ: Conversations from Visions to Thoughts]. Istanbul: Kabalcı.

Türkoğlu, Nurçay, Öztürk, Mehmet and Aymaz, Göksel (2004) *Kentte Sinema, Sinemada Kent* [Cinema in the City, City in Cinema]. Istanbul: Yeni Hayat Yayıncılık.

Vidler, Anthony (2000) *Warped Space: Art, Architecture, and Anxiety in Modern Culture*. Cambridge, MA: MIT Press.

Wharton, Annabel Jane (2001). *Building the Cold War: Hilton International Hotels and Modern Architecture*. Chicago, IL: University of Chicago.

Wood, Jason (2004) *A Quick Chat with Nuri Bilge Ceylan*. Available at http://www.kamera.co.uk/interviews/a_quick_chat_with_nuri_bilge_ceylan.php. Accessed 12 April 2010.

Acknowledgments

I thank Nezar AlSayyad and Deniz Göktürk for nurturing my interest in the relationship between cinema and the city. A preliminary version of this chapter was presented at the Visualizing the City conference held at the University of Manchester in 2005. In the process of developing my ideas, I have benefited from Chris Berry's feedback on an earlier version, and Deniz Göktürk's detailed comments on the final draft. Halit Refiğ whom we lost in 2009 was very sympathetic towards researchers; I was lucky to conduct an interview with him in 2002. Gülper Refiğ graciously granted permission to use video frame captures from *Birds of Exile*, a digitized copy of which Ertem Göreç kindly shared with me. NBC Film was generous to give permission to use publicity film stills and DVD captures of *Distant*.

Chapter 9

Istanbul Convertible: A *Magic Carpet Ride* through Genres

Deniz Bayrakdar and Elif Akçalı

The film *Magic Carpet Ride* (*Organize İşler*, literally 'Organized Business'), a hit at the Turkish box-office in 2005, effectively fulfils the ride promised in its English title:[1] out of approximately thirty scenes, thirteen are composed of helicam shots, showing Istanbul from the air. Nearly ten minutes of the total film are devoted to aerial vision. The camera flies over the Bosphorus from almost every possible angle; it zooms in and out of the historical peninsula, the city centre and the new suburbs, producing spectacular images. Istanbul was never photographed like this, seen from extreme high angles in sunlight and at night. These shots link the various layers of the film. They serve as an invisible eye (I) that rides between different levels of the city, as Aladdin on his magic carpet. This omniscient eye is ours; throughout the film it enables the spectator to master the city of Istanbul and access connections beneath the surface. On these long flights over the city, the cinematography lends the spectator authority. Along with the visual excesses of the helicam shots, the soundtrack, featuring pop icons Brooklyn Funk Essentials, Hüsnü Şenlendirici, Laço Tayfa and Nil Karaibrahimgil, adds a self-reflexive quality to the viewing experience. Brooklyn Funk Essentials' 'Magic Karpet Ride' used over these aerial shots in the film's opening credits introduces the spectator to the ride: 'We have the ticket for a magic carpet ride, non-stop to the other side' (figure 9.1). This flight over the Bosphorus into the city centre is repeated in reverse order in the final sequence, as the camera rises up, flying from the city centre towards the Golden Horn and to the outer spaces of the Marmara Sea with a view of the Prince Islands, like a simulation of Superman in flight. In the theme song 'Organized Business' (*Organize İşler*), which underscores the final shots of the film, singer Nil Karaibrahimgil confesses that she can be neither inside nor outside this 'organized business' that 'commits a crime' against Istanbul; she can neither fly nor escape; she is neither Clark Kent nor Superman.

The past couple of decades have seen a rapid growth in Turkish domestic film production[2] and consequently an increase in the number of films that tell stories in and about Istanbul, the most popular cinematic stage for dramas, gangster films and comedies throughout the history of Turkish cinema.[3] Istanbul has scene is also significant in terms of marking the film's fusion of different genres. Samet's situation is seen in a humourous light, but his benevelont character resembles the good-hearted male protagonists from classic melodramas. Asım could also just as well belong in a drama if he was not caricaturized. As the film progresses, we notice the blending of uncommon themes, characters and incidents. *Magic Carpet Ride* is a comedy that reverses the characteristics of some stereotyped characters and carries traces of dramas and gangster films.

The Aerial Gaze

From the mid-1990s onwards, a substantial number of Istanbul films portrayed the city in new ways while recalling and recycling well-known Istanbul images from earlier films of the 1950s and 1960s. Set in central neighbourhoods such as Cihangir, Tarlabaşı and Beyoğlu, *The Bandit* (*Eşkiya*, Yavuz Turgul, 1996) marked a turning point at the box-office. The film tells a story of violence and vengeance and depicts the difficulty of survival for newcomers from Eastern Turkey, the poor and the marginal in the city. *The Bandit* was one of the first films to use extreme high angle images of the city and the sea shot from rooftops. The city viewed from above became more popular in the following years, as films set out to capture the expansive and contingent totality of the city. *All About Mustafa* (*Mustafa Hakkında Her Şey*, Çağan Irmak, 2003) chose the suburban outskirts and forests as its primary sites. Locations that previously tended to be used as romantic hideaways for lovers, turned into spaces of terror in *All About Mustafa*. *Istanbul Tales* (*Anlat Istanbul*, Selim Demirdelen *et al*., 2004) showcased some of the most recognized cinematic landmarks in the city such as Haydarpaşa Station and the old Galata Bridge, and assigned a story to each. The stories are modelled on well-known fairytales by the Brothers Grimm. There is, of course, the historical figure of Hezarfen Ahmet Çelebi, a Turkish Icarus whose lifelong efforts in seventeenth-century Ottoman Istanbul were devoted to flying over the city with self-made wings. His quasi-mythlogical story was turned into the period film *Istanbul Under My Wings* (*Istanbul Kanatlarımın* Altında, Mustafa Altıoklar) in 1996, featuring ample aerial views of Istanbul. Magic Carpet Ride offers a penultimate realization of this flying fantasy.

At the same time as Turkish cinema witnessed a major increase in film production and ticket sales, efforts to market Istanbul as a global city both to the rest of the world and its own inhabitants intensified. This obviously meant creating attractive images of Istanbul as well as offering experiences to enjoy these images in reality. 'Dinner in the Sky' and 'Looking at Istanbul from 236 metres' from the Sapphire Tower are both city extravaganzas.[7] Views of the city from these heights offer a sense of liberation to those Istanbullites who are squeezed in traffic jams every day or isolated in gated communities. A 'Magic Carpet Ride' could be the desire of everyone who feels restricted or exhausted by spatial confinement in the city. *Magic Carpet Ride* offers the spectator a virtual tour of Istanbul from the air. However, the film also ironically complements the omnisicient aerial gaze by zooming in on significant interactions at street level, revealing unlikely proximities in the circulation of economic and cultural capital in the city.

Istanbul as Convertible Spectacle

Istanbul is a marketing and sales product in *Magic Carpet Ride*. The tour over the city, which hands the power of Superman to the spectator, is experienced by buying a ticket for the film. The cost of a ticket buys a ride over Istanbul, thereby conferring an authority over the city and the rights to enter otherwise inaccessible sites, from a den of thieves in the inner city to gardens of gated communities on the outskirts of the city. *Magic Carpet Ride* enables the reconstruction and recycling of Istanbul's images, sounds, stories, characters and genres into a spectacular world. The film features a glossy visual aesthetic; its art direction and style are reminiscent of advertising, employing colour correction filters, continuous steady-cam and helicam shots, saturated colours, and fast-paced editing. Its production process is also highly unusual, as Yılmaz Erdoğan carries out nearly all the main functions in the film. Originally a stage actor, he wrote the screenplay, directed and produced the film, and played the lead role as Asım, the head of the gang of thieves. *Magic Carpet Ride* ranked ninth among the films released since 1990 in Turkey with a total of 2,552,550 viewers (Akıncı, 2006. p. 42).[8] The film was distributed by *Kenda* (short for '*Kendin Dağıt*', literally 'Distribute Yourself'), a film distribution company started in 2004 by director-producers Yılmaz Erdoğan (BKM) and Sinan Çetin (Plato Film) in collaboration with AFM cinemas. In other words, the production and distribution of this domestic block-buster is a one-man show.

Magic Carpet Ride's true protagonist is Istanbul; in fact, this is a tagline that Erdoğan often repeated[9] before the film's release. It is a film that devotes a considerable amount of time to Istanbul at the risk of upsetting the rhythm of the narrative. Furthermore, the hybridity of genres and characters exemplifies a shift in the contemporary urban order. However, a further issue that this film raises for discussion is the possible economic motivation behind such stylistic choices. The screen time that *Magic Carpet Ride* devotes to desirable shots of Istanbul suggests that the city itself is the main subject of this film, contributing to this economic aspect. *Magic Carpet Ride*'s overarching genre is comedy, but it borrows from many other genres to construct its narrative and style. The film's story develops from the chance meeting of thief Asım and comedian Samet impersonating Superman. This moment is emblematic of the fusion of crime and comedy: in Istanbul, the gangster performs both. Superman Samet has the opportunity to perform his mediocre stand-up show in the gangster world by swindling people. Samet is introduced to the business of stealing cars through Asım's narration of his past as a young man, as the film cuts to Asım's story as an innocent novice.

Figure 9.3. Asım and his girlfriend in a convertible car, which is converted into a sofa. (*Magic Carpet Ride*. © BKM 2005, Scene 9)

This part resembles the mise-en-scène of 1960s Turkish comedies. Crime and comedy in the film are further mixed with romance.

Asım's wife has left him for a dentist to whom Asım sends his men for dental surgery. The ex-wife and her new dentist husband live in one of the luxurious residential areas on the outskirts of Istanbul, while remaining aloof from the crime and dirt of the city's rotten centre occupied by the likes of Asım. Asım longs for his ex-wife and his daughter as he secretly observes them enjoying themselves in their garden. Asım's new girlfriend also leaves him after entering a beauty contest. She tells him that she is going to leave him, as they sit in a classic American 'convertible car' on the balcony (a car without tyres, ironically immobilized). Asım's Hawaiian shirt reminds one of the costumes of characters from cult road movies, with the exception that he himself cannot move anywhere. Asım may be head of a gang of petty thieves but he is a looser in his private life and as miserable as Superman Samet. He, too, engages in comical acts to compensate for unfulfilled longings for his ex-wife and child. Superman Samet's engagement with Asım's business brings him luck in his romantic affairs which then gives him confidence, by the end of the film, to turn back to comedy with a renewed sense of self-esteem.

Superman Samet's first customer/victim is a young girl (Umut, literally 'hope'), the daughter of two academics. Umut represents the upper middle class in the film. As a recent graduate in English literature, she unsuspectingly takes a job as a public relations secretary with a big-time mafioso. After her car is stolen, she decides to ask for help from this Mafia boss and finds herself in a story much like that of *House of Games* (David Mamet, 1987). As the daughter of intellectuals whose professional and familial roles are 'transgendered' according to global professional shifts (her mother teaches quantum physics, while her father is a sociology professor), she symbolizes the integration of natural and social sciences. She stands for the new language of the city, which is the power of black money converted into and legalized through the attraction of the spectacle. The inversion of stereotypical gender roles underlines the change in social and cultural identities. Yusuf, the father, is given the role of a more passive persona at home and in the family, and his logic is always questioned and underestimated by the strong wife/mother.

The film introduces us to Umut's family through the image of the failed father, Yusuf. At the outset, the camera shows an extremely crowded street in which we briefly witness someone snatching a woman's purse. It zooms into a deserted bookstore and pauses on a disillusioned and unhappy man sitting quietly at a table. Yusuf has written a book called *Societies with a Weak Sense of Unity and the European Union*, and this is his book-signing day. Even though the book does not raise any interest in Turkey, it wins him an award of twenty thousand euros from the European Union, which is announced in a letter that surprisingly arrives by regular mail. After the scene in the supermarket where Yusuf is criticized for acting like a boy by 'the queen mother/wife/the master of the atoms and molecules' the couple arrive home where she mocks him continously about his lack of logic and ridicules him because of his attempts to act as the serious father or pedantic bureaucrat trying to uphold standards, as he checks the expiry dates on food packages. The daughter Umut comes home, and a neighbour delivers the letter with the cheque sent by the European Union. The daughter is excited about the long-awaited success of her father. Until this point, the film shows the stereotyped academic father as boring, unsuccessful, and detached from the public. His book does not find any readers in Turkey; only Europe supports science and knowledge. Furthermore, his reward of twenty thousand Euros seems a very small amount compared to the money that the thieves are stealing, but it is exactly the amount that converts itself into a second-hand jeep for Umut which they call 'the white pigeon'. On their first drive, a police officer stops them and declares that the white pigeon is in fact a stolen car! Umut and her mother take on the task of recovering their money from the crooked sellers while Yusuf advises they wait for the results of the official investigation – he never steps out of his failed, cowardly and slow character. In the end, it is Umut and her mother who manage to get their money back – although they do so illegally.

Yusuf's impotence, when contrasted with Asım, suggests a further contrast between Europe and Turkey. Asım/Yılmaz Erdoğan is the central character in the story; this is his world and his film. Director Yılmaz Erdoğan finds uninteresting that which Europe finds the need to reward; his answer to Yusuf's domestically failed book is this film and the vision of a networked Istanbul that it presents. Significantly, Yılmaz Erdoğan's film, which was a box-office hit in Turkey, did not circulate internationally or achieve acclaim at festivals abroad. As the director of the film, he makes fun of the intellectual who is successful in Europe but ignored in Turkey. In a sense, Erdoğan gets his revenge by presenting Yusuf as a failure in his domestic affairs at home and in Turkey. Yusuf who is rewarded by Europe and who is connected to it via books and letters can be seen as representing festival film directors with less box-office success in Turkey who are critically acclaimed in Europe. Yılmaz Erdoğan and his brother Mustafa Erdoğan have both been entrepreneurial organizers of spectacular events in Turkey for domestic audiences.[10]

Cem Yılmaz, who plays the role of the Mafia boss Müslüm, is a prominent media star, actor and director who began his career as a caricaturist and stand-up comedian. He is seemingly on a track parallel to Yılmaz Erdoğan in the entertainment business. They are both rivals and colleagues and prominent figures in the Istanbul entertainment world and previously collaborated in the box-office hit *Vizontele* (Yılmaz Erdoğan, Ömer Faruk Sorak, 2000). Cem Yılmaz and Yılmaz Erdoğan's films are domestic successes without exception, and both are very skilful in marketing their strong media personalities. Their roles in the film are therefore cast in a very clever way, reflecting their real-life status in the entertainment/spectacle business: Müslüm (Cem Yılmaz) the higher-ranked Mafia boss, and Asım (Yılmaz Erdoğan) the ordinary swindler. In the end, Asım and his gang are captured by security cameras in Müslüm's ranch in the new suburban areas outside Istanbul. The confrontation scenes between Müslüm and Asım are organized in such a way that although violence is involved, there is a playfulness between them. The film has multiple happy endings. The daughter Umut gets her father Yusuf's money back from the swindlers with the help of the Mafia boss. The swindlers forswear 'organized business'. Asım's wife leaves her second husband to come back to him with their daughter. Asım decides to turn into a proper 'real estate manager'. Umut brings Samet flowers in the club where he, who could at the beginning of the film neither fly nor make people laugh, finds his way as a stand-up as a comedian who parodies his *alter ego* as Superman. Finally, as in any classic film noir, morality and justice are restored. We again look at the city from above while the theme song plays: 'I'm neither Clark Kent nor Superman; this is an organized business'.

Re-orienting Istanbul

With awe, horror or indifferent familiarity, we are witnessing Istanbul change rapidly in terms of its spatial relations and imaginary geography, as the city has undergone rapid restructuring over the past two decades. Its skies are pierced by ever-taller and multiplying towers, housing developments, offices, and residential apartments, as well as colossal luxury hotels. Its cityscape is crowded by shopping malls, restaurants, cafes and nightclubs in ever increasing numbers. Its arts calendar is getting busier every year, with evermore music and film festivals, exhibitions and activities in the newly opened museums, as well as the Istanbul Biennial. As Istanbul transforms, it makes space for antitheses. The condition of blurred opposites in *Magic Carpet Ride* can be observed in the way settings are used. Theft and fraud are preconditions for living in Istanbul, and its most beautiful locations are owned by underground figures. The mafia headquarters are located in big suburban houses surrounded by green gardens, equestrian paths and swimming pools. They are introduced by the camera flying above filming from different angles; the images look as though they were taken from a commercial for these types of residences. There are people picknicking on the grass, swimming in the pool and wandering around. This peaceful environment contrasts starkly with the criminal intent of its owners. The dimly lit, dirty and hidden locations of the underground are substituted by luxury homes in the gated communities of the *nouveau riche*.

Since the 1980s, successive governments and the elite have pursued Istanbul's global city project, aiming to gain a significant place in the world economy by mobilizing the city's history, geopolitical and cultural values (Öktem, 2005, p. 63). Changes in the actors of the real estate market in particular have had an enormous impact on urban restructuring, especially during the 2000s (Bartu and Kolluoğlu, 2008, p. 16). *Magic Carpet Ride* is especially acute at underlining the phenomenon of real estate speculation. Even the ending of the film refers to real estate as the profession of the future, as gangster Asım decides to become a real estate agent. In his contribution to this book, Çağlar Keyder describes the successful marketing process of Istanbul '[that] proceeded along expected lines: the historical riches of the city as well as its night-life and culinary diversity were (and are) highlighted, along with dozens of music, art, and film festivals, new museums and exhibits' (see the opening paragraph of Chapter 1). Parallel to this marketing process, *Magic Carpet Ride* captures these different forms of culture and entertainment in various shots such as that of a belly dancer and musicians in a narrow street in Beyoğlu, of a trendy and modern nightclub by the Bosphorus or a night shot of the Bosphorus Bridge with fireworks (figure 9.4).

Another important issue in the film is cultural transformation and the preferences of the new middle class for gentrified spaces. The actors in this process are similar to those of Manhattan: 'artists, investors, real estate agencies, manufacturers and politicians' (Şen, 2005, p. 145). Şen defines the process itself as 'an organized experience'[11] (*Ibid.*, p. 145), recalling the Turkish title of the film,

Figure 9.4. A helicam shot of Istanbul by night. (*Magic Carpet Ride*. © BKM 2005, Scene 4)

Organize İşler (Organized Business). Asım and his gang inhabit the city centre and in a sense steal the city from the new middle class who want to transform these spaces. On the other hand, the spaces outside the city, like the Mafia boss Müslüm's gated residential area, build a differentiated sub-urbanization compared to mid-century middle-class suburbs such as Levent which was often featured in domestic films of the 1950s and 1960s. The new breed of gated suburbs are usually accessed only by highways in private cars. They are designed specifically to be shielded from the general public with gates and walls. Their security and other services are provided by private companies. As such, the new breed of suburbs satisfy a particular understanding of social distinction that emerged due to Istanbul's integration into the network of global cities.

As Asım's gang is brought to Müslüm's residence, we see him playing golf, women swimming in a pool and Asım riding a horse. This space resembles the advertisements of new residences in Istanbul that offer 'a distinguished life style'. Müslüm brings Asım's gang into his kitschy office to show the closed-circuit video system and the huge plasma monitors. He tells the terrified gang members in a 'godfatherly' tone that video is a great innovation and plays the recording of Asım and his men secretly entering his residence. Müslüm rewinds this several times to show the swindlers his supremacy and distinction as the one who owns the gaze. This is actually a form of cultural capital, as is made abundantly clear in the next scene in the stable where he hits the gang with golf balls. Before commencing this torture, he speaks very quietly of his love of golf and how he would like to spend time every day playing golf. However, he confesses that small things, like the need for torture, distract him from his sports activities. He asks, 'am I expecting too much?' and the miserably tortured gang members show empathy for his wish. In this scene, the video and the closed-circuit surveillance symbolize the 'profound security need', while the private swimming pool, horse riding and golf, which serve as distinctive sports and cultural activities – and hence as cultural capital – underline the harsh 'walls' that divide the city spaces according to class.

There is a spatial and social shrinking in the lives of the residents due to the constraints on mobility in the 'off worlds' (Bartu and Kolluoğlu, 2008, p. 23), which can be seen in the isolated suburban residential areas where we are shown, for instance, the image of Asım's ex-wife's newly formed family shot with telephoto lenses. She also lives in one of those villas in the suburbs similar to Müslüm's with her dentist husband. She has taken her daughter away from Asım and away from illegal businesses to bring her to an uncorrupted space while, ironically, Müslüm may be their neighbour.

Money and wealth are indeed recurrent themes throughout the film, and they are used to hint at the contrast in interests of different groups, as well as their overlapping and diminishing borders. In other words, *Magic Carpet Ride* depicts an Istanbul that is operated with money; morality is irrelevant as long as you have the financial means to survive. The characters' momentary slip into cruelty can therefore be forgiven. For instance, the portrayal of Asım's headquarters emphasizes the different titles that Yılmaz Erdoğan has for this film. After the two-minute credit sequence at the film's opening, composed of aerial shots of Istanbul, the camera zooms from very high above into a small spot in the city. This is an abandoned parking lot, squatted by Asım's gang and used as a car recyling yard as well as living quarters. The end of the film mirrors the beginning; the camera zooms out from the very same spot up to the sky to show some of the most tourist locations of Istanbul in sunset for another two minutes. The gang's home base is framed as the core of the city or a root from which all the activity sprouts. A parallel can be drawn here to Beşiktaş Kültür Merkezi (BKM), Erdoğan's production company and theatre ensemble in real life. Erdoğan and his collaborators are the ones responsible for this fictional Istanbul. This root, therefore, stands as much for the source of money as for creativity. Erdoğan defines the origin of his funding chain as 'the spectator's pocket': his career as an actor in a television series allowed him to invest the money he earned in theatre; his popular TV image boosted the ticket sales of his theatre productions, which in turn let him save money to shoot movies (Erdoğan, 2007, p. 184).

Magic Carpet Ride exposes the money spent on this production through visual images. Yılmaz Erdoğan is the 'image agent' of the film, and he is, to put it plainly, marketing Istanbul. Many shots of the city are independent of the film's narrative; they come between sequences without necessarily adding any meaning to the stories or driving them forward. However, Istanbul is a strong underlying element in the film; the sense that this story belongs above all to Istanbul is emphasized through the overwhelming images of the city. The Istanbul that *Magic Carpet Ride* constructs is a beautiful city and its faults are concealed. Visually, we witness this by watching the city in perfectly filtered and colour-corrected minute-long scenes; thematically, the cruelties of characters are forgotten both by their amiability and the conventions of the comedy genre. Furthermore, Istanbul in this film is owned by power, wealth and fraud – a constellation which the film does not seem to criticize, but rather celebrates.

Conclusion

Magic Carpet Ride exploits Istanbul images. The film does not only project transformations of the city, it also uses its images as a marketing product, thus giving them a value beyond the world of film. As Mazierska and Rascaroli (2003, p. 22) argue, 'the contemporary city points towards difference, individuality, and pleasure. The city too becomes a product, which has to be beautified and promoted accordingly among the citizens/customers'. In this sense, Istanbul's recent promotion and marketing as a global city finds its reflection in this film. *Magic Carpet Ride* fulfils the spectator's expectations through magnificent travel above and around the city, as well as between different genres. The city becomes the centre of attraction, the peak of the spectacle that comforts the spectator with the sense of having exploited Istanbul to its extremes. The film presents Istanbul as a convertible commodity that has an exchange value gained through the price of a ticket.

While the branding and marketing of Istanbul as a global city serves the purpose of attracting foreign investment and tourists from abroad, the presentation of Istanbul as spectacle in *Magic Carpet Ride* addresses primarily a domestic audience. In that sense, the film reflects an inward looking tendency in the spectacularization of the city, mirroring many cultural initiatives that are promoted under the umbrella of Istanbul 2010 European Capital of Culture. There seems to be as much a need to market Istanbul to its inhabitants as to foreign tourists and investors.

Istanbul in *Magic Carpet Ride* is thus a setting and a protagonist as well as a spectacle that offers a mix of sensations ranging from desire to confusion and shock, realized through the combination of different genres. The story not only takes place in Istanbul, but Istanbul takes place in this story. The city occupies screen time similar to a character, and the story suggests that the incidents are pertinent to this city only. This is a chaotic city in which survival and progress necessitate fraud. The camera captures Istanbul for the spectators who acquire the authority to perceive the city and themselves in different modes. This multiplicity thrives on the hybridity of genres emulated in this magic camera ride. By watching Samet making fun of Superman Samet, spectators view the enactment of social roles on the city stage in an ironic light.

Notes

1. The correct translation of the original title is *Organized Business*. The film owes its English title to a song on the soundtrack by Brooklyn Funk Essentials called 'Magic Karpet Ride'.

2. A quick look at the yearly box-office figures will illustrate this shift. See the websites 'Sinematürk', http://www.sinematurk.com/gise.php?action=goToBoxOfficePage&type=2&year=&firma=&pagenum=1 or 'Box Office Türkiye', http://www.boxofficeturkiye.com/yillik/. Accessed 2 February 2010.

3. Agâh Özgüç claims that between 1922 and 2004 seventy-one films were made with titles that contain either Istanbul's or its neighbourhoods' names (see Özgüç, 2005, p. 377).

4. According to John Urry (2002, p. 78), 'Tourism has always involved spectacle' and the advance of mobile technologies has shifted the nature of vision from static to mobile. 'There are a variety of tourist glances, the capturing of sights in passing from a railway carriage, through the car widescreen, the steamship porthole or the camcorder viewfinder' (*Ibid.*, pp. 152–153). If the static tourist gaze was captured through the still camera, this mobile gaze finds its match in the helicam shots in this film.

5. Both Nil Karaibrahimgil's song 'Organized Business' and Brooklyn Funk Essentials' 'Magic Karpet Ride' mix traditional Turkish rhythms with modern beats.

6. The Twin Towers were the iconic symbol for New York, a city which, according to Umberto Eco (1999), 'reveals itself like Istanbul'. New York has probably had an influence on the imagination and design of many cities like Istanbul, which has been developing and marketing regions similar to Manhattan.

7. Dinner in the Sky is an event that took place in May 2008 in Istanbul where people were 'hosted

at a table suspended at a height of 50 metres' in Ortaköy (see http://www.dinnerinthesky.com/dits_dinner/dinner.php. Accessed 2 February 2010). Still under construction, Sapphire Tower will be the tallest building in Europe and promises to offer spectacular views of Istanbul from its rooftop.

8. The film ranked first during the year it was released and currently ranks tenth at the all-time box-office list according to the website Sinematürk: http://www.sinematurk.com/gise.php?action=goToBoxOfficePage&type=2&year=&firma= (accessed 2 February 2010). 2006 is also the first year, after a long period, in which the cumulative box-office gross of domestic productions exceeded that of foreign films.

9. In one his interviews, Erdoğan explains that on the first page of his script is written, 'Istanbul is one of our protagonists. She should be shot and made to perform in the way she deserves (our translation)' (Okyay, 2005). Erdoğan claims that 'it is not only about Istanbul's beauty. The city contributes to everything that takes place in it, just like a character'.

10. Mustafa Erdoğan's dance shows played in the largest performance centres, Anatolian Fire (*Anadolu Ateşi*) and Troy (*Troya*) in particular, have been recent blockbuster successes in Turkey, but internationally only successful in Western European cities with a large Turkish migrant population.

11. Şen borrows this term from Bourdieu (1986).

References

Akıncı, Tolga (2006) Fragman Gişe Raporu. *Film +*, April, p. 42.

Bartu, Ayfer, Kolluoğlu, Candan Biray (2008) Emerging spaces of neoliberalism: a gated town and a public housing project in Istanbul. *New Perspectives on Turkey*, no. 39, pp. 5–46.

Bourdieu, Pierre (1986) *Distinction: A Social Critique of the Judgement of Taste* (trans. Richard Nice). London: Routledge.

Eco, Umberto (1999) Istanbul as unity and trinity (trans. Seyra Faralyalı). Atlas Special issue. Available at http://www.wan-press.org/article3190.html. Accessed 2 February 2010.

Erdoğan, Yılmaz (2007) Yapımcı Yönetmenler I [Producer Directors I], in Bayrakdar, Deniz (ed.) *Türk Film Araştırmalarında Yeni Yönelimler* [New Directions in Turkish Film Studies VII]. Istanbul: Bağlam Yayınları, pp. 177–206.

Mazierska, Ewa and Rascaroli, Laura (2003) *From Moscow to Madrid: Postmodern Cities, European Cinema*. London: I.B. Tauris.

Öktem, Binnur (2005) Küresel Kent Söyleminin Kentsel Mekanı Dönüşürmedeki Rolü; Büyükdere-Maslak Aksı [The Role of the Global City Discourse in Transforming the Urban Space; The Büyükdere-Maslak Axis], in Kurtuluş, Hatice (ed.) *İstanbul'da Kentsel Ayrışma: Mekânsal Dönüşümün Çeşitli Boyutları* [Urban Disintegration in Istanbul: Various Aspects of Spatial Transformation], pp. 25–76. Istanbul: Bağlam Yayınları.

Okyay, Sevin (2005) Bu söyleşiyi Erdoğan organize etti [It was Erdoğan who organized this interview], ntvmsnbc, 21 December. Available online at http://arsiv.ntvmsnbc.com/news/354795.asp. Accessed 11 October 2009.

Özgüç, Agâh (2005) *Türlerle Türk Sineması* [Turkish Cinema in Genres]. Istanbul: Dünya Yayıncılık.

Şen, Besime (2005) Soylulaştırma: Kentsel Mekânda Yeni Ayrıştırma Biçimi [Gentrification: New Form of Exclusionion in Urban Space], in Kurtuluþ, Hatice (ed.) *İstanbul'da Kentsel Ayrışma: Mekânsal Dönüşümün Çeşitli Boyutları* [Urban Disintegration in Istanbul: Various Aspects of Spatial Transformation]. Istanbul: Bağlam Yayınları, pp. 127–159.

Urry, John (2002) *The Tourist Gaze*. London: Sage.

Chapter 10

Projecting Polyphony: Moving Images, Travelling Sounds[1]

Deniz Göktürk

Locations on the Move

In one of the earliest cinematic documents of Istanbul, the camera pans across a bridge. This also happens to be one of the pioneer travelling shots taken with a mobile camera. Less than a minute long, the film depicts the Galata Bridge from a boat moving across the Golden Horn. *Constantinople – Panorama de la Corne d'Or* was filmed in 1897 by Alexandre Promio, the Lumière Brothers' most widely travelled operator, who had also put a camera on a gondola in Venice one year earlier.

The view of the historic silhouette from the water was a common perspective on Constantinople for nineteenth-century European travellers, some of whom would never leave their ship to avoid spoiling the fantastic spectacle through exposure to the squalor of the streets (Schiffer, 1999, p. 151). What is new in Promio's cinematic panorama from the water is the focus on modern transportation technology. Ignoring the famous historic sights at the tip of the peninsula (Topkap*ı* Palace, Hagia Sophia and the Blue Mosque remain outside the frame), the camera lingers instead on the big wheel of a steam ship in the foreground and then moves across the bridge while the Süleymaniye Mosque remains aloof in the background. Part of the attraction might have been the moving bridge itself, which had been rebuilt on twenty-four pontoons in 1875 to ensure increasing access for ships to the port up the Golden Horn.

The nexus between cinematography and transportation technology captured in this short filmic document is characteristic of early cinema, where travel films, also called 'actualities', were a popular genre. Moving pictures set locations on the move and brought 'the whole world within reach'; in fact, they not only offered *Ersatz* journeys to the armchair traveller, but in many ways also enticed and promoted tourism (Gunning, 2006). We can imagine an elegant audience up the hill north of the Golden Horn in the Cosmos Cinema on the Grande Rue de Pera (see Nezih Erdoğan, Chapter 7) watching this travelling perspective on their own city – alongside vistas from other European cities – with a sense of amazement. Meanwhile, visitors to cinemas in Paris, London and Berlin also

Figure 10.1. Lumière Brothers' *Constantinople – Panorama de la Corne d'Or*. DVD capture. (*Source: The Lumière Brothers' First Films*. Kino Video 2003)

saw this brief glimpse of bridge traffic in Istanbul in an assemblage of 'foreign views' and 'virtual voyages' (Ruoff, 2006). As locations became portable and cities merged into a cinematic geography, the gaze within and the gaze from outside became increasingly intertwined. However, the places depicted would have been met with different degrees of recognition and emotional resonance by audiences in different cinemas.

It is important to remember that early cinema audiences, especially before World War I, did not constitute a nationally contained 'imagined community' (Anderson, 1991), but rather a community of spectators who were networked through a fascination with modern technology, movement and spectacle. As films circulated internationally, cinema 'brought the world to the world', as Bertrand Tavernier states in his narration on *The Lumière Brothers' First Films* (1998). These world spectators enjoyed access to events and locations on the move and potentially felt connected to viewers elsewhere who were engaged in simultaneous viewing.

One of these avid cinema spectators might have been the young Siegfried Kracauer. In 1925, when cinema had long shifted from a compilation of short attractions to the more immersive long feature format, Kracauer wrote:

> The more the world shrinks thanks to automobiles, films, and airplanes, the more the concept of the exotic in turn also becomes relativized. Though at present the exotic may still cling to the pyramids and the Golden Horn, someday it will designate any spot in the world

> whatsoever, to the extent that the spot appears unusual from the perspective of any other point in the world. This relativizing of the exotic goes hand in hand with its banishment from reality – so that sooner or later the romantically inclined will have to agitate for the establishment of fenced-in nature preserves, isolated fairy-tale realms in which people will still be able to hope for experiences that today even Calcutta is hardly able to provide. (Kracauer, 1995, p. 65–66)

Architect and film critic, Kracauer highlighted that moving pictures were just as crucial as transportation technologies for the experience of expanding modernity in a shrinking world; in fact, he anticipated what cultural geographer David Harvey (1990) would later term 'time-space compression'. Cinema and other spectacles from panoramas to world exhibitions served the 'tourist gaze' (Urry, 1990) of the armchair traveller who, in Kracauer's account, was still firmly grounded in Central Europe and projected his desire for things authentic onto 'exotic' places, such as the Pyramids, Calcutta, and the Golden Horn. That desire, however, was already shot through with the realization that commodification of places relies on recognizable uniqueness while, paradoxically, erasing specificity in the spread and expanse of tradable resemblance. As David Harvey points out, capitalist globalization requires that 'the Picasso has to have a money value as does the Monet, the Manet, the aboriginal art, the archaeological artefacts, the historic buildings, the ancient monuments, the Buddhist temples, and the experience of rafting down the Colorado, being in Istanbul or on top of Mount Everest.' (Harvey, 2001, p. 396). The narrative of total homogenization that theorists of 'Disneyfication' propose (Zukin, 1996; Bryman, 2004) needs to be complicated by considering specific inflections and appropriations. Nonetheless, questions about the mediated 'production of locality' (Appadurai, 1996, p. 178) in the context of global circulation remain salient for the analysis of moving city images today.

Over a century after the Lumière Brothers' panorama of the Golden Horn, *Crossing the Bridge: The Sound of Istanbul* (2005), a cinematic portrait of Istanbul which adds music to moving images, focuses once again on bridges. The American DVD version of Fatih Akın's feature-length documentary opens with a view of the Manhattan skyline through clouds, the Brooklyn Bridge emerging in the foreground and dissolving into the distribution company's name: Strand Releasing. A helicopter shot of the Bosphorus Bridge follows, then a cut to a dancing belly, back to the Bosphorus Bridge, followed by break dancers performing on a street. Condensed into less than a minute, this pre-menu trailer promises a wide-ranging cross-section of popular tunes and acts. Eurovision winner Sertap Erener's cover version of Madonna's 'Music[2] (makes the people come together)' further accentuates fusion and amalgamates a montage sequence of several performers featured in the film, emphasizing the community-building quality of music, which 'mixes the bourgeoisie and the rebel'.

The DVD trailer presents a slightly condensed version of the theatrical trailer, which is included as an extra on the DVD and streams online on platforms such as the Internet Movie Database, Netflix and the video sharing site YouTube, where many scenes from the film have been uploaded as well. In many ways, these trailers and other video clips are reminiscent of early cinema programmes, which comprised a series of short attractions such as the Lumières' vista of the Golden Horn as well as filmed stage acts. Digital media provide access to deterritorialized spectators with the added benefits of choice and repeatable home viewing (Mulvey, 2006). In fact, a focus on media archaeology reveals many correspondences between 'old' and 'new' media (Elsaesser, 2005). If early cinema was instrumental in setting locations on the move and bringing 'the world to the world', the World Wide Web connects the spectator to the world from his or her home station and potentially turns the spectator into the producer of such clips, re-edited with subtitles, a new soundtrack and/or comments.[3]

The film *Crossing the Bridge: The Sound of Istanbul* was the second feature produced by Akın's Hamburg-based company Corazón International in 2005. The film was shot in Istanbul and marketed in the United States by a company which distributes international art house cinema; this company uses the Brooklyn Bridge as emblem, but is located in Santa Monica, California, not Manhattan. Along with its trailers and digital editions, the film poignantly exemplifies the global circulation of cultural forms and the mediated production of local specificity.

Fatih Akın's discovery of Istanbul occurred at a time when Istanbul was on its way to becoming the third most visited city in Europe after London and Paris (the seventh most visited city in the world, with over 7 million visitors per year)[3] and in the wake of Istanbul's selection as a European Capital of Culture. Initiatives to promote Istanbul's candidacy took shape in July 2000, and on 13 November 2006 Istanbul was officially announced a European Capital of Culture for 2010 (see Carola Hein and Oğuz Öner, chapters 14 and 15). *Crossing the Bridge* opened at the Cannes Film Festival on 12 May 2005 and in the US and UK in 2006. The film captures a junction of German and Turkish aspirations within broader frameworks of European integration and globalization.

Any imagination of a city – whether from within or from outside – feeds on images that are tinted with the desires of actors and spectators. What kinds of mediation have taken place leading up to 2010, when Istanbul is celebrated as a European Capital of Culture? Who has been staging and projecting the city to whom, how, and why? How has the bridge been mobilized as an image and metaphor in attempts to stage Istanbul as an open, diverse and polyphonic city?

This chapter will address these questions by focusing on Akın's documentary. It situates the film in the context of debates on Turkey's geopolitical 'bridging' role in the EU accession process and the wake of Istanbul's nomination as a European Capital of Culture, focusing on the role of participating mediators who channel and amplify the music of Istanbul into global circuits, projecting a utopian vision of the city as a model for pluralist polyphony. The film's engagement with Roma performers gives rise to questions about mediations of the exotic in world music and cinema. The chapter concludes with reflections about spectatorship, participation and representation in times of digital media that transcend established conventions of documentary cinema.

The Bridge as Metaphor

In an interview with the German weekly magazine *Der Spiegel,* director Fatih Akın explains:

> Istanbul is the real capital of Turkey and reflects the social situation of the country pretty well. With this film I wanted to pursue the question of whether Turkey is European enough for the European Union. And I thought that music would be an interesting way to answer this question. (Dürr and Wellershof, 2005)

The interviewer conjectures that *Crossing the Bridge* 'does not give an answer'. Akın replies (echoing a statement made in the film by the saxophonist Richard Hamer from the band Orient Expressions):

> Because there are no clear-cut answers. In the course of shooting the film I have learned that the situation in Turkey is far too complex for that... The West ... does not end in Greece, and the East does not stretch from Istanbul to China. Those are imaginary boundaries. (*Ibid.*)

This distancing from common place East/West oppositions suggests that the bridge metaphor in the film's title is employed as an antithesis to the self-fulfilling prophecy about a 'clash of civilizations' (Huntington, 1993). Much along the same lines Turkish politicians have used the metaphor of a 'bridge across continents' or a 'bridge across civilizations' in their speeches and the iconic 'image of the Ortaköy Mosque in the foreground and Bosphorus Bridge in the background ... as a photo backdrop' in visual stagings of state visits in the post-Cold War era, thus '(re)producing Turkey's exceptionalism' (Yanık, 2009, p. 532). While 'culture clash' has gained high currency as the mantra governing world politics, especially over the past decade, elites across the political spectrum in Turkey – from socialist and Republican-nationalist to neoliberal Islamist – have drawn in their national self-fashioning on an eclectic mix of influences and aspirations, thereby lending popular appeal to the bridge between Europe and Asia as an image of elegant synthesis.

The bridge metaphor has also abounded in advance publicity material promoting Istanbul as a European Capital of Culture, a welcome opportunity to counteract the East/West divide through cultural policy and aesthetic interventions. The 'Istanbul 2010 European Capital of Culture Master Plan', published on an earlier version of the Istanbul 2010 website, emphatically framed Istanbul as a cosmopolitan model for coexistence, emphasizing the city's unique location, which connects Europe and Asia by literal and figurative bridges:

Figure 10.2. *Crossing the Bridge*, promotional image. (*Courtesy of* Intervista/ Corazón International)

> The second determinant shaping the Master Plan is Istanbul's rich cultural heritage. As the only city in the world that bridges two continents, Istanbul in all its guises has always been a crossroads of cultures.[5]

Meanwhile, scholars have been pointing out the limitations of the bridge metaphor, suggesting that it has arrested Turkey in ambivalent indeterminancy between 'backwardness and progress' for over a century (Ahıska, 2003, p. 353). A bridge by definition implies a stable construction connecting two fixed shores. When evoked in a metaphoric sense, it posits distinct, self-contained continents or cultures, Asia *vis-à-vis* Europe or the East *vis-à-vis* the West.

As the sociologist Georg Simmel pointed out in his 1909 essay 'The Bridge and the Door', relating and separating always precondition each other; indeed, we can only relate things that we perceive as separate.

> Things must first be separate in order to be together... The bridge becomes an aesthetic value not only because it in reality achieves the inter-relation of what is separate, and because it achieves practical purposes, it becomes an aesthetic value because it makes the interrelation immediately visible. The bridge encourages the eye to inter-relate parts of the landscape just as in practical reality it encourages bodies to relate with one another. (Simmel, 1994 (1909), p. 408)

Simmel's relativist epistemology poignantly speaks to current discussions of cultural contact. His essay about sites and figures of transition – the bridge and the door – and their aesthetic value carries seeds of a dynamic conceptualization of representation as interaction.

The logic of distinct and self-contained cultures, which is inherent in the rhetoric of both ethnonationalist purity as well as multicultural plurality, often merely replicates and reinforces a sense of identity as a trap (Sen, 2006). Regimes of nationalism as well as multiculturalism tend to be underwritten with ethnocultural identifications about belonging to a community of indigenous people who have some organic attachment to their land. This model does not allow for multiple affiliations. The festival and exhibition parades in world cities and the global competition in showcasing diversity tend to perpetuate this additive logic. In Istanbul, the celebration of pluralism is furthermore infused with nostalgia:

> Throughout history ... Istanbul has been home to countless societies and cultures. Yet this 'beautiful harmony', which is embedded in the city's foundations and entwined in the branches of its family tree, is not just a pleasant memento from a bygone era. Istanbul retains still its rich cosmopolitan character, sometimes concealing and sometimes revealing the evidence of its unrivalled physical and cultural legacy. The city is a living example of the much sought-after meeting of civilisations – something so desperately missing in the modern world that the search for it seems almost utopian.[6]

A major irony lies in the fact that non-Muslim groups such as the Rum (Greeks) and the Armenians who had lived in the city for centuries have mostly left (Kirişçi, 2008), but their heritage, especially their former buildings and neighbourhoods, are claimed today in a kind of retro-chic fashion. Istanbul 2010 features commemorative exhibitions on the work of Greek and Armenian architects.[7] Meanwhile, new migrants to the city (Kurds from southeastern Turkey and neighbouring countries, refugees from Iraq and Afghanistan, migrants from Africa and the former Soviet Union) have remained largely invisible in Istanbul's celebratory pictures of cosmopolitan diversity.[8] The question arises of which multilingual voices are incorporated into spectacles of selective additive multiculturalism. This focus on staging and projecting polyphony brings us back to the use of music in the films of Fatih Akın, particulary in his sound portrait of Istanbul *Crossing the Bridge*. Following the proposition to conceive of *keyif* as the prevailing mood of Istanbul over *hüzün* (Engin Işın, Chapter 2), I would like to propose that *Crossing the Bridge* offers a counter perspective to those nostalgic projects, Orhan Pamuk's melancholy-laden Istanbul memoir, or Nuri Bilge Ceylan's staging of the displaced migrant in the city (İpek Türeli, Chapter 8). Clearly, this documentary is less interested in things past, in history and nostalgia, than in highlighting present-day dynamism and a polyphonic utopia fuelled by migration.

Mediators on Sound Bridges

In a 'global culture industry' (Roberts, 1998), questions arise as to who projects cosmopolitan consciousness, who are subjects and objects of the camera, who determines ownership and authenticity, what circulates and what does not, and who is addressed as audience. Mediators engaged in processes of cultural transfer in *Crossing the Bridge* – the film's director as well as featured travelling performers and producers of music – will be discussed in more detail in what follows.

In his films, Hamburg-based director Fatih Akın has demonstrated a successive approach to Istanbul, contributing his share to putting the city on the map of contemporary European and world cinema. Music is a driving force for Akın's mobile imagination. In *Head-On* (2004), music sets in with the establishing shot: a medium-long shot of an orchestra of six, facing the camera. The musicians, dressed in black suits with bow ties, are seated in a row of chairs, at the centre of which a female singer stands in a red dress. The musicians are located on the northern shore of the Golden Horn with a view across the water of the Süleymaniye Mosque and the fire observation tower in Beyazıt (first built in 1749, today located on the grounds of Istanbul University), architectural landmarks which might be read as emblematic for the proximity of religious and secular spheres in Turkish society. Boats are arriving and departing on the opposite shore. Luxurious red carpets, arranged asymmetrically on top of each other as in a carpet shop, fully cover the ground upon which the orchestra is

Figure 10.3. Fatih Akın discovers a site of technical reproduction on a rainy day in Istanbul. Promotional image. (*Courtesy of* Intervista/Corazón International)

set and appear to extend beyond the frame. The carpets create a stage for the ensemble's open-air performance and unsettle the opposition between indoor and outdoor, thereby suggesting a conflation of public and private spaces, a fusion between theatrical stages and 'real' locations.

This enigmatic tableau, showing the musicians on the shore of the Golden Horn, is repeated five more times throughout *Head-On*, with the exact same framing and staging of the orchestra, although in changing daylight. Short clarinet solos by Selim Sesler function as 'sound bridges', conveying temporal and spatial transition between Istanbul and Hamburg in a highly condensed fashion. The brief appearances of the orchestra signal ironic interruptions, a different level of awareness, inviting the spectator to step out of the illusionist enactment of the protagonists' *amour fou*. The direct address of the musicians introduces a different mode of performance by actors who stand in no relation to the dramatic fiction. In the orchestral interludes, the recurring frontal arrangement of the orchestra playing diegetic music breaks up conventions of mimetic realism. The repeated interruption of the linear narrative by the same tableau with significant variations reminds the viewer of the staged nature of the film as spectacle. In this sense *Head-On* evolves as an 'ironic melodrama' in the spirit of the novel and film *Ağır Roman* (1997) – the fusion of a heavy story and a slow Romani melody (Göktürk, 2008*b*).

Akın's subsequent documentary feature *Crossing the Bridge* serves as a companion piece to the musical interludes in *Head-On*, giving centre stage

Figure 10.4. Interlude in *Head-On*: Singing on the shore of the Golden Horn. İdil Üner standing in the middle, Selim Sesler playing the clarinet to her left. Film still (*Courtesy of* Wüste/Corazón)

to musicians and the sound of Istanbul. The film captures the music of the city through the narration and mediation of Alexander Hacke, a member of the experimental band Die Einstürzenden Neubauten. Hacke had previously collaborated with Akın on the sound recording for *Head-On*. In *Crossing the Bridge*, he acts in front of the camera as a bass player and sound engineer who sets out to explore the sound of the city.

Unlike the more recent documentary *In Berlin* (2009), where Alexander Hacke serves as a participating tour guide in his native city, *Crossing the Bridge* is presented as a travelogue that experiences the city through the eyes of a stranger, explaining difference to his audience. Despite the irritation that his presence and commentary causes for some spectators of *Crossing the Bridge*,[9] I would like to suggest that it is actually a distinguishing feature of this documentary to put mediators like Hacke visibly on the screen. Compared to Wim Wenders' *Buena Vista Social Club* (1999), which also features the mediators – Ry Cooder and his son Joachim Cooder – as actors, *Crossing the Bridge* casts the mediators in a more casual and subdued spirit of collaboration. Unlike Ry Cooder, who positions himself centre stage and foregrounds his own importance in bringing the Cuban musicians to Carnegie Hall and 'making them heard', Hacke is seen gathering and recording on his computer, at times participating in a diverse and vibrant scene that does not depend on him as facilitator.

With his travelling sound studio, Alexander Hacke electronically records musical performances, comments on them in a voice-over that adopts the mode of a travelogue, and at times plays along with the bands. Scenes from a daylong

Figure 10.5. Alexander Hacke with his travelling sound studio in front of the Grand Hotel de Londres. Film still (*Courtesy of* Intervista/Corazón International)

jam session with Baba Zula's psychedelic rock band on a boat on the Bosphorus frame this series of performances at the beginning and end of the film. The documentary presents an eclectic cross-section of performers. The grunge band Duman with Seattle connections; the Replicas, who point out that they grew up following the taste of Western teenagers and only discovered local music later on; and Erkin Koray, the rebellious pioneer of Turkish rock who still fills a stadium with an enthusiastic young audience. Then, in Üsküdar (on the Asian side of Istanbul), the rapper Ceza and his clan, who are introduced by Hacke as representatives of 'the black music of Istanbul', promote a serious, non-violent, anti-gangsta approach to rap. They are followed by Istanbul Style Breakers, who perform on the street and insist that they are not imitating the Americans, but want to 'make Turkey's name heard here'. They dance to a rap adaptation of Orhan Veli's famous poem 'İstanbul'u dinliyorum gözlerim kapalı' (I am listening to Istanbul with my eyes closed). Meanwhile, Ceza's father, an Eric Clapton and Jimi Hendrix fan, endorses rap as 'the music that Turkey needs today'. The film successfully interweaves the performances, occasionally assembling some of the musicians and having them comment on each other. Preparations are underway on an open-air stage, which Ceza shares with young Romani musicians and Mercan Dede, accompanied by a young American woman whirling on stage like a dervish.

The musical performances are interspersed with shots from daily life on the streets, aerial shots of the city, and some archival material such as a scene from *İstanbul Geceleri/Istanbul Nights* (1950) and clips from Orhan Gencebay's films. Long-time screen hero Gencebay, a virtuoso bağlama player and star of *arabesk* music, is cast as an icon of migration introducing folk music to an urban setting (Stokes, 1992). He explains that his 'urban *saz* style' took inspiration from Egyptian music, once again complicating notions of pure tradition with references to sonic hybridity.

Kurdish street musicians from the group Siyasiyabend, displaced from their villages in southeastern Turkey, explain why they prefer to play and interact on the streets rather than in concert halls. They believe in the transformational power of music and see the street as a level playing field, which brings people together irrespective of social class – the homeless and the professionals carrying their laptops. This puts a different spin on Sertap's cover version of Madonna's 'Music', which runs over the credits of the film. However, the street musicians also clearly state that the street has no memory; as long as there are no recordings, voices on the street will not be archived. Despite their resistance to commercialization, these street musicians claim their place within the Turkish nation and also present themselves as articulate actors who play 'a role in the game of Europeanization'. 'Street music is one of the values of Europe', they say, hence the police can push them off the main street but ultimately have to tolerate them. Even if these street singers did not get a share of the revenue, the film adds their voices to the cultural memoryscape of Istanbul in its pre-European Cultural Capital days. The young female singer Nur Ceylan, too, peforms on the street in a neighbourhood populated by poor Kurdish migrants. Meanwhile, Aynur Doğan sings in Kurdish and powerfully fills the space of an old bathhouse with her voice. The producer of her CDs from Kalan Müzik Hasan Saltık, himself an important mediator and archivist of diverse and multilingual music in Turkey, explains about the ban on broadcasting in minority languages. As he puts it, the ban was lifted in the 1990s, but the politics of broadcasting only liberalized more recently, in the course of the EU accession process. With these statements, the film makes a strong point for an inclusive and polyphonic vision of Turkish society. The inclusion of these street musicians, in particular, points in the direction of a participatory aesthetic in this utopian vision of pluralist polyphony, which is held up as a model not only for Turkey but also for Germany and Europe.

Most prominently, *Crossing the Bridge* introduces the performers of the interludes in *Head-On* as the Roma clarinettist Selim Sesler and his orchestra, consisting of friends and family members. His son Bülent Sesler plays the kanun, a kind of zither. They are filmed in their living room, chatting and performing together with the Canadian singer Brenna MacCrimmon who talks about retrieving these 'Roma songs' from old records found in Bulgarian villages under layers of dust while on tour with Muammer Ketencoğlu, compiling a selection to perform with Selim Sesler, and producing a CD with Selim Sesler (MacCrimmon, 1998). In the same year, she also organized a Canadian tour for the group.

Figure 10.6. Brenna MacCrimmon, a nightingale on the Bosphorus. (*Courtesy of* Intervista/ Corazón International)

Crossing the Bridge pays belated tribute to Brenna MacCrimmon, who had not been credited in *Head-On*, although all the interlude songs and arrangements were chosen from her compilation, but rerecorded with actress İdil Üner as singer. As Brenna MacCrimmon points out in the conversation in Selim Sesler's living room, the Roma musicians told her that they had previously not really valued the music they played, until she came along as a stranger from afar who took a keen interest in their songs and learned to perform them. Her articulation in the Turkish songs is so good that she can easily pass for a native, maybe with a slight Rumeli accent. She explains that the mix of sorrow and joy in these songs also appeals to a North American audience who cannot understand the words.

Crossing the Bridge travels to Keşan to explore this connection – the only trip that the film takes outside of Istanbul. Roma are numerous in this town near the Greek border; in *Crossing the Bridge*, Hacke claims that they constitute about two-thirds of the town's population. Roma people in this border zone were initially relocated in the population exchanges of 1923/1924 from the area around Thessalonica. *Ağır Roman* and other kinds of Romani melodies are characterized by a 9/8 rhythm and typically performed by improvisation (*doğaçlama*) in a *fasıl* (drinking and playing session) (Seeman, 1998; Duygulu, 2006). Selim Sesler plays centre stage at such a *fasıl* in Keşan. On the bus, he explains the difference between Turkish and Romani music as follows: 'The listener of Romani music can't help himself. He just has to get up and dance. The listener of Turkish

Figure 10.7. Selim Sesler at a *fasıl* in Keşan. (*Courtesy of* Intervista/Corazón International)

classical music sits and listens'. The folkloric practice of Romani music is explored as Selim Sesler and Alexander Hacke both dance at a wedding in Keşan. Thracian 'gypsy music' incorporates a rich mix of Turkish, Greek, Romani, and Bulgarian influences. This music signals cross-Balkan mobility and the fusion of multiethnic influences at the core of what is commonly considered Turkish music or culture. The medley of music in an improvisational *fasıl*, as in the performance of Selim Sesler and the other musicians in the pub in Keşan, embraces Balkanization as fusion in a positive sense and transcends scenarios of ethnocultural tribalism. Released in the same year as *Crossing the Bridge*, Yılmaz Erdoğan's *Organize ışler/Magic Carpet Ride* (2005) also uses Balkan tunes played by clarinettist Hüsnü Şenlendirici (see Deniz Bayrakdar and Elif Akçalı, Chapter 8).

'Gypsies' have generally tended to fulfil the projective needs of the majority (Iordanova, 2001, pp. 213–232). Are 'Gypsies' once again exoticized in Akın's films as vagrant figures and perpetual migrants, as a stand-in for the German Turks who have transcended ethnic identification? As David Malvinni maps out, music has been one of the markers of 'Gypsy' identity ever since Franz Liszt praised the pure musicality, the depth of emotion, and the virtuosity of improvisation in the Gypsy fiddler in his *Des Bohemiens et leur Musique en Hongrie* (The Bohemians and their music in Hungary) (1859). Ideas of 'Gypsiness' as natural musicality were inflected early on with discourses of nationalism. While Liszt claimed Gypsy music as Hungarian music, Béla Bartók contested these claims of authenticity by pointing out that Gypsy music originated in urban cafés and, unlike Hungarian peasant music, could not be considered true folk music (Malvinni, 2004). Adorno (2002) picked up on Bartók's critique in his essay 'On Jazz', written in 1936:

> The relationship between jazz and black people is similar to that between salon music and the wandering fiddle players whom it so firmly believes it has transcended – the gypsies. According to Bartók, the gypsies are supplied with this music by the cities; like commodity consumption itself, the manufacture [*Herstellung*] of jazz is also an urban phenomenon, and the skin of the black man functions as much as a coloristic effect as does the silver of the saxophone. (Adorno, 2002 (1936), p. 477)

Adorno's essay is less concerned with racial stereotyping than with the deconstruction of folkloristic romanticization and an interrogation of assumptions about what is commonly considered natural, primitive and traditional.

Akın and his crew might be inclined to second Adorno in his resistance to staging ethnic and national identity in folkloristic guise. They included the Roma performers in *Head-On* and *Crossing the Bridge* in a celebration of diversity, emphasizing the multiethnic mix within Turkish music. Akın stages Selim Sesler and his orchestra not as exotic, folkloristic figures, but as distinctly urban musicians in costume who appear to most spectators at first glance to be indistinguishable from performers of classical Turkish music. In fact, Selim Sesler also performs in the orchestra accompanying 86 year-old Müzeyyen Senar, legendary star of Turkish classical music, singing 'Haydar Haydar'. This intermingling has a long tradition in musical entertainment. Turkish music is gypsy music and *vice versa*. It is precisely this convergence of intertwined traditions that challenges any nation-based definition of music and, more generally, of culture. In a non-exoticist way, the Roma performers thus become desirable figures in a play on stereotypes and multiple ironies of identification. Perhaps it has become an appealing trajectory for filmmakers and actors to inscribe themselves into 'the uniquely transnational phenomenon of the Gypsy films' (Iordanova, 2003).

Alexander Hacke and Brenna MacCrimmon play prominent roles as participant mediators in *Crossing the Bridge*. In fact the film as a whole provides an international stage for some previously unknown musicians alongside big stars who were only known to a Turkish audience. Subsequently, the music of many of these performers became readily available on CDs in stores across Istanbul and globally on the Internet. Like Alexander Hacke and Brenna MacCrimmon, Fatih Akın himself is a mediator who valorizes and amplifies local culture in global circuits.[10] At the film's end, however, as Alexander Hacke packs his bags to leave Istanbul, he admits: 'I could not unlock the magic of this city. I merely scratched the surface'.

Sites of Participation

Istanbul's cinematic appeal transcends Europe. Most recently, Indian cinema, the largest film industry in the world, has discovered Istanbul as a location with Mani Ratnam's *Guru* (2007) and Apoorva Lakhia's *Mission Istaanbul* (2008). In 2008, the Ministry for Culture and Tourism in collaboration with the Turkish Foundation of Cinema and Audiovisual Culture (TURSAK) published a 183-page book in English, marketing locations to international film production companies, featuring glossy photographs of natural treasures and heritage sites across Turkey, including many buildings, models of which can be found in the architectural theme park Miniaturk (İpek Türeli, Chapter 6). The book emphasizes diversity in relation to the coexistence and mutual tolerance of three faiths (Judaism, Christianity and Islam). Views of Istanbul are scarce in this glossy volume. The section titled 'Urban' only comprises six photographs: fishing boats in Ortaköy with the mosque and the Bosphorus Bridge in the background; Aya Irini and the Topkapı Palace from the air; the tram in Beyoğlu; the roof of the Topkapı Palace; a fishermen's café, photographed by Ara Güler; and, finally, the Galata Bridge from the air.

Although *Crossing the Bridge* includes some stunning views of Istanbul from the water and the air, the film clearly offers a different vision of life than in the glossy book published by the Ministry for Culture and Tourism focused on promoting heritage sites and film locations. *Crossing the Bridge* promotes a less nostalgic take on diversity by featuring Kurdish and Romani performers and putting travelling mediators on stage. The film opens up an alternative archive: the archive of rock, rap and street music. Despite these differences, Akın's films are readily incorporated by the Ministry for Culture and Tourism in their showcasing of eleven films 'made in Turkey', which include *Gegen die Wand* (the German title!) and *The Edge of Heaven* (English title!), both German productions. The list also includes three films by Nuri Bilge Ceylan (*Distant*, *Climates*, *Three Monkeys*) as well as Yeşim Ustaoğlu's *Journey to the Sun*, all realized as co-productions with funding from Eurimages (TÜRSAK, 2008, pp. 112–113). Clearly, the boundaries of national cinema have become porous; the inside and outside perspectives are intertwined; official self-representation is mediated through European funding and production channels and geared to an international market.

Meanwhile, the call for inclusive polyphony and the new popularity of Balkan tunes on the world music stage do not prevent minority populations from falling prey to cleansing and urban regeneration efforts. It is worth mentioning here that 8–10 million Roma constitute the largest minority in Europe, but due to their lack of territorial affiliation they rarely figure in the nation-bound rhetoric about immigration, integration, and multiculturalism. Their reality bears little resemblance to romanticized visions of 'Gypsiness'. The traditional Romani neighbourhood of Sulukule near the Theodosian city walls of Istanbul – one of the oldest settlements of Roma in the world – was recently demolished

Technical University, Ankara, 9 July 2007, and have been previously published (Göktürk, 2008*a* and 2008*b*). Earlier versions of this material were presented as invited lectures and conference contributions at the University of California Santa Barbara, Istanbul Bilgi University, Oxford Brookes University, the University of British Columbia in Vancouver, and Indiana University. I would like to thank organizers and participants at all these occasions for engaged discussions. Conversations with Sonia Tamar Seeman in Santa Barbara and Brenna MacCrimmon in Berkeley have been particularly enlightening and inspirational. Thanks are also due to Nicholas Walter Baer, Aleks Göllü, Levent Soysal, and İpek Türeli for comments and corrections.

2. *Music*. Written and produced by Madonna and Mirwais Ahmadzai. Maverick, Warner Bros. Released 19 September 2000. The album *Music* sold over 11 million copies worldwide. The hit song samples elements from the song 'Never Young Again' by Mirwais Ahmadzai.

3. Ironically, the video sharing site YouTube has been closed in Turkey by a court ruling since 7 March 2007. However, users continue to find ways of circumventing the ban.

4. The 10 Most Visited Cities in the World. Vacation Ideas. Available at http://www.vacationideas.me/travel-tips/top-10-most-visited-cities-in-the-world. Accessed 31 March 2010.

5. Istanbul 2010 European Capital of Culture Master Plan, 20 November 2007. Available at http://www.istanbul2010.org/doc/MasterPlan_eng.pdf, p. 1. Accessed 8 July 2007.

6. Available at http://www.istanbul2010.org/?p=5&lang=eng. Accessed 8 July 2007.

7. Available at http://www.en.istanbul2010.org/PROJE/GP_518598 and http://www.en.istanbul2010.org/PROJE/GP_518641. Accessed 31 March 2010. The film projects for Istanbul 2010 are also looking backward to heritage to find diversity: *Istanbul Do Redo Undo: Waters, Streets, Faces*, an archive compilation 'which will be looking at Istanbul of 100 years ago with a thrill of discovery rather than a sense of nostalgia', compiled by Nezih Erdogan with a team at Bilgi University; *Colors of Cultural Diversity*, a 'documentary project aiming at demonstrating the richness added to Istanbul by diverse cultures that have been coexisting for centuries ... especially artistic contributions of the Greek, the Jewish, Armenians, Assyrians and gypsies to the cultural richness of Istanbul'; *Holy Days in Istanbul*, a documentary on festivals celebrated by various religious communities in the city (available at http://www.en.istanbul2010.org/PROJELERVEBASVURULAR/projearama/index.htm?kategori=SINEM, accessed 13 April 2010); and Hüseyin Karabey's *Unutma Beni/Forget Me Not*, a proposed international collaboration about 'shared memory' regarding Istanbul's cosmopolitan heritage with directors Harutyun Khachatryan (Armenia), Hany Abu-Assad (Palestine), Jafar Panahi (Iran), Stergios Niziris (Greece), Aida Begic (Bornia), Omar Shakrawi (Palestine/Denemark), Mani Maserrat-agah (Iran/Sweden), Stefan Arsenijevic (Serbia). Email communication by Hüseyin Karabey, 21 December 2009.

8. Esra Ersen's video *Brothers and Sisters* (2003) http://www.ok-centrum.at/english/presse/esra_ersen.html (accessed 13 April 2010) and Berke Baş's documentary film *In Transit* (2004) http://www.arteeast.org/cinemaeast/filmfest_05/fest05-films/intransit.html (accessed 13 April 2010) offer glimpses into the lives of clandestine migrants in Istanbul. On irregular migration to Turkey see also Ahmet İçduygu, International Organization for Migration (IOM), Irregular Migration in Turkey, February 2003, Migration Research Series, No. 12. Available at: http://www.unhcr.org/refworld/docid/415c6e904.html. Accessed 31 March 2010. Also Yükseker, Hatice and Brewer, Kelly. The Unending Migration Process: African Refugees and Illegal Migrants in Istanbul Waiting to Leave for Europe. Paper presented at the annual meeting of the American Sociological Association, Montreal Convention Center, Montreal, Quebec, Canada, 10 August 2006. Available at http://www.allacademic.com/meta/p105362_index.html. Accessed 31 March 2010.

9. In a review of the film published in the *San Francisco Chronicle* on 14 July 2006, for example, Jonathan Curiel wrote: '*Crossing the Bridge* has two main flaws: the emphasis it puts on German bassist Alexander Hacke, the film's ostensible narrator, who shows up in too many scenes, and the fact that it doesn't identify many of the film's performers until the very end. Even so, *Crossing the Bridge* is satisfying to watch, especially with its postcard shots of Istanbul, one of the world's most beautiful cities'. Available at http://www.sfgate.com/cgi-bin/article.cgi?f=/c/a/2006/07/14/DDG9KJU2BI1.DTL#flick. Accessed 31 March 2010.

10. The worldwide box office revenue for *Crossing the Bridge* is listed on Box Office Mojo as $555,754, considerably lower than for Akın's dramatic features *Head-On* ($11,030,861) and *The Edge of Heaven* ($17,804,565), but still relatively high for a documentary. The film remains in circulation on DVD through the 'long tail' of Netflix and Amazon. (Anderson, 2004)

11. Fatih Pınar and Neşe Ozan: Sulukule'de Kentsel Dönüşüme Yakalanan Hayatlar http://bianet.org/bianet/siyaset/115273-sulukule-de-kentsel-donusume-yakalanan-hayatlar. Accessed 31 March 2010.

12. A clip can be seen on YouTube at http://www.youtube.com/watch?v=CRr3FdAwIMM. Accessed 31 March 2010.

References

Adorno, Theodor W. (2002) On Jazz. (Original 1936), in *Essays on Music* (translated by Richard Leppert). Berkeley, CA: University of California Press.

Ahıska, Meltem (2003) Occidentalism: the historical fantasy of the modern. *South Atlantic Quarterly*, **102**(2/3), pp. 351–379.

Akın, Fatih (2004) *Head-On/ Gegen die Wand/ Duvara Karşı*. Germany/Turkey. Dir. Fatih Akın. DVD (US). Dist. Santa Monica, CA: Strand Releasing, 2005.

Akın, Fatih (2005) *Crossing the Bridge: The Sound of Istanbul / Istanbul Hatırası*. Dir. Fatih Akın. DVD (US). Prod. Corazon International and Intervista Digital Media. Dist. Santa Monica, CA: Strand Releasing, 2006.

Anderson, Benedict (1991) *Imagined Communities: Reflections on the Origin and Spread of Nationalism*. London: Verso.

Anderson, Chris (2004) The long tail. *Wired Magazine*, 12 October. Available at: http://www.wired.com/wired/archive/12.10/tail.html. Accessed 18 April 2010.

Appadurai, Arjun (1996) *Modernity at Large: Cultural Dimensions of Globalization*. Minneapolis, MN: University of Minnesota Press.

Bryman, Alan E. (2004) *The Disneyization of Society*. London: Sage.

Christensen, Miyase and Nezih Erdoğan (eds.) (2008) *Shifting Landscapes: Film and Media in European Context*. Cambridge: Cambridge Scholars Press.

Dürr, Anke and Wellershoff, Marianne (2005) Unsere Zeit ist zu unpolitisch. Interview with Fatih Akın. *Der Spiegel*, 6 June. Available at http://www.spiegel.de/spiegel/print/d-40630230.html. Accessed 31 March 2010.

Duygulu, Melih (2006) *Türkiye'de Çingene Müziği: Batı Grubu Romanlarında Müzik Kültürü* [Gypsy Music in Turkey: Music Culture Among the Western Roma]. Istanbul: Pan.

Elsaesser, Thomas (2005) *European Cinema: Face to Face with Hollywood*. Amsterdam: Amsterdam University Press.

Erzen, Jale N. (ed.) (2008) *Congress Book: Panels, Plenaries, Artists' Presentations. XVIIth International Congress of Aesthetics*. Ankara: Sanart.

Göktürk, Deniz (2002) Beyond paternalism: Turkish German traffic in Cinema, in Bergfelder, Tim, Carter, Erica and Göktürk, Deniz (eds.) *The German Cinema Book*. London: BFI.

Göktürk, Deniz (2008*a*) Sound bridges and travelling tunes, Erzen, Jale N. (ed.) (2008) *Congress Book: Panels, Plenaries, Artists' Presentations. XVIIth International Congress of Aesthetics*. Ankara: Sanart, pp. 423–435

Göktürk, Deniz (2008*b*) Sound bridges: transnational mobility as ironic melodrama, in Christensen, Miyase and Erdoğan, Nezih (eds.) *Shifting Landscapes: Film and Media in European Context*. Cambridge: Cambridge Scholars Press, pp. 153–171.

Gunning, Tom (2006) The whole world within reach: travel images without borders, in Ruoff, Jeffrey (ed.) *Virtual Voyages: Cinema and Travel*. Durham, NC: Duke University Press, pp. 25–41.

Harvey, David (1990) *The Condition of Postmodernity: An Enquiry into the Origins of Cultural Change*. Oxford: Blackwell.

Harvey, David (2001) The art of rent: globalization and the commodification of culture, in *Spaces of Capital*. New York: Routledge, pp. 394–411.

Huntington, Samuel P. (1993) The clash of civilizations? *Foreign Affairs*, **72**(3), pp. 22–50.

Iordanova, Dina (2001) *Cinema of Flames: Balkan Film, Culture, and the Media*. London: BFI.

Iordanova, Dina (2003) Editorial to the Special Issue on 'Romanies and Cinematic Representation'. *Framework*, **44**(2), pp. 5–15.

Kirişçi, Kemal (2008) Migration and Turkey: the dynamics of state, society and politics, in Kasaba, Reşat (ed.) *Turkey in the Modern World* (The Cambridge History of Turkey, Vol. 4). Cambridge: Cambridge University Press, pp. 175–198

Kracauer, Siegfried (1995) Travel and Dance. *The Mass Ornament* (translated by Thomas Y. Levin). Cambridge, MA: Harvard University Press, pp. 65–73.

The Lumière Brothers' First Films. DVD. New York: Kino on Video, 1998.

MacCrimmon, Brenna and Sesler, Selim (1999) *Karşılama*. Green Goat Recordings/Kalan Müzik CD 113.

Malvinni, David (2004) *The Gypsy Caravan: From Real Roma to Imaginary Gypsies in Western Music and Film*. London: Routledge.

Mulvey, Laura (2006) *Death 24x a Second: Stillness and the Moving Image*. London: Reaktion.

Roberts, Martin (1998) Baraka: world cinema and global culture industry. *Cinema Journal*, **37**(3), pp. 62–82.

Ruoff, Jeffrey (ed.) (2006) *Virtual Voyages: Cinema and Travel*. Durham, NC: Duke University Press.

Schiffer, Reinhold (1999) *Oriental Panorama: British Travellers in 19th Century Turkey*. Amsterdam: Rodopi.

Seeman, Sonia Tamar and Sesler, Selim (1999) *Keşan'a giden yollar/Roads to Keşan: Regional and Roma Music of Thrace*. Kalan Müzik CD 154. Traditional Crossroads CD 80702-6001-2.

Sen, Amartya (2006) *Identity and Violence: The Illusion of Destiny*. New York: Norton.

Simmel, Georg (1994) The bridge and the door (Translated and introduction by Michael Kaern). *Qualitative Sociology*, **17**(4), pp. 397–413.

Stokes, Martin (1992) *The Arabesk Debate: Music and Musicians in Modern Turkey*. Oxford: Clarendon Press.

TÜRSAK (Turkish Foundation of Cinema and Audiovisual Culture) (2008) *Filming Guide Turkey*. Istanbul: Mas Matbaacılık

Urry, John (1990) *The Tourist Gaze: Leisure and Travel in Contemporary Societies*. London: Sage.

Yanık, Lerna K. (2009) The metamorphosis of metaphors of vision: 'bridging' Turkey's location, role and identity after the end of the Cold War. *Geopolitics*, **14**(3), pp. 531–549.

Zukin, Sharon (1996) *The Cultures of Cities*. Oxford: Wiley-Blackwell.

Part IV. Art in the City

Like many other cities, Istanbul has turned to culture and the arts in the endeavour to reinvent itself as a world city. Processes of deindustrialization and urban restructuring go hand in hand with the emergence of museum and gallery spaces – frequently in re-purposed industrial buildings. Since the late 1970s, Istanbul has slowly but surely developed its ties to other metropolitan art scenes. Festivals and art events in the city have grown more internationally connected in their programming, publicity and audience appeal. Europe still dominates this 'international' horizon.

As one of the key players in Istanbul's incorporation into global circuits of the arts, the Istanbul Art and Culture Foundation (İstanbul Kültür Sanat Vakfi, (İKSV) has endowed festivals of classical music, jazz, theatre, film, and exhibitions, most prominently, since 1987, the Istanbul Biennial. İKSV is also proactive in promoting Turkey in Europe, collaborating with the government in organizing cultural events to showcase artists and artworks from Turkey. The foundation has acted as a coordinator to Istanbul's application to the European Capital of Culture Programme and its administrators continue to play an important role at the 2010 Istanbul Agency.

Focusing on the Biennial, the three chapters in this part interrogate the social function of art as spectacle for citizens and tourists. Continuing some strands of discussion from the previous Part, the chapters here also address questions of location and circulation with regard to image production, spectatorship, and participation. Experimental documentary cinema has migrated into multiscreen installations in exhibition spaces, reconfiguring the nexus between art and society. Conversations with and between practitioners of the art scene and mediators who have been active in staging art shows form the core of this Part.

'Optimism Reconsidered', an interview with Hou Hanru, the curator of Tenth Istanbul Biennial, conjures from his motto that 'optimism is not only possible, but necessary'. In the interview, as in the Biennial he curated, Hou approaches 'The Third World' as a global project that has mobilized states and masses on the road from decolonization and independence to modernization and globalization, creating its own crises and success stories along the way. To re-examine 'the promise of modernity', Hou chose exhibition venues in Istanbul's contested modernist edifices. For Hou, contemporary art, itself a product of modernization and modernity, relates to 'explosive urban expansion' as a source of inspiration to mobilize democratizing creativity and imagination – hence, his call for optimism.

In 'Art in Istanbul: Contemporary Spectacles and History Revisited,' Jale Erzen, scholar of art and herself a practitioner of the Istanbul Biennial like Hou Hanru, takes a more cautionary road towards the prospects of contemporary art and biennials. Focusing on a variety of art shows that thematically address Istanbul in and outside Turkey including the Biennial, Erzen draws our attention to the obscuring of the local and national heritage of modern art through the contemporary practice of biennials staged by globetrotting artists and curators. She sees potential in the 'postmodernist gaze' in so far as it exploits the counter-modernist discourses in the condition of Istanbul, but detects hints of orientalist, exoticizing impulses as biennials concern themselves solely with so-called marginal lives and spaces. Erzen is disaffected with the exhaustive focus on experiment, marginality, and festivity, which has produced its own capitalist system of production, aesthetic monopoly, and audience.

Banu Karaca's chapter 'The Politics of Urban Arts Events: Comparing Istanbul and Berlin', closes the Part by expanding her analytical horizon to Berlin, a city which is rapidly becoming next door to Istanbul's art community. Drawing on observations from the biennials of Istanbul and Berlin, she examines the proliferation of urban art events and traces parallels and divergences in the discourses and practices that account for the 'spectacularization' of art in the contemporary metropolitan order. In both cities, she finds that contemporary art events are often presented to engender 'democratized' access to culture and hence an extension of the public sphere. On the other hand, she contends that urban art spectacles operate as vital economic development strategies in cityscapes that are increasingly characterized by social segmentation and gentrification, and thus may actually feed into exclusionary practices in the urban realm. Karaca's comparative approach that reads Berlin and Istanbul in conjunction, was confirmed recently when the director of the first Berlin Biennial in 1998, Jens Hoffmann, Director of the CCA Wattis Institute for Contemporary Art in San Francisco, was appointed curator of the Twelfth Istanbul Biennial in 2011. Clearly, there will be further intersections to analyze in the future – in the convergence of European Capital of Culture and world city aspirations.

Chapter 11

Optimism Reconsidered

Curator Hou Hanru interviewed by Nilgün Bayraktar

The Istanbul Biennial has been organized since 1987 by the Istanbul Culture and Art Foundation (İKSV).[1] As one of the forerunners of the increasing number of non-Western biennials (São Paulo, Havana, Tapei, Cairo, and Tirana, to name but a few) it has become an influential contemporary art event. The Tenth Istanbul Biennial took place from 8 September to 4 November 2007 under the title 'Not Only Possible, But Also Necessary: Optimism in the Age of Global War'. Curated by Hou Hanru, the Tenth Biennial explored the tensions, contradictions, and dynamism created by alternative modernities in a 'New World Order dominated by liberal capitalist economic forces and their related values' (Hanru, 2007, p. 21). Abandoning the traditionally used tourist circuit of historic sites, the Tenth Biennial pervaded a wide range of central and peripheral urban zones of Istanbul and addressed the relationship between artistic production and social change. Hou Hanru's curatorial strategies highlighted the cultural, political, and artistic currents that run through contemporary art, from non-conventional exhibition sites to site-specific public art commissions, from multi-screen installations to public talks and events taking place as part of the biennial event. Hou Hanru curated numerous other exhibitions, including the Tenth Lyon Biennial in 2009, the Chinese Pavilion at the 2007 Venice Biennial, the Second Guangzhou Triennale from 2004 to 2006, the Third Tirana Biennial in 2005, Paris's Nuit Blanche in 2004, and the Gwangju Biennial in 2002.

This interview, conducted in San Francisco in November 2009, aims to unpack the curatorial process, how the Tenth Istanbul Biennial ventured into the complexity and multiplicity of Istanbul's urban reality, and how various artistic works by ninety-six international artists and artist groups with more than 150 projects engaged with the city. It also raises questions about transnational art networks in which the Istanbul Biennial participates and the role of cultural events in the staging of Istanbul as a global city.

◆ ◆ ◆

Hou Hanru: I have been organizing exhibitions and writing on art for many years, but curating biennials has been particularly significant and interesting for me because every biennial is part of the globalization process itself. At the same time, every biennial should be able to make waves in a specific context. In doing so, a biennial becomes not only a part of the global circle, but also a part of the effort to generate diversity, multiplicity, and complexity, enriching both the local and the global cultural scene. My work is very much related to this effort. I have had opportunities to do projects such as the Venice Biennial, the Shanghai Biennial, the Istanbul Biennial, and the Lyon Biennial. Essentially I am very interested in how a biennial is connected to the city, to the site at which it takes place. Curating a biennial is also interesting in terms of trying to produce a new exhibition format each time – a format that has a physical connection with the place itself.

Nilgün Bayraktar: Is curating biennials a way of learning about a new city and becoming part of a new cultural and artistic scene for you?

Hou Hanru: Exactly. For example, the Istanbul Biennial was an opportunity for people to learn about their city through the mobilization of an event. At the beginning of the preparation process I did research on the city, on how art activities have been happening in the city in a particular way and how a particular setting of the city can provide a unique condition for people to produce art or to look at art from their own place. So this is why I chose some specific sites such as Atatürk Cultural Centre (Atatürk Kültür Merkezi, AKM)[2] (figures 11.1, 11.2, 11.3) on Taksim Square and Istanbul Textile Traders' Market (Istanbul Manifaturacılar Çarşısı, İMÇ) (figures 11.4, 11.5, 11.6) in Unkapanı.[3]

Figure 11.1. Atatürk Cultural Centre. (*Photo*: Sercan Taycan)

Figure 11.2. Atatürk Cultural Centre. Interior of the main lobby. (*Photo*: Sercan Taycan)

Figure 11.3. 'Mon grand récit: Weep into stones…' by Lee Bul. (*Photo*: Sercan Taycan)

Figure 11.4. Istanbul Textile Traders' Market. (*Photo*: Sercan Taycan)

Figure 11.5. Istanbul Textile Traders' Market. (*Photo*: Sercan Taycan)

Figure 11.6. Istanbul Textile Traders' Market. (*Photo*: Sercan Taycan)

Nilgün Bayraktar: Why do you think you were invited to curate the Istanbul Biennial?

Hou Hanru: The mechanism in inviting a curator works in this way: the members of the advisory board of the Biennial in the Istanbul Foundation for Culture and Arts (İKSV) propose different names and discuss the most relevant person to curate the upcoming Biennial. Of course this process is based on personal connections, on knowing who is doing what. I think the reason they chose me is because some people on the committee had been following what I had been doing in the past from other exhibitions and from my writings in various publications.

Nilgün Bayraktar: Why did you decide to accept the offer?

Hou Hanru: The Istanbul Biennial has always been one of the most attractive events for me. I have always wanted to do it because of its particular agenda and also its location. It has been a very important event in terms of its quality as well. Curating the Istanbul Biennial is one of the most significant things I have done.

Nilgün Bayraktar: What is the process of developing an agenda for the Biennial like? How much time did you spend in Istanbul? Who were your interlocutors?

Hou Hanru: I was invited right at the end of the last Biennial (curated by Charles Esche and Vasıf Kortun in 2005). Therefore I had two years to work on the Tenth Biennial. This was very productive because usually a curator has only a few months to work on a project. For instance, the Lyon Biennial provides six months for preparation. For two years I went to Istanbul every month, sometimes for a few days and sometimes for longer periods to do research and work with the team to build up the project. The Biennial team of İKSV is very good. I spent two months residing in Istanbul before the opening of the Biennial.

Nilgün Bayraktar: How did you develop the conceptual framework for the 2007 Biennial? Your introductory text to the catalogue (Hanru, 2007) provoked reaction in some circles in Istanbul. For instance, several academics from Marmara University published a counter-press release in response to your conceptual framework.[4]

Hou Hanru: It is true that the issue I tried to bring up did touch on some essential aspects of Turkish society. My text asked the question of where it is going. Clearly, my questions were about the old-fashioned modernization project of the republicans. Turkey is stuck between the desire to become European and finding a new identity in this big climate of the Middle East. One has to develop a critical view on this from an historical perspective and also from the perspective of what is going on today. I was aware that bringing up these issues would inevitably touch on the nerves of some people who want to preserve a certain form of privilege. Even if they are not in the most privileged position themselves, they are in a privileged spirit, which is fine. I am pretty happy about the fact that there were such public responses to the Biennial, because that proves that an exhibition project can have a social impact too, which is not easy to create.

Nilgün Bayraktar: How did you gain access to the local reaction and ensuing debate?

Hou Hanru: The reaction happened at the same time as the opening. Therefore, I was extremely busy. When the counter-press release came out, I had already left Istanbul. So I was at a distance from all the debates. Actually they (the group of local academic artists organized by the then Dean of the Art Faculty at Marmara University) tried to organize a meeting with me when I was away. I counter-proposed having a meeting with them when I went back to Istanbul. I was ready to go to Marmara University. But for some reason they did not want further conversation. However, when I arrived in my Istanbul office, I found piles of books on my desk – books sent by those people. At least five different biographies of Atatürk and different writings on Turkey. Of course, I had read some of them before for research. It was really interesting how they tried to communicate with me. I still keep those books in my collection. Most of them are in English. A couple were in Turkish despite the fact that I do not read or speak Turkish!

Nilgün Bayraktar: How did you come up with the title 'Not Only Possible, But Also Necessary: Optimism in the Age of Global War' for the 2007 Biennial? Why 'global war', for instance?

Hou Hanru: 'Global war' refers to not only physical war, but also economic war. Basically it is a war of different modernization projects. I would not call it a 'clash of civilizations' like Samuel Huntington. Rather, it has to do with different projects that each society can generate regarding what a modern world should be. That is what I referred to as alternative modernities. Turkey is of course one of the most intense examples. On the one hand, the Ottoman background generates a radically Westernized society. On the other hand, this society embodies a hybrid model. This issue of alternative modernities has a very contemporary significance: contemporary society is not only facing the challenge of new fundamentalisms of all kinds, but also the collapse of utopias. However, new utopias are constantly being generated as well. It is interesting to ask what kinds of utopia are actually happening today. We are living in the age of global war. Yet the only way to deal with this situation is by creating a new form of hope.

Nilgün Bayraktar: Is this why you chose the word 'optimism' rather than utopia?

Hou Hanru: Yes. Optimism is an attitude. It is not a project in itself. This implies a possibility for people to navigate various difficulties with hope. Of course this is ironic as well. We are living in a world in which we ought to have a sense of humour.

Nilgün Bayraktar: What was the process of selecting the sites such as AKM, İMÇ, ANTREPO,[5] and KAHEM?[6]

Hou Hanru: I learned a lot about these sites in the process of preparing the show. I talked to members of the local community, as well as professionals, architects, and artists. The first site I wanted to use was İMÇ (figures 11.7 and 11.8). Many Istanbulites do not know about this place because they would never go there. İMÇ is not the place of local intellectuals and artists. No one would go there to buy things. I used to take a cab from the airport to Taksim. Every time I took this trip, I saw this building, and I thought that it was very interesting. Then I decided to do research on it. I asked my team to take me there. They explained that this is the textile market. I tried to collect all the drawings and books about the design, the history, and the transformation of that place from the 1950s onwards. And then we visited the architects and the union that runs the building.

I found that this is actually something that embodies a tiny universe of the

Figure 11.7. 'ReverseDirection: Counter-Services' by Burak Delier. (*Photo*: Sercan Taycan)

Figure 11.8. 'Informal Economy Vendors' by Julio César Morales. (*Photo*: Sercan Taycan)

history of Istanbul. A kind of microcosm that made us think about the history of the utopian aspect of the modernization project as reflected in the architectural design of the building. How has this building been transformed and used over the past fifty years? And how does the building reflect the societal changes that are happening today? How has the population inside changed? How have the shops changed? What kinds of commodities have been sold there? It is very interesting! In the meantime I was working on the project called 'World Factory' in San Francisco.[7] I used that project as a laboratory and extended it to the building, integrating its various elements into that shopping area itself. İMÇ was the first place I was compelled to use for the Biennial.

Another building was of course the AKM. It is very clear: when you arrive in Taksim, you cannot miss this building. It is a monument. You can tell from the first glance that there is a very interesting history behind this building – the history of how socialists or the nation state celebrated their ideal in this place and how this building has gone through a form of decadence. People would easily forget such a history! And I learned that this place was about to be taken over by the city government to make room for a new development, financed by private developers to become a more 'contemporary' cultural and commercial complex. Obviously, it is a typical gentrification. Similar cases happen all around the world today. For instance, in East Berlin, the Palace of the Republic was demolished. There are many cases like this in China also. Even in Paris, the Communist Party Quarter has been regularly rented out for fashion shows. In fact, all these things embody the kind of globalization happening today: various ideals about a more just, more democratic society are being taken over. I visited AKM, learned about the history of the building through talking to local people and through research. Having that building as a venue in the Biennial was a very complicated issue because nobody knew what was going to happen to the building the following year. Moreover, there was an ongoing public debate about the building, and this debate was further intensified because of the Biennial. AKM and İMÇ are modernist buildings that have reflected a political and an economic destiny simultaneously.

Another venue I had in mind was the urban space at large. Istanbul is a place in which people do not sleep. Therefore I decided to do a show that would take place 24 hours a day. This attempt to extend the Biennial to diverse urban spaces was also about bringing contemporary art as well as the discussion around the idea of modernized society to the different parts of the city – parts that have no access to such public events and discussions. For this 24-hour project, we worked with younger curators from Turkey.

This project was actually based on another project I did for the Nuit Blanche in Paris three years before the Biennial. I called it 'Image Dazibao'. 'Dazibao' is a kind of poster, or public announcement, with large letters. During the Cultural Revolution in China in the 1960s and 1970s, the 'lower classes' were encouraged to post their critiques of the elite in the form of 'dazibao', turning street corners into a public forum. It could be defined as one of the most democratic forums. In this regard, 'Image Dazibao' was a multimedia forum for screening on the street. Both regular people and artists contributed to this project in Paris.

For the Istanbul Biennial, I invited other curators to do the 'Nightcomers' project. In the evenings, the programme was projected in public spaces in different parts of the city, taking the Biennial to every corner of the city. 'Nightcomers' journeyed through different city spaces during the three months of the Biennial.

Related to this project was the exhibition in the Antrepo, in which I tried to emulate a street situation (figure 11.9). I created not a nice museum space but rather a living space. For instance, in the 'Dream House' project, you could spend the night in Antrepo. The visitors could sleep on a huge bed and could watch the Bosphorus view. Therefore, the Antrepo project was not an exhibition of objects but a site of activities: a site for people to come to experience something extraordinary that they do not have in their everyday lives.

Nilgün Bayraktar: The artworks in Antrepo dealt with global war much more directly (figures 11.10, 11.11, 11.12, 11.13). For instance: Multiplicity's piece 'Road Map' on the territories of Israel and Palestine; Michael Rakowitz's piece

Figure 11.9. Antrepo Building. (*Photo*: Sercan Taycan)

'The invisible enemy should not exist' on the artefacts stolen from the National Museum of Iraq, Baghdad, in the aftermath of US invasion in April 2003; Rainer Ganahl's 'Silenced Voices: Bicycling Istanbul's Topography of 21 Murdered Journalists'.

Hou Hanru: Also, there was a project on the question of Armenia, a dialogue between Kutlug Ataman and Atom Egoyan. The works in Antrepo touched on

Figure 11.10. 'Lines of Lessons' by Hamra Abbas. (*Photo*: Sercan Taycan)

Figure 11.11. 'Construction Site' by Huang Yong Ping. (*Photo*: Sercan Taycan)

Figure 11.12. 'What?' by Extramücadele. (*Photo*: Sercan Taycan)

Figure 11.13. 'From Worldwide to International' by Yan Pei Ming. (*Photo*: Sercan Taycan)

different aspects of society. In their condensed form, Antrepo projects tried to tell people that they are living in a city in which there are problems with history, various geopolitical conflicts, and tension around migration, but also utopian projects. This is why I had people like Rem Koolhaas, who produced projects with visions for the future. Additionally, there was a place to rest ('Dream House'). Altogether, these diverse elements constituted a living city.

Nilgün Bayraktar: In your text in the Biennial catalogue, you criticized the top-down modernization project of the early Republican era. At the same time, you chose modernist buildings such as AKM and İMÇ, which were built after the Republican era, and for which Turkey's current government, under the so-called Justice and Development Party (AKP), has had demolition plans. Do you think the Biennial helped save those modernist buildings?

Hou Hanru: It is true that these buildings embody a top-down model. However, I was interested in exploring how people have been using them, and how the confrontation between the needs of the bottom of the society and the top-down power system can generate a new energy. I am definitely not interested in preserving a building as it is. A building is not a dead object that should be preserved. I am also not interested in authenticity or turning a city into a museum. What is significant is how a building can be reactivated – how it can be injected with a new energy and new possibilities. Yet these buildings are also part of a certain heritage. In that sense, they should be preserved in a dynamic way as a part of the changing social life itself. A city should have a dynamic memory. Unfortunately, people tend to forget their recent history, because it is too close to them and it is very much related to their own interests. For instance, there are politicians who want to erase the recent history. They would celebrate Hagia Sophia as part of the cultural heritage of the city. However, they would not want people to understand the actual national identity or any other identity that comes from the recent history. This is very political and this is how the power system works: erasing the memory of the recent and fixing people on a remote imagination that constitutes a manipulable image, or 'tradition', or 'identity'. And this is how a dictatorship works. Talking about the project of modernization is actually talking about democracy itself. What is important in a democracy is that every individual participates in the making of the history – a right that is very often taken away by the power system. In this regard, by putting forward such a conflictual situation, the Biennial was an attempt to expose to people the dynamic that constructs our existence.

Nilgün Bayraktar: What are some of the ways in which you aimed to establish diverse relationships between the works and the sites?

Hou Hanru: In each building I developed a specific vision, a strategy and an understanding of the relation between our work and the space. AKM's concept was about the paradox of utopia. For AKM, I tried to build up a situation in which all the works organically related to the building as if they had already been there. I tried to render works as integrated as possible into the building in terms of the format and the content. There were a few works commenting on the history of the building. There were works on how utopias failed: for example, Armenian artist Vahram Aghasyan's piece 'Ghost City' (a piece on an unfinished modernist construction in the aftermath of the 1988 earthquake in Armenia) shows the failure in the utopian aspect of society in relation to political change as well as natural change – the earthquake. Moreover, there were works connected to the building architecturally. For instance, Nina Fisher and Maroan El Sani's work on the library (the old Bibliotheque Nationale de France) in Paris and their work on the headquarters of the communist party building in Paris; the work by Daniel Faust on the UN building; the Austrian artist Markus Krottendorfer's 'Hotel Rossija', which is based on the last days of the Hotel Moscow.

For AKM, I tried to create a reactivated memory that integrates different kinds of historical situation. For the İMÇ, I tried to create an exhibition in a shopping format. The exhibition was almost displayed in the form of shops. So İMÇ did not have huge installation projects, but rather narratives and images being displayed in various spaces in the complex. The experience of viewing İMÇ was like going from one shop, a normal shop, to an exhibition shop. That actually raised very interesting questions, because people came to İMÇ expecting to see a gallery exhibition. Instead they discovered this place and they met the local people. People from the art world especially would never have a chance or the desire to talk to someone who sells tea or kebabs or folklore music. I think such an encounter would be a very important lesson for the art world.

Nilgün Bayraktar: Could you unpack this notion of the elitist art world a little more?

Hou Hanru: What is interesting is to see the hidden desires of people who want to embrace different worlds. For instance, a shop owner, who was also an artist, decided to give her shop to a group of young artists to have their exhibition there. Later on, that shop became a gallery. Now there is a gallery in the İMÇ!

Nilgün Bayraktar: İMÇ works brought up issues around social activism. Most of the works confronted capitalism directly. Could you talk about the works in this building?

Hou Hanru: As I said, I started with the project 'World Factory', which dealt with the question of how the Third World has been transformed, in the process of globalization, into factories to produce consumer goods for the First World market. For instance, the way in which Turkey has developed over the past fifty years and in which Istanbul has expanded is part of this process. Today, Mexico, China, and many other places are also going through a similar process. Indeed, many artists are dealing with these issues today. This is why I came up with this project called 'World Factory', and we used İMÇ as a laboratory to explore this and then extend it to Istanbul. We collaborated with people who work in urbanism and architecture – people who strive to create alternative forms of public space. This is essentially about issuing proposals to improve the quality of public space. We also looked at very direct confrontational situations such as gentrification. We had three Turkish filmmakers, Ege Berensel, Serhat H. Yalçınkaya, and Banu Orat, who did a documentary titled 'TST – The Hill Doesn't Chant Anymore' on a village being gentrified. We had people like Ramazan Bayrakoğlu, who talked to the people in the shops in the İMÇ about how the Turkish textile industry is being marginalized because of Chinese production, which is now taking over the industry. In Antrepo we showed a piece by Wong Hoy-Cheong, who spent two to three months working with the kids in Sulukule[8] during the 2007 protests against state-led gentrification and produced the film 'Oh Sulukule, Darling Sulukule'. During the Biennial, we organized panels with architects and people living in the city to talk about what is happening in Sulukule. In KAHEM, we had the 'Emergency Biennial', curated by Evelyne Jouanno, which was also a very provocative project. We had a debate about human rights in Turkey: we invited a Kurdish filmmaker, Hüseyin Karabey, and other activists to discuss human rights issues in Turkey.

Nilgün Bayraktar: In Antrepo, most of the works dealt with border crossing, hybridization, and migration. Such themes have also been appropriated by the culture industry, especially when we think of the commodification of diversity or cultural difference.

Hou Hanru: Of course it is a matter of fact that somehow major corporations are taking over. But there is also a very dynamic resistance movement that continuously produces difference. It is true that we all live in this liberal capitalist system. But I am not so pessimistic. Indeed, I would like to see the complexity in it, as well as the zones in which different systems wrestle or try to negotiate with each other. Yet KOÇ Holding is a sponsor of the Biennial. Ironically though, most of the artists do projects against a corporate mentality. This is the reality. The important thing is not providing one solution to that conflict, but rather maintaining or encouraging situations in which this complexity can become more dynamic.

Nilgün Bayraktar: 'Nightcomers' is very interesting, for it made the Biennial accessible to people outside of the art world. Do you know what kind of people watched those videos?

Hou Hanru: Unfortunately, I could not follow the whole process, but the team went to the different parts of the city. At each screening they gathered a big crowd – sometimes hundreds of people. Yet the important thing was not only how many people were there. Rather, this project was also a gesture for new artistic and cultural initiatives in the urban space. Indeed, artists from Istanbul, from the 1990s generation, already started this process of living in a neighbourhood and sharing space with local people as in the projects of the artist collective Oda Projesi.[9] We used the Biennial to extend such projects. And hopefully this will go on. Yet there is also the question of access. How can you create the opportunity for people to access these projects?

Nilgün Bayraktar: There were many artists from China and Taiwan in the Biennial. Do you see parallels between these countries and Turkey in terms of this notion of a modernization project? How did your background inform the Biennial – for instance during the process of selecting the artists?

Hou Hanru: The experience of living in different parts of the world has been very important in that sense. Indeed, many artists and intellectuals today have similar experiences. Being multiply located also paves the way for a historical reflection on how non-Western societies, following processes of de-colonization and independence, have tried to reinvent their own modern world in their different contexts. These non-Western societies also share the idea of creating something that will be equal to the West. There has been a very important potential energy behind that. Still today, the question is about how to embrace a new idea of a world constituted by diversity and a multiplicity of modernities, rather than a world with one centre or dominant philosophy. I think the Biennial reflected on this process. Hopefully we can recharge the art world itself. Perhaps there is a new task that the art world can take over.

Nilgün Bayraktar: Modernization in Turkey is very much related to secularization. How did the Biennial address religion and the mobilization of Islam since the 1980s?

Hou Hanru: Personally I am not a specialist in the interpretation of religion. But I believe it is important to recognize that the recent 'return' of Islamic influences should be understood in the context of political and geopolitical changes for the last decades. Somehow, the reason is more social, political and even economic than religious. This is very similar to the ways in which the so-called radical Islam is 'invented'. These are often contemporary products. Taliban is a contemporary invention. It is invented in a very specific geopolitical situation; a kind of vernacular religious resistance was created and used by local and international powers to prevent the so-called communist expansion. Therefore, I do not see religion as an isolated thing. This is why I did not put any specific response to the question of religion in the Biennial. However, we addressed the question of secularization, which points to certain religious issues. Secularization and religion are connected to each other in multiple ways. I prefer to look at religion from a historical perspective rather than simply essentializing it.

Nilgün Bayraktar: Thinking about modernization as a term, many scholars have tried to understand post-1980s Turkey through the lens of postmodernism. But your text brought up this issue of a top-down modernization process. I am wondering how you situate modernism in relation to postmodernism.

Hou Hanru: Actually the claim that there are multiple modernities could be taken as a postmodernist attitude. Yet it is not a postmodernist attitude that tries to create a nostalgic reconstruction of the past as a justification of a certain political stance in the present. That was what happened in architecture, and it was about how to reconstruct a conservative paradise. This is what I oppose. I understand modernities as a plural concept that implies a common desire of human society to evolve into a better way of existence. You might call it progress, to use a very old-fashioned word that people hate so much. There are different ways to achieve progress and different visions of what progress is. Yet what is important is not to look at the final image that progress would produce but to maintain the dynamic that can make us continuously creative and alert to the danger of being limited in our activities. For me that is essential. Modernity is not a fixed concept but a process. It is a concept like identity that evolves all the time. Therefore, I do not try to situate the notion of modernity between modernism and postmodernism. I think they are completely different things. What is important is to see these processes as part of a dynamic system.

Nilgün Bayraktar: How do you think we can create resistance to dominant forces appropriating diversity?

Hou Hanru: This is why in my catalogue text for the Biennial (Hanru, 2007) I referred to people like Tony Negri and Michael Hardt, who use the concept of the multitude. Discussing the multitude has to do with finding another way to create a social contract based on a new understanding of the common. I think it is time to talk about how to re-imagine a new common. It is a very idealist dream. But we can talk about a solid ground to achieve a major part of this social contract. Why not?

Nilgün Bayraktar: What is the role of the Istanbul Biennial in the transnational art network and its relative position *vis à vis* other biennials?

Hou Hanru: Istanbul is among the most important ones. It is a leading biennial in the field because it is one of the first biennials that happened outside of the West.

At the time, it was one of the cornerstones that actually produced a whole culture of biennial. This culture is not only the Biennial event but a system that replaced the conventional institutional art framework. The culture of biennial was created through some key points in the 1980s and especially the 1990s, and Istanbul has been one of the central points in this process. Also, over time the Instanbul Biennial proved that the quality of its projects is among the highest. Now it makes even more sense that the last few editions of the biennial have increasingly engaged in some major political and geo-political discussions. This is not very often the case with other biennials. Therefore, the Istanbul Biennial is not merely a spectacle or a showcase for the best artists. It is much, much more than that.

It is very interesting that the biennial itself can become not only a very important global event, but also a permanent laboratory in itself. Just as Istanbul is a huge laboratory for human life, so that spirit has to be preserved, but also developed.

Nilgün Bayraktar: Do you think the biennial paves the way for Istanbul to become a 'global city'?

Hou Hanru: Maybe it is part of the process. I think the biennial, along with all other big festivals, has generated a very significant art community in the city. That is really important. Somehow that actually forced the whole society to embrace the idea that culture plays a role in making a position for a city in the world. What is crucial is not only how good one event is, but also the fact that people have this consciousness that Istanbul, as a global city, has a major role in the globalization process itself.

Notes

1. İKSV was founded in 1973, on the fiftieth anniversary of the foundation of the Turkish Republic, by local businessmen under the leadership of Dr. Nejat F. Eczacıbaşı of the pharmaceutical corporation with the same name, with the goal of organizing an international arts festival in Istanbul. The initial activities of the Foundation centred mainly on classical music. In 1983 a film festival and in 1989 a theatre festival were launched, followed by a jazz festival in 1994. Today IKSV organizes five festivals in Istanbul and various collaborative projects in the field of culture and art in major European cities ('Şimdi Now' in Berlin in 2004; 'Şimdi Stuttgart' in 2005; 'Turkey Now' in Amsterdam in 2007; and Rotterdam in 2008).

2. Atatürk Cultural Centre (Atatürk Kültür Merkezi, AKM) was first proposed in the mid-1930s as a performance venue for opera, theatre, classical music and fine arts, for Taksim Square. After several designs and multiple halts in the construction, architect Hayati Tabanlıoğlu designed the final version. AKM opened its doors in 1969. Following a fire in 1970, the building was closed down and opened again in 1978. In addition to the art galleries located on its top floor, AKM has two halls for opera, ballet, theatre classical music performances and concerts (Tabanlıoğlu, 1979).

3. Istanbul Textile Traders' Market (İstanbul Manifaturacılar Çarşısı, İMÇ) accommodates nearly a thousand shops in six blocks that stretch along one side of the Atatürk Boulevard between the Valens Aqueduct (Bozdoğan) and Unkapanı. The complex was designed in 1959 by architects Metin Hepgüler, Doğan Tekeli and Sami Sisa, as the result of an architectural competition, and built between 1960 and 1968 (Yırtıcı, 2003).

4. A few weeks after the opening of the Biennial, the Dean of the Art Faculty at Marmara University (Istanbul), Nazan Erkmen, along with 131 academics, made a statement to the press, criticizing the curatorial statement of Hou Hanru in the Biennial catalogue for 'talking about the Republic and the reformations of Atatürk as a top-down elitist project' (NTV-MSNBC, 2007).

5. Antrepo no. 3 is one of four former customs warehouses along the bank of Salıpazarı Harbour in Tophane with a privileged panoramic view of the Bosphorus and the historic peninsula. The first three of these warehouses were proposed as part of a site plan developed by architect Sedad Hakkı Eldem for the Turkish Maritime Bank in 1958. Since the warehouses became redundant in the late 1990s, they have been used for exhibitions such as the Istanbul Biennial. The fourth warehouse that was added later was converted in 2004 into the Istanbul Modern Museum.

6. Kadıköy Halk Eğitim Merkezi – KAHEM (Kadıköy Public Education Centre) was designed in 1938 as a People's House by architect Rüknettin Güney. People's Houses were launched by the People's Republican Party (CHP) in the 1930s as progressive institutions of social education with the stated objective of teaching people about the ideals of the Kemalist revolution. They were closed down when the Democrat Party came to power in 1950. Several years after closure as People's House, the Kadıköy building was reopened as Kadıköy Public Education Centre and has served in that capacity since then (Bozdoğan, 2001).

7. For more information, see the online exhibition available at http://activeweb.sfai.edu/galleries/waltermcbean/worldfactory.aspx. Accessed 9 January 2010.

8. Sulukule, an old settlement located within Istanbul's historic peninsula, in Fatih municipality, has historically accommodated many Roma citizens. The Roma community living in Sulukule had made a living through performing music and dance in their own neighbourhood until 1992. In 1992, the community's entertainment businesses were closed down, causing serious socio-economic problems in the area. In 2009, the neighbourhood was demolished by the local government to displace the Roma and is to be replaced by a new housing scheme.

9. Oda Projesi is an artist collaborative initiated by three women artists. In 1997, the three artists rented an apartment intended as a shared artists' studio in the neighbourhood of Galata, Istanbul. From 2000 to 2005, Oda Projesi carried out projects that explored the possibilities emerging out of their shared lived experiences with their neighbours in Galata.

References

10th Istanbul Biennial. Available at http://iksv.org/bienal10/english. Accessed 8 January 2010.

Bozdoğan, Sibel (2001) *Modernism and Nation-Building. Turkish Architectural Culture in the Early Republic*. Seattle, WA: University of Washington Press, pp. 94–98.

Hanru, Hou (2007) Not Only Possible, But Also Necessary: Optimism in the Age of Global War? in Ayvaz, İlkay Baliç (ed.) *10th International Istanbul Biennial Catalogue*. Istanbul: İKSV and Yapı Kredi Publications.

Istanbul Foundation for Culture and Arts (İKSV) (2005) Biennial Venues. Available at http://www.iksv.org/bienal/english/bienal.asp?cid=8. Accessed 8 December 2009.

Istanbul Foundation for Culture and Arts (İKSV) (nd) 10th Istanbul Biennial Special Projects: Kadıköy Halk Eğitim Merkezi (KAHEM). Available at http://www.iksv.org/bienal10/english/detail.asp?cid=12&ac=kahem. Accessed 8 December 2009.

NTV-MSNBC (2007) 10. İstanbul Bienali'nde Kemalizm Tartışması (Kemalism Debate at the Istanbul Biennial). Available at http://arsiv.ntvmsnbc.com/news/421065.asp. Accessed 8 December 2009.

Tabanlıoğlu, Hayati (1979) *Atatürk Kültür Merkezi Kitabı* [Book of the Ataturk Cultural Centre]. Istanbul: Apa Ofset.

Yırtıcı, Hakkı (2003) İstanbul Manifaturacılar Çarşısı. *XXI*, **4**, pp. 74–83.

Chapter 12

Art in Istanbul: Contemporary Spectacles and History Revisited

Jale Erzen

This chapter proposes two basic arguments concerning the artistic life of Istanbul. One is the issue of Westernization, which I claim has created the main thread of continuity between the historical and the contemporary. However, the use of the West as a model has also been the ground for controversies and polemics in art and cultural discourses. Although the chapter will not deal with these polemics, the arguments will show that controversies continue. The other argument is based on the claim that Istanbul's idolatry of itself has generally constituted the basic reason and content of its artistic production. In the twenty-first century, Istanbul has become one of the world's popular cultural curiosities, not only because it provides a very rich spectacle of contemporary artistic events, but also because in its urban heritage it has stored a cultural memory of two thousand years which stimulates its cultural ambience. Here I propose to show how the very claims of 'contemporary', still use the layered historical presences of the city to redeem an aura for art.

These arguments are based on my personal experiences rather than on academic presuppositions. As a practicing artist, art historian and editor of an art journal (*Boyut*) I was also involved in the Third Istanbul Biennial and in the consultant committee of the succeeding Biennial. The Association of Aesthetics and Visual Culture, Sanart, of which I was the founder and president, has been organizing international symposia since 1992, and investigating transformations in art and aesthetics, both in Turkey and abroad. However, it is my childhood and adolescence in Istanbul through the 1950s and 1960s, which has provided me with certain insights into the character of the city and the changes it has lived through.

The Urban Aesthetic

Through the ages Istanbul has perpetuated its particular qualities in a synaesthetic way, appealing to all the senses, through spectacle, poetry and art, reproducing and representing, hence, regenerating itself. Thus, to view Istanbul by losing oneself in its intoxicating multiplicity may be the right approach, truthful to its social aesthetics. This hedonism lies at the base of a lot of art done in Istanbul and is part of Istanbul's charm to which a large portion of the world's artistic and intellectual public is drawn. As with most spectacles, Istanbul with its mythical past and imaginary present, is a dream machine in much the same way as the Orient has been.

The qualities which draw international crowds today have to do not only with the natural aesthetic attractions of the city such as the historic monuments, the rich flora and the beauty of its layout, not to mention art events, but also with a certain disorder and diversity both in the social makeup and behaviour, and in the physical qualities of the urban texture. The symbiosis of the West and the Orient in Istanbul largely makes up its general aesthetics today. Therefore it is not farfetched to claim that part of the attraction is due to the city's safeguarding a lot of its Oriental character. Today, the content of this attraction is not so much the historical monuments and symbolism, as it had been for the French Orientalist painter J.L. Gerome (1824–1904), but has shifted focus onto the 'other' as the margins of society either in terms of class or culture. As examples we can cite some art shows of the last decade such as 'Meltem of Istanbul' (2006), curated by Necmi Sönmez, an exhibition of contemporary German and Turkish art at the Elgiz Museum of Contemporary Art in Istanbul; the 2005 Berlin Show 'Urban Realities: Fokus Istanbul' at Künstlerhaus Bethanien in Berlin; and 'Stadtansichten Istanbul' in 2004 at the ifa-Galerie Stuttgart.[1] All these shows, two of which took place in Germany, not only showed Turkish art, but displayed the city and its marginal people. Some of the essays and photographic works produced for these shows deal with penury, prostitution and trans-gender performances, with the claim that in so doing they redeem the right to exist for all diversity and provide a critique of intolerance.

The contemporary artistic scene of Istanbul could be treated as a global reality of the present, where any artist anywhere in the world can have autonomy, as is often claimed by curators of contemporary art. This presupposition ignores the relation to power politics or to the market and assumes that artists are independent and selected primarily for quality. The interpretation that I choose, would be to treat the artistic scene of Istanbul as a continuation of the developments that Istanbul lived in history, with their political and economic implications still largely valid today. Istanbul's actual presence on the world scene is the fruit of global economy, postmodern critique, and the production of difference neutralized through capitalist means. As we shall see, the new approaches in art along with the publication of several radical art journals started in the 1980s, the period of political opening of Turkey to the international market. As Sibel Yardımcı explains, it is ironic that the development in contemporary art, implying freedom of expression, came contemporaneously with the 1980 military coup that enforced a severe repression on the intellectual realm (Yardımcı 2007,

p. 4). The coup's goal was actually to put an end to the leftist and liberal thought and expression that had been evolving in the intellectual, artistic and academic fields since the late 1960s. However, still in the early 1980s, state bureaucrats were so far behind in any artistic awareness that for them art, except when it dealt with nudity and sex, could not be a threat. Thus, in the years following the coup, although many universities were under scrutiny, one could say that art had a certain freedom. It is interesting that the army general who was responsible for the coup and who immediately became the Republic's President had the painting of a nude taken down, after his visit of a large national painting exhibition. The same person, after his retirement, has been painting nudes and popular screen artists, and selling them for exorbitant prices for charity.

Behind the visual manifestation of Istanbul, narcissistically adored, dressed up, reflected in paintings, in mirrors and in its own waters, there had always been the hybrid identity that held a fascination. Süleyman the Magnificent's pride, as he commissioned from the Venetian jewellers a four layered tiara, was based on the double or triple citizenry of the city, of which he was the sovereign: Christians, Muslims, and the immigrant – Sephardic Jews (Necipoğlu, 1989). In much of today's obsession about Istanbul's identity, this pluralistic aspect still plays a role, in fact it even increased in the late twentieth century with national and international migrations, notwithstanding the fact that many of the old local minority families of Greeks, Jews and Armenians had left. The Istanbul art scene, as was very evident in the 2005 Istanbul Biennial, feeds greatly on this fact (Istanbul Biennial 2005, catalogue).

In Istanbul, identities are preserved and the pluralistic make-up of the society is always visible as a cultural fact. This is because rather than producing its own potpourri culture, Istanbul becomes a land where every different culture wants to persist, to claim its own space and time, to speak its own language and play its own music. What is special about the history, the topography and the geography of the city is that in Istanbul space and time are never homogenized, leaving liberty to self-expression. Possibly, in contrast to many other cities which are made of diverse cultures, Istanbul can keep these differences, due to its topography. All through the ages, particular ethnic groups inhabited different districts of Istanbul. In spite of the impossibility of keeping districts apart, in present conditions, immigrants from different parts of Turkey usually inhabit neighbouring areas, for economic solidarity.

One important issue in creating a focus on Istanbul was the rise of postmodernist critiques, especially around the 1980s. While Ankara was considered the seat of modernist Turkey, which never quite fulfilled its promises, Istanbul remained the city of diversities, the city that did not and could not become an example of modernity. This missed opportunity, and the nostalgic power of the past, present everywhere is, according to Orhan Pamuk, the 2007 Nobel Laureate, one of the attractions of Istanbul, in a sense, its truthful face (Pamuk, 2003). Istanbul, which never could become homogenized, became the postmodern city par excellence, the city where all the various languages were spoken, but never fully understood by each other; where otherness forces itself to all efforts of unity.

I remember as a child, walking with my grandfather on Beyoğlu, Pera, where he had his house on Eski Çiçekçi Sokak (which is still standing), he would signal to me the banners that read 'Citizen, speak Turkish'. I remember how at night the nostalgic music of the nearby nightclubs kept me awake. You would always hear tangos and waltzes round the streets. Most vendors were Greek, especially in Beyoğlu where the fashionable boutiques and pastry shops were located. One night I woke up to a blazing light in my room, a famous store, Lyon, was burning. Emblematically, this fire signalled the end of the Greek culture in Beyoğlu, and later, on 6–7 September 1955, when stores belonging to Greeks were vandalized, the expulsion became even more tangible. Beyoğlu underwent a period of abandonment and decay, until populist culture invaded every corner with plastic goods and *kebab* smoke. Istanbul is a phoenix of a city unequalled anywhere, both a ruin and a city of riches.

Spectacles of the Past

Istanbul owes its attraction not only to its unique geographical position, which creates a constant flow of international encounters, but also to having been the capital of two world empires, both of which used it as a stage for their show of power and of glory. Presaging the mediatic atmosphere of today's Istanbul, Justinian's Constantinople vied for publicity and public spectacle. Justinian's Byzantine architectural heritage still adorns the city with churches and city walls that are outstanding examples of urban architecture, as the famous Haghia Sophia from the sixth century (figure 12.1). The show of power through public celebrations and holidays was not missing in the Ottoman capital either. The important fact is that both in the Christian and Muslim worlds, the elite as well as the poor, were at all times both actors and spectators, as one can witness in the miniatures of Levni from the eighteenth century (Atıl, 1999). This continues today and may be one of the few real democratic experiences, as well as what transforms the urban spaces into theatrical settings. As the miniatures of the circumcision celebrations of Sultan Ahmet's son illustrate, people are watching parades, but are themselves part of the show. These celebrations and parades were common in the Ottoman Empire, using every possible occasion to create a spectacle, such as the procession of the Sultan and his retinue to the Friday prayer, or the departure of the military for a new campaign. Today, the crowds in Istanbul, around mosques and public spaces, attending the unending festivals throughout the year, become themselves the spectacle. The diversity in the crowds, with people speaking different languages or dialects, creates a dynamic colourful view in many areas of the city, such as Eyüp, the traditional religious district or Beyoğlu, the more modern area.

Figure 12.1. Hagia Sophia, The Royal Fountain of Sultan Ahmed III, Istanbul Peddler. (*Photo*: Jale Erzen, 2000)

Even if notions of 'art' did not exist or were different from today, one can still talk about similarities in the interest for art between the present and the historic past of Istanbul. Conqueror Sultan Mehmet II's fine taste and judgement can be attested in his choices of architecture and art. One can talk about a pioneering effort in his interest in Western art. It is believed that he invited the renowned Filarete to build his imperial architectural complex; his portraits, one of which is by the Venetian painter Gentile Bellini, display an aesthetic that is difficult to find in any of the later Ottoman imperial portraits. The one attributed to Şiblizade Ahmet stands out in its elegance and its symbolism of the Sultan as an aesthete (figure 12.2). After the Conqueror who, by inviting back the Greek population and adorning the city with palaces, mosques and commercial complexes, gave the city its initial Ottoman character, almost every sultan and great vizier added to the glory of the city undertaking ambitious architectural projects.

Istanbul's many religious complexes, which constitute the city's basic focal points in terms of how the urban structures have been built around them through the ages, have become today the most popular social focal points, both for the common citizen and for the tourists. The present tendency of the political party in power towards creating a religious society implies strong ties with the pre-modern past of Istanbul. Many of these religious complexes that had been left without care have all been restored since the Justice and Development Party (AKP) came to power both in the Istanbul Metropolitan Municipality and in central government. Many headquarters of important institutions have also been moved to Istanbul, implying a desire on the part of the government to reconstitute Istanbul as the capital, as it was during the Ottoman period.

Figure 12.2. The portrait of Mehmed the Conqueror by Şiblizade Ahmed (*c.* 1480 AD). (*Source*: Topkapı Palace Museum, Miniature: 39 x 27 cm, H: 2153, Folio: 10a, with the permission of the Topkapı Palace Museum)

Many of Istanbul's Westernized aspects have their precedent in the eighteenth century when, with the realization of Ottoman inferiority to Europe in military matters, and with the fascination with Parisian lifestyles, Westernization and its artistic approach began to take root. Women began to appear on the social scene, at least in recreational contexts, and a life of pompous spectacles for the capital began to take form. In contrast to the expansionism of the Ottoman state until that time, a pacifist leisure loving court began to form. The miniaturist Levni's aforementioned book of the circumcision celebrations is a lively document of how Istanbul began to turn into a narcissistic mirror of itself. The poet Nedim of Sultan Ahmet III's entourage, wrote in one of his many *gazels* (poetry for garden rituals) admiring the city, 'Holy Paradise! Is it under or above the city of Istanbul?'. A doctoral dissertation on poetry honouring the city, has elaborated on poets active between 1453 and 1730 who in different ways formed brotherhoods to devote themselves to the beauties of the city which they considered as a garden. Much of their poetry was collected under the title 'Şehrengiz' implying formidable city (Çalış, 2004). From the sixteenth to nineteenth centuries, foreign diplomats and travellers created albums of the city's panoramas, the most notable one being the panoramic view of Istanbul by Melchior Lorichs, now in the Leiden University library (Mango and Yerasimos, 1999). A group of European painters, who worked mostly in the eighteenth century creating landscapes of the Bosphorus, came to be known as the 'Bosphorus Painters' (Boppe 1989). The picnics and festivals on the shores of the Golden Horn became the subject matter of a lot of pictures executed by Ottoman artists, which may be considered the first examples of a Westernized style of painting (figure 12.3). As well as the large figurative pictures, different in formal and spatial composition from the miniatures, there began to appear portrayals of people in their typical costumes, which are called 'market' paintings. Landscapes often depicting the views of Istanbul began to be seen as wall decorations in mosques and private villas (Renda, 1977). Until the early twentieth century, celebrations and parades continued to add to the city's spectacles. One important procession was the annual caravan transportation of gifts to Mecca, when decorated camels loaded with gifts left the palace (Finkel 2005, illustration 36).

In a lot of visual art produced in Istanbul the subject has been the city itself. The first painters to emerge from the military academies, as a result of reforms that brought military engineers teaching perspective, drawing, and mechanical rendering, use the photographs of Istanbul's parks and palaces as their inspiration. The use of the photograph and the new concern with linear perspective created a new understanding of representation which led to the development of a Western

Figure 12.3. Performances in Golden Horn; Naval Celebration organized for Sultan Ahmed III by Levni. (*Source*: Miniature: 18.5 x 27.5 cm. Topkapı Palace Museum, Folio: 96b, with the permission of the Topkapı Palace Museum)

approach to painting. In the late nineteenth century Istanbul became the centre for all artistic activity not only in Turkey but in the whole Middle East. As Turkish artists emerged, European artists also came to work and exhibit in Istanbul, mostly in galleries sponsored by non-Muslims, in the European district of the city, the fashionable Pera. Amongst the artists who studied in France, Osman Hamdi, a student of Gustav Boulanger, founded the Fine Arts Academy as well as the Archaeological Museum of Istanbul. The Fine Arts Academy, was until the late 1970s the core institution of art teaching and production in Turkey. Viewed from such a historical perspective, the actual art scene in Istanbul can be seen as a further development of a kind of Westernization which has to do with the global market and greater political involvement with Europe in the twentieth century.

The New Art Scene and the Biennials

In the second half of the twentieth century, the questioning of modernity brought about a vital interest in cultural differences. Places that had had no interest for the artistic world began to hold biennials, to invite artists from the West to hold workshops, and the number of artists from non-European countries to study in the West grew enormously. Beginning with the 1980s, Turkey's opening to the liberal market, art education and study abroad, especially in the United States, became fashionable. Art became a prestigious occupation for the elite and for the intellectuals. It also began to be a kind of arena for social commentary. Many new exhibition series began to form. For the centennial of the foundation of the Fine Arts Academy in 1983, 'New Tendencies Exhibition', and the 'Young Talents' exhibitions were organized, showing interest in non-conventional media. The opening of new galleries like Urart, Nev, Siyah Beyaz followed, and Southeby's interest in Turkish art can be counted amongst new trends.

The new consumerist economy of the 1980s created a kind of insatiable hunger amongst business companies to reach out to any possibility of making money and of creating prestige. Banks and industry were the patrons of Istanbul, drawing artistic resources to the city. As a new possibility for investment, elitist galleries began to invite Turkish painters from abroad to create a new market. Fashionable clothes stores like Vakko, or banks, began to open art galleries. A totally private enterprise, *Boyut* art journal (of which I was the editor) began to appear in 1980 and continued until 1984, when the new economic competition created monopolies in the distribution of journals and printed media.[2] Other new journals such as *Gergedan*, or *Kalın* were often involved in *avant-garde* art or subversive criticism, as an approach to bring to the understanding of art a critical and questioning stance.

Artists who had been engaged in the first private university in Turkey, Bilkent which was opened in Ankara in 1984, moved to Istanbul as soon as other private universities founded by big business concerns began to follow in Istanbul. Many of these universities opened art departments, employing artists who were flooding to Istanbul from all parts of Turkey, attracted by the promise of fame and money. Art which had until then been considered only as a practice concerned with beauty and elite culture, began to assume new meanings as a critique, a provocation or an intellectual challenge. I believe that the role of several younger artists like Bedri Baykam and Mehmet Güleryüz, who had studied in France and worked in the United States, was important in creating this new attitude; their works were informal, often untidy, concerned with erotic and political content. Their attitude helped shake off the dust of the 'bureaucratic' approach of the faculty of the Fine Arts Academy. The new art journals, the artists arriving from Europe, exhibitions concerned with 'new' art approaches, created an atmosphere ripe for an explosion of contemporary art. The Istanbul Biennial began to take form in this atmosphere.

The Istanbul Biennial, which began in the late 1980s, was different in character in its pioneering years. The key issue then was who were the Turkish artists worthy of representing Turkish art. Another important development was the foundation of the Turkish branch of the International Artists Association, which put Turkish artists in direct contact with their European colleagues. In the competitive, commercial atmosphere of Istanbul, artists vied for public commissions and for recognition. In time, the fierce competitive atmosphere became stifling for the Istanbul Biennial. After the Third Biennial, the Istanbul Culture and Art Foundation (İstanbul Kültür Sanat Vakfı, İKSV) invited foreign curators.

The Istanbul Biennial has become one of the important events of the global art milieu. Yet, without doubt, part of the success is due to its urban backdrop. The development of cultural tourism in the twenty-first century and investment in art and leisure in Istanbul are major grounds for the attraction that the Biennial has assumed. Yet, the question remains as to whether the Biennial and the Istanbul Modern Museum, both active under the auspices of the İKSV, related to the Eczacıbaşı family involved in pharmaceutical production, have made important contributions to the general quality of art and art appreciation in Turkey.

Although Turkish culture has constantly revised itself according to the Western model, like its language, it has certain aesthetic peculiarities which are due to its rich historical background. Similar examples may be found Indian and Mexican art to mention just two. Indian artists of the Baroda school, no matter how much they looked to the West, produced an art idiom that had a definite identity. A close understanding of Turkish art in its development period shows that this is also true. However, if one looks at Turkish painting from the point of view of European or American art, which are also different from each other, this kind of comparison will underestimate the inherent qualities of Turkish art, especially its painting and sculpture. These qualities have to do with the concepts of space and time that gave special formal approaches to traditional visual arts all the way from the Asian to the Ottoman past. The shift in spatial concepts with the use of linear perspective have not altogether changed these visual preferences. On the other hand, unlike Western art, in the Orient, and especially in Islamic cultures, the human figure never attained a major symbolic and formal priority. Therefore, the approach to the human figure, and the formal treatment of it has been very different from that in Western art. This difference has not been altogether lost with the development of modern art in Turkey.

Besides these basic differences, the visual traditions of Ottoman and Turkish art always tended towards a decorative form, rather than favouring critical and philosophical implications. Therefore, to view modern Turkish art with the historically developed Western visual values will always create a misunderstanding as well as misjudgement. Many naive looking modern Turkish paintings or sculptures contain values that could be understood only if we have a good formal understanding of the historical development of Turkish art. As with many cultural developments in Turkey, the shift from the former understanding of art to the contemporary happened in one leap, leaving no time for natural critical and interpretative processes to develop for its evaluation.

The emphasis on contemporary art practice and media such as video, installations and performance, at the Biennial and other exhibitions, obliterated the evaluation of earlier Turkish art on its own terms and undermined aesthetic appreciation in the country's art public. The first two Istanbul Biennials both aimed to show Turkish art with reference to its particular values, on a common platform with international examples. At the time, many European works of painting and sculpture were also exhibited. It is important to remember that, no matter how unconventional much of the art shown might be, the Venice Biennale or the yearly Whitney show of American art never completely abandon painting and sculpture. One of the responsibilities of museum exhibitions and biennials or triennales all over the world also seems to be to create opportunities for the re-evaluation of older art. On the Turkish art scene painting and sculpture ceased to exist as an actual practice because a great majority of important exhibitions tended to show only contemporary media art. Artworks produced in Turkey before the 1980s suddenly became 'antiques' whose value was determined at auction. Many serious artists who work in traditional media or with aesthetic concerns, such as Melike Abasıyanık Kurtiç, Seyhun Topuz, or İsmail Ateş do not appear on the scene.

This sudden shift, caused mostly by international curators who have no knowledge or real understanding of Turkish art, has destroyed the potential of a new positive evaluation. Art fairs and galleries which do not show contemporary art have turned into cheap markets of popular production. The Istanbul Modern Museum (İstanbul Modern), whose collection is mostly borrowed from the Ankara State Museum and the University Museum of Istanbul, stages periodic exhibitions also mostly on contemporary art. There is a great need to re-assess Turkish cultural history via its modern art. İstanbul Modern could step up to this task, but instead it chooses to function primarily as a 'gallery' of contemporary art. The museum's collection of Turkish painters is meagre in comparison with what could be expected of a museum of art. Thus it cannot fulfil its important mission of informing the public about Turkish art and its history.

I believe these are grave deficiencies in a world where globalization erases the specificity of cultural values. On the other hand, the Istanbul Biennial is increasingly appearing as an *ad hoc* show of carelessly collected and exhibited works of minor quality which are often repetitious. In comparison to the seeming arbitrariness and disorderliness of the 2009 Biennial displays, which were referring to what one would see on the streets or in the quotidian, the reality in the streets of Istanbul has always been more poignant and related to political and cultural conditions, and stores more surprises than the art exhibited. Sibel Yardımcı, looking at the Istanbul art scene from a sociological perspective, has also voiced valid criticisms along with her uneasiness about the 'privatization of culture' in the absence of any substantial state support and the taking over of all art events by private holdings and foundations.

> Festivals ... base themselves on the aesthetic taste and the criteria of legitimate art that these institutions define in the cultural capitals of Europe or the United States, then only to influence art markets globally... Most of the time, Istanbul Festivals fail to develop their own language ... overlooking their own specificity, and relying on international curators in the hope that their names or practices would bring recognition and acclaim. (Yardımcı, 2007, p. 5)

The interest in the notion of difference in creating new political utopias is well justified as a humanitarian mission of art. Although art's role in manifesting the 'other' is very important, the way Istanbul's 'other' was repeatedly offered as a spectacle seemed to imply exploitation. The gaze into the private life of the poor was taken as something rather normal or as a kind of social engagement. The use of Istanbul's penury as a subject for art may not immediately involve ethical considerations, however, one has to think twice before indulging in the pleasures of watching misery as an art experience. These kinds of exposure show that many artists of the Biennial are removed from common realities and that they view the world from an elevated economic and cognitive position (Istanbul Biennial 2005, catalogue).

The Oda Project which was several times featured in the group of events accompanying the Biennials was one such exposure, with a patronizing and protectionist attitude towards the lower income individuals of the city. As Yardımcı (2007, p. 6) stresses, the covert censorship applied by the sponsors often softens political criticism. Politically motivated art thus turns into mere spectacle. Alternative art spaces created for those excluded from the Biennial and from other institutionalized exhibitions sometimes offer interesting critiques, but often the distinction between art and non art become blurred in the intention of social engagement, and the pursuit of aesthetization borders on kitsch.

Venues and Politics

In almost all the Istanbul Biennials the principle art focus has been the city itself with its sacred, profane, historical places as well as its squatter shacks. Monuments become the background for artists' fantasies: Buren in the Süleymaniye, Kounellis at St. Irene, Haghia Sophia partitioned into artists' private show cases. This indicates two things: that the city spaces no longer have established status or identities, or that space of any kind can be transformed in unexpected ways. This gives Istanbul a fluid character in its constant metamorphosis. Marginality and transgression have been amongst the important issues for artists, especially since the 1980s. Decontextualization has turned Istanbul into a neutral space where art pretended to be the signifier of urban phenomena. The use and transformation of certain historical spaces and monuments as Biennial venues, and the application of contemporary art onto historic monuments, such as French abstract artist, Daniel Buren's, application of stripes to the columns of the Süleymaniye Mosque, which is one of the most valued and meaningful edifices of the city, in my opinion is an effacing of memory and neutralizes publicly cherished values. The approach to the mnemonic values of a city as though they did not exist, or as though a contemporary work could revitalize them, might be considered mere cynicism, following certain postmodernist deconstructivist intentions.

On the other hand, it is obvious that political concerns have played a major role in the artistic events of Istanbul and their venues. Before art, architecture and international cultural events became the obvious prestige builders, neither Istanbul Metropolitan Municipality nor the political party which later gained power (Justice and Development Party – AKP) had anything to do with art. In fact, even today, the most severe cultural critique that the AKP voices is against modern Western culture, as the source of corruption in Turkey. However, the same party has shown a different attitude towards international art events. As AKP aims to maintain its liberal face in Europe and the US, support for internationally visible art exhibitions serves as an important vehicle for that purpose. One of the party's important agendas was to take Turkey into the European Union. The cultural issue became important in this relation. The new face-lift for Istanbul, with the art projects attracting Western attention served two functions.

With the foundation of the Republic, Modernity's denial of the past had become the tool to reconstruct a new time and space for modern Turkey. For many historians such as Kinzer (2001, pp. 10–13), present social polemics are rooted in this forced oblivion. Ankara became the capital. Istanbul had belonged to the past, its history was preserved in museums. For the AKP, Istanbul was the past to be revived for its Ottoman glory. For some post-modernist thinkers as well, it was now time to reject and to criticize the Modernist era, the Republican past, and to create the continuity with the religious past, as an indication of non-elitist democracy. The revival of Istanbul could serve also the close connection of political power with commercial power. AKP's leaders used Istanbul as a show case for their important meetings with foreign delegations and toyed with moving the headquarters of government institutions such as the Central Bank of Turkey to Istanbul – however, this has not happened yet. Istanbul became the supermarket for many fashionable foreign firms who claimed to have more clients here than in many European cities. Foreign luxury goods had great sales opportunities in Istanbul. Art seemed to be amongst these luxury goods. The city of festivals, of luxury boutiques, galleries and museums, Istanbul was drawing cultural tourism. Since the mid-1990s, it has become the great leisure and recreation city. It is interesting to note in this connection that, ironically, 80 per cent of the Istanbul Modern Museum's visitors are tourists.

It is also worthwhile to remember the controversy that was created over the choice of the Biennial's venue after the Third Biennial, between the greater Istanbul Metropolitan Municipality which owned the Feshane[3] (hat factory) building intended for the Biennial and the Eczacıbaşı corporation which had sponsored the restoration and which also was also behind the İSKV which launched the Biennials. The Third Istanbul Biennial was held at the re-designed Ottoman hat-factory Feshane, adapted to its new use by Gae Aulenti, the architect who oversaw the design conversion of Paris's Gare d'Orsay to today's Musée d'Orsay. The question that concerned the future Biennials as well as the Istanbul Modern Museum which was to reside here was what percentage of the museum the Municipality should own, and how much say it should have in its direction. The odds were that with rising fundamentalism and Islamic political power, the Municipality would fall into the hands of the Islamicist Party, the AKP. This is what happened, which proved to be the end of the use of Feshane for artistic purposes.

Istanbul, on the whole, feeds on its glorious past in many ways. It is fascinating to listen to a concert in the Hagia Irene, or to see an exhibition at Hagia Sophia or at the Topkapı Palace Museum. These are extraordinary spaces. As the photograph of Ferhat Özgür, the contemporary artist who took part in the 2007 Biennial, symbolically shows, no art can equal the glory of these spaces (figure 12.4). On the other hand, for a city which claims to be part of the contemporary world, the production of new spaces for art is a responsibility. The only example in this sense is Santral, the huge exhibition space that was built by the Bilgi University in the grounds of the Ottoman-era electric power plant which was given for use to the university by the AKP leader, Erdoğan, the current Prime Minister. This new museum and gallery annexed to the old power plant, which supplied the city with electricity from 1911 to 1983, is a remarkable architectural complex which synchronizes the aesthetics of the machine with contemporary design.[4] The central exhibition spaces of the Istanbul Biennial, and the building of İstanbul Modern, which is converted entrepôt beside the Bosphorus, have also been allotted several of these buildings by the AKP's leader. With these examples and that of the Feshane, it is obvious that, even though a certain growth of interest and production has been visible in the art scene since the 1980s, there has

Figure 12.4. 2007 İstanbul Biennial by Ferhat Özgür, from the 'Jump/Sıçra' Series, 2004. Photograph, 99 x 75 cm. (By courtesy of the Artist)

not been any clear policy on the part of the AKP in cultural and artistic matters, beyond its short-term interests. To understand the rapid growth of artistic investments we should also not forget the role of the billionaire investor George Soros as a sponsor in some of the recent artistic happenings of Istanbul.

As art became an important asset in the economic and social field, with Turkey's adoption of a liberal economy in the 1980s, many holdings belonging to the rich families of Istanbul began to invest in art collections. By the new millennium, several private museums had opened in Istanbul. The Sabancı Museum, several Koç family museums such as the Technology Museum and the Pera, and the Kadir Has Museum can be counted among the important ones. Many private collections are awaiting the right occasion to found their museums. Local auction houses such as Portakal or Antik Dekor, which both have their luxurious periodicals, are also very prominent on the art scene. At present, the number of galleries in Istanbul is not less than 350. While only about twenty of these are dealing with major artists, clearly art has become a lucrative business. This is also an indication that the residents of Istanbul are developing a bourgeois culture, notwithstanding the Islamist affiliations of many of them.

Contemporary art, with its seemingly free approaches, gives the impression of freedom and democracy. Such an impression is important in the way art has become a social interest in Istanbul. Like many other visual effects and spectacles, often this does not go beyond being a general impression. According to several artists there is severe censorship applied by the sponsors of the Biennial and of other international shows. Recently, a group of young artists voiced their opposition to the international business relationships of many sponsors of the 2009 Biennial, claiming that while the theme was based on political critiques, some of the support came from business dealing with arms trade. Both the 'Resistanbul' and the 'Adrennale' opposition groups' main protest was that the Biennial had increasingly become a business deal, for the sponsors and their star artists to show off.[5]

Conclusion

I have tried to argue that Istanbul's present artistic richness is part of its traditional concern with spectacle and self-idolatry, and that in different periods of its past Istanbul's sovereigns have shown interest in the arts to reflect their glory. The present Turkish government, in spite of its ongoing critique of modern Western culture and its Islamist affiliation, has supported art events and institutions, as a show of its cultural tolerance, and because of its relations with big business in the city, which is behind many of the artistic events. The claim that the neoliberal attitude of the present government defies real Islamist affiliation is a misinterpretation, in view of the fact that almost all states of the Gulf and the Arabian Peninsula, such as Dubai, Abu Dhabi and Kuwait have in the recent decades sponsored various manifestations of Western culture and have favoured a liberal economy. We see that presently neoliberalism and neo-conservatism can be employed together in different fields. I have also argued that Istanbul's interest in contemporary art since the 1980s can be seen as a continuation of the project of Westernization that began in the eighteenth century and which cannot be completed. One of my criticisms has been that the complete takeover of contemporary art in the present Istanbul artistic milieu has certain negative side effects such as the oblivion of older Turkish art which needs to be re-evaluated and promoted for its particular values, with comprehensive shows and criticism.

Istanbul claims to be the European city of Turkey, diverse and open to the world. The city has grown more than tenfold over the last half century, and in recent decades informal housing has increased to more than 50 per cent of the dwellings of the city (Tekeli *et al.*, 1994). Today vast areas are inhabited by Anatolian migrants who have few facilities. Over time these areas are opened to development and contractors move in to provide ghettos for the rich. In reaching even the richest neighbourhoods of Istanbul, one goes through such *gecekondu* settlements. A criticism that was emphasized throughout the chapter was that most of the content of contemporary art in Istanbul has been the city itself and that a sort of 'Orientalizing' gaze has exaggeratingly shown penury and injustice claiming to make political critique. Besides ethical implications, often the real scenes on the street seemed to make more sense than the art videos and that sometimes even the chaos of the city could have more aesthetic content and totality than the art about it.

Although post-modern discourse has often shown the world of spectacle and fantasy as the only antidote to the servitude of the capitalist system, and the investment in the implosion of diverse realities through globalization, in the production of simulacra and hyper-realities by way of festivals and festivities, has increasingly become a politics of social exclusion depending on status, income and power. This focus on play, experiment and festivity has produced its own capitalist system of production. What feeds this economy is the entertainment industry which has also produced its monopoly and its art lovers. Yet, for the wider public of Istanbul neither the museums nor the Biennial are accessible.

Unmindful of the rest of Turkey, and not really a European city, Istanbul is a narcissus dreaming its own spectacle. There is a fascinating aspect to this, especially if one is not a citizen but drops into its labyrinth temporarily. One may, in the end, lose oneself in this labyrinth, maybe only because the streets go nowhere, transportation does not move and one is stuck and stranded. But on the whole, Istanbul is watched from a sort of distance in time and in space. This may also be because one watches it as a work of art, and hence time and space are transformed. For many artists Istanbul itself is a work of art whether seen for its geography or architecture, or seen simply from the bird's eye view,

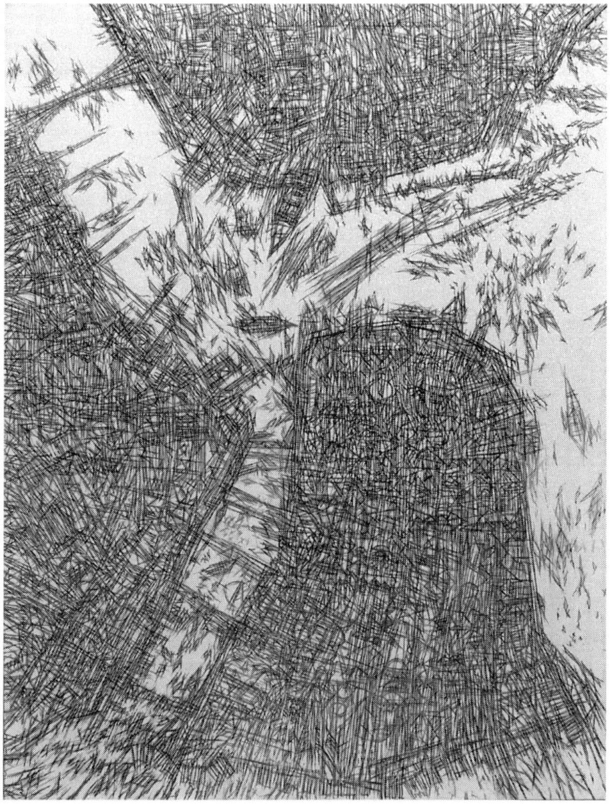

Figure 12.5. Bird's Eye View Istanbul (Kuşbakışı İstanbul), 2009, by Devrim Erbil, 160 x 120 cm, Oil on Canvas. (By courtesy of the Artist)

as the Istanbulite painter Devrim Erbil has seen it throughout his long career (figure 12.5). In his many mappings of Istanbul, the feeling of fluidity, of haste and of an all-over action-painting like quality implies that no matter how much we try to bring to light Istanbul's character, its millennia-old layers will never be completely penetrated. This is one of the great fascinations of the city. But Istanbul is also watched with a somewhat Orientalist gaze, and with the platitude that anyway, in the postmodern world, there are no pure or genuine meanings that would correctly orient the city.

Notes

1. *Meltem of Istanbul*, Emerging German and Turkish Artists from the Elgiz Collection, Project 4L Elgiz Museum of Contemporary Art, Istanbul, 2006, curator Necmi Sönmez; *Stadtansichten Istanbul*, ifa Galerie Berlin, ifa Galerie Stuttgart, Maurer Druck und Verlag, 2004, curator Vasıf Kortun; *Urbane Realitaten*: Fokus Istanbul, Martin Gropius Bau Berlin, Künstlerhaus Bethanien, 2005.

2. *Yeni Boyut Plastik Sanatlar Dergisi* and *Dimensions*, edited by Jale Erzen between 1980 and 1984 in Ankara.

3. Feshane was the Ottoman factory on the Golden Horn, producing the red felt hats that the Ottoman men wore in the nineteenth century.

4. For images of this museum complex see http://www.santralistanbul.org/. Accessed 8 January 2010.

5. The website of these groups is: http://resistanbul.wordpress.com/2009/09/04/conceptual-framework-of-direnal-istanbul-resistance. Accessed 31 December 2009.

References

Atıl, Esin (1999) *Levni ve Surname* (Levni and the Surname). Istanbul: Koçbank.

Boppe, Auguste (1989) *Les Orientalistes*. 8. *Les peintres du Bosphore au XVIIIe siècle*. Paris: ACR Edition.

Çalış, B.D. (2004) Ideal and Real Spaces of Ottoman Imagination – Continuation and Change in Ottoman Rituals of Poetry – Istanbul 1453–1730. Doctoral Dissertation, Faculty of Architecture, Middle East Technical University, Ankara.

Erzen, Jale (ed.) (1980–1984) *Yeni Boyut Plastik Sanatlar Dergisi and Dimensions*. Ankara.

Esche, Charles and Kortun, Vasıf (2005) *The 2005 Istanbul Biennale Catalogue*. Istanbul: Istanbul Kültür Sanat Vakfı.

Finkel, Caroline (2005) *Osman's Dream: The Story of the Ottoman Empire, 1300–1923*. New York: Basic Books.

Ifa-Galerie Stuttgart (2004) *Stadtansichten Istanbul*. Stuttgart: Ifa-Galerie.

Kinzer, Stephen (2001) *Crescent and Star: Turkey Between Two Worlds*. New York: Farrar, Straus and Giroux.

Mango, Cyril A. and Yerasimos, Stéphane (1999) *Melchior Lorichs' Panorama of Istanbul 1559*. Bern: Ertuğ and Kocabıyık.

Necipoğlu, Gülru (1989) Süleyman the Magnificent and the representation of power in the context of Ottoman-Habsburg-Papal rivalry. *Art Bulletin*, **71**(3), pp. 401–427.

Pamuk, Orhan (2003) *İstanbul: Hatıralar ve Şehir* [Istanbul: Memories of a City]. Istanbul: YKY.

Renda, Günsel (1977) *Batılılaşma Döneminde Türk Resim Sanatı: 1700–1850* [Turkish Art of Painting of the Westernization Period: 1700–1850]. Ankara: Hacettepe Üniversitesi Yayınları.

Tekeli, İlhan *et al.* (1993–1995) *İstanbul Ansiklopedisi: Dünden Bugüne* [Encyclopaedia of Istanbul: From Yesterday to the Present], Vol. 6. Istanbul: Ministry of Culture of Turkey and History Foundation of Turkey.

Yardımcı, Sibel (2005) *Kentsel Değişim ve Festivalizm: Küreselleşen İstanbul'da Bienal.* [Urban Transformation and Festivalism: The biennial in *a* globalizing Istanbul]. Istanbul: İletişim Publishing.

Yardımcı, Sibel (2007) Festivalising Difference: Privatization of Culture and Symbolic Exclusion in Istanbul. European University Institute RSCAS 2007/35. Working Paper. Available at: http://cadmus.eui.eu/dspace/bitstream/1814/7670/1/RSCAS_2007_35.pdf. Accessed 2 January 2010.

Chapter 13

The Politics of Urban Art Events: Comparing Istanbul and Berlin

Banu Karaca

The Spectacle is capital accumulated to the point where it becomes image

Debord, 1994, p. 24

Enter the (Urban) Arts Spectacle

A recent article in the *New York Times* entitled '31 Places to Go in 2010' ranked Istanbul in nineteenth place by referencing its 'steadily growing' contemporary art scene as 'one of the most innovative in the world' (Emmrich, 2010). Istanbul's increasing presence on the global stage has been advanced in no small part through the Istanbul Biennial. Inaugurated in 1987, its rise to prominence in the 2000s falls into what has been called the 'biennial decade' beginning in the second half of the 1990s (Obrist, 2007, p. 360), during which the proliferation and increase in visibility of biennials worldwide reflected global economic restructuring processes, and particularly the role of art and cultural production in the positioning of urban centres within these processes. This chapter takes the Istanbul Biennial of 2005 as a lens to interrogate large-scale urban arts spectacles. It does so by comparing the case of Istanbul with that of Berlin (2006), another location that has drawn much attention through its burgeoning art scene.

As the biennial format has become '[a]n increasingly popular institutional structure for the staging of large-scale exhibitions' (Wu, 2009, p. 108), the parallels and divergences between the two examples may serve to understand the discourses and practices that account for the 'spectacularization' of art in the contemporary metropolitan order. The emphasis here is not on the frequently claimed mass appeal that these spectacles might provide for respective audiences, but on the normative framework that arts spectacles posit for articulating desirable and viable connections between art and the economy from the viewpoint of urban revitalization and city marketing strategies. As such this chapter deals with three sets of interrelated convergences that presently manifest themselves in the spectacularization of art: that between economic interests, cultural policy, and artistic and curatorial practices; the format of the biennial in different locations;

and, finally, that of the cityscapes that host them. The respective instalments of the Istanbul and Berlin Biennials are interesting in this regard, not least because the host cities played an important role in the perception of both events.

The following is thus an interrogation of the artistic, political and economic variables that contribute to the spectacularization of art – at a time when art spectacles emerge as the preferred form of presenting art in the urban setting. I use the notion of 'preferred' here not to indicate artistic (or curatorial) predilections – although participation in biennials and international art fairs has become an indispensable stepping-stone in artistic careers – but those of sponsors and municipalities who consider these events lucrative with regard to the cultural and economic capital they produce. While in Istanbul as in Berlin, art events and festivals, like biennials, are often presented as 'democratizing' access to the arts when compared to 'conventional' art formats, I posit that in cityscapes increasingly characterized by income polarization, gentrification, and social segmentation, urban art spectacles have become a vital economic development strategy, and may actually feed into exclusionary practices in the urban realm.

Apart from the above-mentioned institutional structure that biennials tend to create and rely on, they are also productive entry points in this discussion because of:

(*a*) their pronounced international character;

(*b*) their status as self-declared fora of artistic experimentation away from market pressures;

(*c*) their formation as events that claim both globality and site-specific uniqueness, a duality that characterizes the commodification of culture in other areas as well (Harvey, 2001); and

(*d*) the increasing number of biennials launched in geo-political locations such as Sao Paulo, Havana, Dakar, Gwangju, Taipei, Busan, Cairo, Centinje, Tirana and Athens which have challenged (Kortun and Hanru, 2003) their ascribed status as 'artistic peripheries' (Bydler, 2004), or – as some might argue – have aimed to join the Western establishment.

Surely, one way of establishing the ascription 'spectacle' to this increasingly prevalent manifestation is the reference made by interlocutors and the literature to the descent of biennials from world fairs; in particular, this can be traced in the longest running biennial of Venice, originating in 1895. Another, however, is their very characterization as 'events'. Never 'just' an exhibition, these events are designed to deliver an encompassing experience denoted by an array of accompanying programmes in order to 'contextualize' either the artworks included or the overall concept of the show. These additional elements range from academic lectures (in order to provide a theoretical background), films (offering a 'different, more popular and accessible medium of visualization'), to the notorious DJ-dance party, opening and closing cocktails (highlighting the fun, entertaining aspects of the whole endeavour that also serve as professional networking platforms), and the arsenal of related (if 'tasteful') merchandise. To approach urban art events such as large-scale exhibitions, biennials and performance festivals from this perspective might help to break away from prevalent romanticized modernist conceptualizations of art in general, to highlight some of the current shifts in the workings of the art world, and to contextualize the larger social, political and economic dimensions that contemporary art inhabits today.

Following the global economic restructuring measures that marked much of the 1980s, a notable convergence emerged between economic interests, cultural policy and artistic practice, three realms generally conceptualized as pursuing opposing interests and operating logics. Together they have formed the basis for a type of city marketing that is intimately connected to, and founded on, discourses of 'cultural entrepreneurship' (Langley and Abruzzo, 1986; McLaughlin, 1991), and policies centred on the 'creative industries' (Florida, 1992). Such policies have gained currency in the realm of arts planning and management, and have reconfigured international artworlds based on a (re)conceptualization of art as a vital urban economic development strategy within which artists and other cultural workers are increasingly understood as service providers for global metropolises (Karaca, 2000).

Neither the spectacularization nor the enterprising of art is new. Indeed the politics of spectacular capitalist display – and its critique – are inscribed in the trajectory of modernity – as not least the engagement with the topic by Guy Debord and the *Situationist International* of half a century ago demonstrates. Yet, the specificities through which this enterprising is configured, the discourses it brings together and the strategies it entails have reassembled. Although spearheaded by different sets of actors or interest groups, who seemingly occupy different positions within the international art world and are heirs to divergent cultural policy trajectories, this development is strikingly similar in Istanbul and Berlin.

Proposing that parallels arise only in part from congruent economic, social and political processes, I see both locations as locked into an internationalized circulation of ideas pertaining to the uses and pliability of art – and especially of art spectacles for urban renewal and city marketing. Urban art spectacles are seen to enhance the city's representational prestige and to provide incentives for (foreign) investment, as well as a draw for tourism. Together these facets have led to an increasingly normative understanding of how art events should be presented, and which locations, such as Istanbul, that were long understood as peripheral, are actually constitutive for and expressive of overall developments. As Levent Soysal (2005, p. 263) suggests, arts events of this kind are 'neither local

Figure 13.1. 'I keep an eye on' by Ines Tartler (Berlin, 2006), an intervention by the artist on Augusstrasse which was not officially part of the Fourth Berlin Biennial (*Photo:* Banu Karaca).

Figure 13.2. 'Bu da geçer yahu' ('This too shall pass') by Fuat Şahinler, Murat Şahinler, Ayten Başoğlu and Yakup Çetinkaya (Istanbul, 2005) was realized as part of the Second Pedestrian Exhibition curated by Fulya Erdemci and Emre Baykal which coincided with the Ninth Istanbul Biennial. It has become quite common in Istanbul and Berlin for arts events and programmes to coincide with biennials. These artworks are just two examples of interventions which were part of such initiatives displayed in public spaces in the vicinity of the biennial locations (*Photo*: Banu Karaca).

nor particular. They borrow from each other, they mimic each other, and they approximate each other. Their discursive and performative vocabulary is gradually but surely resembling each other and becoming the same'.[1] At the same time, city spectacles have to offer local specificity to provide an incentive for people to travel to a particular city in a year-round circuit of art biennials and fairs, thus resulting in an interplay between homogenizing tendencies and struggles for uniqueness in order to remain relevant and profitable (cf. Harvey, 2001).

Biennials: Last Resorts for Artistic Experimentation or Venues for Exquisite Commodification?

The Istanbul and Berlin Biennials took place over a period of six weeks in September 2005 and April 2006, respectively. While the Ninth Istanbul Biennial (IB9), curated by Vasıf Kortun and Charles Esche had as its theme the city of Istanbul itself, the Fourth Berlin Biennial (BB4), curated by Maurizio Cattelan, Massimiliano Gioni, and Ali Subotnik, borrowed its title from Steinbeck's novel *Of Mice and Men*. Introducing the concept, or as they insisted, the 'non-concept' the BB4 curators state:

> We believe that art is about the experiencing of possible worlds… So, don't think this is a show about Berlin. Even if it takes place in one of the city's streets, it is not a show about Auguststrasse, or even about Mitte in the year 2006. If we were to be tremendously ambitious, we would say our Berlin is just a landscape of the mind, it only exists in our heads: So it could be anywhere – and yet it could only possibly exist here, this street is an archetype or maybe just an example: it is real and yet completely imaginary. (Catellan *et al.*, 2006, p. 23)

In contrast, Kortun and Esche open the catalogue of the Istanbul Biennial with the following statement:

> This biennial is for and about Istanbul. We have sought to address the environment in which the work will be shown and to place art in a dialogue with different aspects and observations of the city itself. This has meant asking artists to stay in the city and produce new work, as well as selecting work that might shed light on the particularities of Istanbul in comparison with cities elsewhere… It will hopefully provoke a new awareness of some common perceptions or a reassessment of the personal clichés that we all carry in our heads. This is what we believe art can do to and for us, and how it can, in its own way, change the world. (Esche and Kortun, 2005, p. 9)

Besides the possibilities that art might entail, these positions seem to provide diametrically opposed starting points. On one hand, of the Berlin Biennial curators present to a seemingly undefined audience the Berlin Biennial not as anchored in the city space, but in 'a landscape of the mind,' that 'could be anywhere,' and yet only exists in Berlin. On the other, we find a 'biennial for and about Istanbul', and, by extension, for and about the region as Kortun and Esche emphasized during numerous panel discussions. Despite stating very different intentions, both biennials exhibited striking parallels in the form in which artworks were presented. Together the elements of intention and presentation illustrate the tensions that David Harvey (2001) and Levent Soysal (2005) map out in their discussions of the commodification of culture, highlighting how local particularity guarantees authenticity while 'global form' ensures market compatibility and competitiveness.

Both biennials used a dispersed set of locations as exhibition venues, favouring sites usually inaccessible to the public which induced art critics and visitors to note that the sites were often more impressive, if not spectacular, than the artworks themselves (Hohmann, 2005, p. 12; Stange, 2006, p. 81). However, the actual staging of the artworks varied considerably. In Berlin, they were shown in different locations along the 920 metre long Auguststrasse in the Mitte district (figure 13.3). Infamous as the city's poorhouse in the eighteenth century, Auguststrasse emerged as an important centre for Jewish life in Berlin and as an iconic site for the Weimar Republic's art scene where Joseph von Sternberg set his film *The Blue Angel* in 1930.[2]

The most evocative location in this context was the Jewish Girls School (built in 1927–1928). Closed down by the National Socialists in 1943, it was renamed Bertolt Brecht High School during the Socialist era, only to be closed down again in the process of German reunification. Having been restituted to the Jewish Central Committee of Germany, the building was the only biennial location

Figure 13.3. Auguststrasse with *KunstWerke*, the institutional host of the Biennial, on the right. (© Fourth Berlin Biennial for Contemporary Art. *Courtesy of* Uwe Walter, 2006)

subject to the tight security measures in place for synagogues, Jewish schools and community centres throughout Germany. Here the work of Polish artist Robert Ku□mirowski, entitled 'Wagon' (2006) (see figure 13.4) – a life-size train wagon, eerily reminiscent of the cattle cars that transported millions of people to concentration camps during the NS regime – received a lot of attention. This work was identified as especially haunting by visitors because of its historical and spatial connotations, and the juxtaposition between the light materials of which it was built and the heaviness of the memories it invoked. The desolate wallpapers, crumbling walls, and the accumulated dust seemed to speak to the curators' assertion that the location had remained unaltered. The arrangement of the site played intentionally into what could be called 'the spectacle of authenticity' – an over-dramatization that in some cases enhanced, but at others overpowered, the artworks.

In Istanbul the locations were spread – at walking distance – throughout the Beyoğlu district (figure 13.5), reaching down to the Antrepo at the Tophane waterfront, making accessible otherwise unused spaces such as the Garibaldi building (founded by the Italian Opera Society in 1863) or an old tobacco depot.[3] While employing a wider set of locations than previous Istanbul Biennials which had centred on prominent historical sites in Sultanahmet, it nonetheless did not break away from concentrating on areas already known for their artistic and entertainment caché. As Charles Esche commented, the organizers and curators had not been 'brave enough' to make a biennial that was for *all* Istanbul, which would have required further inclusion of other and more remote parts of the city.

Among the locations, Deniz Palas featured Michael Blum's 'A tribute to Safiye Behar', a play with reality and historical connotations, similar to Kuśmirowski's 'Wagon' at the Berlin Biennial. Blum's work introduced the audience to a fictional character by the same name and a 'reconstruction' of her office, diaries and photographs (figures 13.6 and 13.7). Behar, the visitor was told, was not

Figure 13.4. Robert Kuśmirowski: Wagon, 2006. (© Fourth Berlin Biennial for Contemporary Art. *Courtesy of* Uwe Walter, 2006)

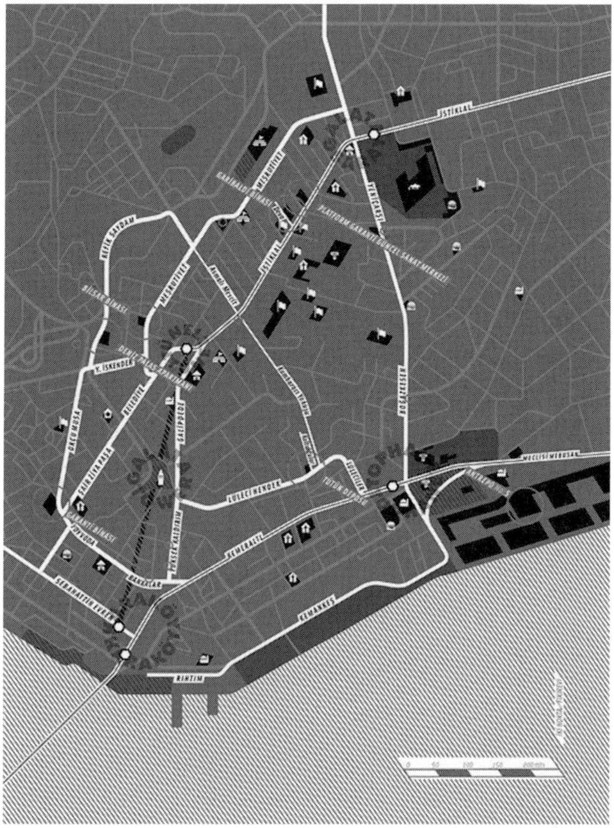

Figure 13.5. Ninth Istanbul Biennial locations, 2005. (*Courtesy of* İKSV)

Figure 13.6. Michael Blum, A Tribute to Safiye Behar, 2005. (*Courtesy of* the artist)

only a Marxist and feminist of Jewish descent, but also a close confidant and, as the displayed correspondence between the two reveal, a lover of Mustafa Kemal (Atatürk), asserting considerable influence on his political thinking, especially with regard to women's rights. The detailed set-up generated animated discussions among visitors and reviewers. While Kuśmirowski reproduced or re-enacted history – a history that is very present in Berlin, Blum's fictionalized history pointed to silences in Turkey's historiography, with Behar's emigration

Figure 13.7. Michael Blum, A Tribute to Safiye Behar, Ninth Istanbul Biennial, 2005. (*Photo*: Banu Karaca)

addressing the repression and expulsion of non-Muslim minorities after the Republic was established: did she leave because of the wealth tax imposed on non-Muslims in the 1940s or the pogroms of 6–7 September 1955? Why had no one ever heard about this Jewish lover, confidant and intellectual collaborator of Atatürk?

Mobilizing the memory of Jews in Europe, both works presented critical interventions that reveal/rediscover the former Jewish presence in both Berlin and Istanbul (and by extension Germany and Turkey). As Andreas Huyssen (2003) argues, these kinds of 'recoveries' are not incompatible with the post-1989 marketing of memory by the cultural industries. Huyssen's assessment that remembering and commodification are not necessarily mutually exclusive (*Ibid.*, pp. 18ff) manifests itself particularly in official, state-initiated 'celebrations of diversity' which have become vital markers in cities' self-representation (Karaca, 2009, pp. 314–359; see also Mary Brink-Danan, Chapter 16). There is, however, another convergence between the history of Jewish absence in Berlin and Istanbul (as with non-Muslims in general in the case of the latter) and the reconfiguration of the urban space, to which I will return below. Namely, that the voids left by the slaughter, expulsion or dispossession of minorities frequently end up being vital spaces for urban regeneration projects often under the vanguard of artists who in face of little available financial means search to occupy vacant spaces in the city.[4]

Throughout the Ninth Istanbul Biennial, different artists' groups and cultural centres organized public discussions, some related to the exhibition and its conceptual framework, others pertaining to current trends in the local and international art world. Issues of sponsorship, power relations within and outside Istanbul's art world, commercial cooptation and the ethics of curating and producing art were discussed in a very controversial and often fervent manner. It was no exception that specific artists and curators were confronted with accusations of being handmaidens of the political and economic establishment. Harsh criticisms were levelled against the organizer of the Biennial, the Istanbul Foundation for Culture and the Arts (IKSV) and its political orientation (Yardımcı, 2005). The accusation that Biennial participants were 'selling out' to corporate sponsors was repeatedly raised in these fora. Some curators and art critics, wary of the influence of big corporations, such as co-sponsor Koç Holding, within the Biennial context, proposed to take the funding and 'bring down the system from inside'.

The issue of commercial cooptation was rarely, if ever, addressed during the Berlin Biennial of 2006.[5] One journalist made an implicit criticism during the opening press conference by interrupting Biennial director Gaby Horn while she was acknowledging the string of sponsors saying: 'This is a press conference, we are interested in the conceptual framework of the exhibition, not the sponsors!' It seemed that the organizers and curators had, for lack of a better word, 'succumbed to', or embraced, commercial interests, maybe in part, because for the first time financing had been guaranteed through a commitment of the corporate sponsors (BMW and Allianz) and the Federal Cultural Foundation (Hautpstadtkulturfond, HKF) for two consecutive biennials. This embrace of the rapprochement between art and commerce might also be interpreted as breaking a taboo in that it explicitly, and not without irony, crossed the boundaries between the 'pristine' realm of art and the domain of the art market.

Yet the lack of need or the refusal to problematize these concerns and the absence of controversial debate on the spatial proximity between commercial galleries and the biennial locations was striking. After all, the Mitte district hosted some of the major galleries of the city. One of my interlocutors who worked as a guide at the Biennial frequently expressed her discomfort with high-profile visitors inquiring about where to purchase certain artists' works. This emphasis on consumption by visitors and organizers alike seemed at odds with the panel discussions at the Young Curators workshop, 'Fast Forward', conducted in the framework of the Berlin Biennial. Under the heading 'Does the World need more Biennials?' the invited senior curators emphatically stressed biennials as the sole remaining space for experimentation in the international art world setting, away from market rationales.

Differential Perception and Converging Formats

Despite the divergences of the discussion accompanying the two biennials, the promotion and reviews of both events were bound to rather streamlined representations of each city. In contrast to reviews of the Berlin event, which focused more strictly on the exhibition itself, the perception of international critics of the Istanbul Biennial often circled around well-rehearsed orientalist themes, i.e. mentioning Istanbul as a 'bridge between East and West' (themes that were taken up in reviewing subsequent Istanbul Biennials) and allusions to the effects of the local cuisine on those unaccustomed to it (Herzog, 2007). Not only obscuring the fact that the Istanbul Biennial, established in 1987, predates Berlin's, inaugurated in 1997, reviewers (e.g. Hohmann, 2005) also tended to deny the 'coevalness' (Fabian, 1983) between the two events by emphasizing or proposing the novelty of Istanbul's arts scene.

Notably, in both instances the host city did not give any (Berlin) or only logistical (e.g. by providing the necessary permits, Istanbul) support. Yet, city officials and many local journalists alike, having highlighted the cosmopolitanism, and by extension the 'modern-ness', that these events establish for their cities, emphasized the media attention and ensuing attractiveness for tourism and investment that the biennials might entail. The biennials thus presented another item in the cities' public relations portfolios along with other urban events of ever higher flying superlatives, such as Fashion Week (Berlin) and Formula One (Istanbul).

While it might not be surprising that affirming the city as a brand to be marketed in order to attract tourism and other types of business was furthered by city officials, I found that art world actors often used the same line of reasoning when trying to establish the significance of their respective art scenes in the global circuit. In Istanbul, this was frequently done in order to counter historical particularities of the national arts scene as well as orientalist (self-)perceptions. In Berlin, this discourse was employed to shift the city's image away from that of the 'poorhouse of Germany' and 'fiscally irresponsible' to a place where trading, 'buying and selling' – in this case of art, can be conducted – as KunstWerke director Gaby Horn emphasized during our conversation (Berlin, May 2006).

That Horn emphasizes the connection between the biennial and the commercial art market situated in the city is especially telling at a conjuncture where the 'Biennial decade' has converged into the 'Art Fair Age' (Barragán, 2008). Biennials and art fairs now cover between them an almost year-round circuit and have become increasingly indistinguishable from each other. The Contemporary Istanbul Art Fair and the Tenth Istanbul Biennial were held contemporaneously in 2007. The director of Berlin's Art Forum, Anne Maier, expressed pride in the fact that the fair is able to provide each stand with a 'gallery-look' by securing the sole remaining vintage hall of Berlin's commercial trade fair complex. Frequently co-sponsored by public institutions for their accompanying educational programmes, art fairs in Istanbul, Berlin and beyond, evidence that the self-representation of the commercial art market and non-profit art spaces increasingly resemble each other.

Varied as they might be from the outset, commercial on the one hand and supposed fora for artistic experimentation on the other, both art fairs and biennials today are subject to similar economic pressures, cater for collectors more than to the general public, take place in comparable settings, and employ discourses around engaging a broader audience through educational programmes. As exhibition openings become the talk of the town, the increasing convergence between the settings in which 'art for sale' and 'art for the public good' (a fickle distinction to begin with if one considers interdependency between private and public arts institutions and collections for instance, Graw, 2008) has manifested itself in the spectacularization of both.

Cultural Entrepreneurship and the Restructuring of Urban Space

In order to understand why these market rationales can be voiced and ultimately executed today with very little protest in the case of Berlin, and such little consequence (if at all) in the case of Istanbul, one has to take into account how intensified global economic restructuring has impacted and articulated itself in the urban realm. The arts spectacle and its surrounding discourses have emerged in an urban setting marked by growing income segmentation (Mollenkopf and Castells, 1991) and by gentrification – a development in which artists are both vanguard actors and negatively affected (Smith, 1996, pp. 3–29; Deutsche and Ryan, 1984; Zukin, 1995).

In Berlin the processes of gentrification have been most visible in Mitte and Prenzlauer Berg, formerly parts of East Berlin. The same is – partially – true for Kreuzberg, a neighbourhood in West Berlin which, because of its proximity to the Wall, became unattractive for its German inhabitants who ventured further west into the city, transforming it into an immigrant quarter from the 1960s onwards. On adjacent sides of the former wall, these districts found themselves catapulted into the geographical centre of the capital of a newly unified Germany. In the early 1990s, many deserted spaces in Mitte and Prenzlauer Berg offered squatting possibilities for 'adventurous pioneers' as well as for makeshift galleries, clubs and performance spaces. While the post-Wall shift is well known, there is little awareness that the very availability of these free spaces for the

> young Berlin scene is indirectly based on the aryanization policies during the Hitler regime: Jewish Berliners were dispossessed, the GDR transferred the apartments into public property, the implosion of the *Realsozialismus* then provided cheap available living spaces. (Kessen, 2004, p. 143)

Attracting not only national but international attention these neighbourhoods were the first to be redeveloped in the early 1990s due to their newly attained cultural caché and central location. In contrast to these eastern districts of Berlin, Kreuzberg, heralded as a space for 'alternative culture' since the 1960s, has proven more difficult to gentrify due to a long history of disinvestment and strong community advocacy organizations.

It is no coincidence that in Istanbul, too, it is the areas of Galata, Tünel, Cihangir and Tarlabaşı in the district of Beyoğlu, home to many arts spaces, that are among the most affected by gentrification (İslam and Behar, 2006; Tan, 2006). The exodus following waves of repression against the remaining non-Muslim minorities who had mainly inhabited these neighbourhoods along with European expatriates in Ottoman and early Republican times left many buildings unoccupied. With their owners unable to claim their former property, the disinvestment in the area rendered it a refuge for Kurds fleeing violence in the south-east as well as immigrants from Africa and other parts of the world stranded at the walls of fortress Europe (e.g. see İçduygu and Kirişci, 2009; Özlem Ünsal and Tuna Kuyucu, Chapter 3). Beyoğlu has been subject to repeated 'urban regeneration' campaigns that have targeted prostitution and drug trafficking as well as alternative lifestyles, and has been claimed by the Istanbulite intelligentsia, art studios and galleries, film and advertising executives followed by cafés and restaurants. With them redevelopment has intensified since the late 1990s, especially as the regulations on foreign investment in Turkey have been softened.

In both Berlin and Istanbul there is a push by policy-makers and city planners to privatize and homogenize public space, expressed in exclusionary practices inscribed in the aesthetics of the architectural environment and in policing techniques (see also, Aksoy, 2008). These intertwined developments have also impacted discussions and changes in the field of arts management and subsequently the self-conception of the artist him/herself. In Berlin, governmental cutbacks in social service expenditure have led to a rollback in state funding for the arts. But as German cultural policy frames support for the arts as a duty to citizens and as imperative for the education of 'critically-minded' human beings, politicians and policy-makers have been quick to portray these budget-cuts as a chance to reform arts institutions and enhance their economic efficacy. In Turkey, where private sponsors and the state assume a rather symbiotic relationship, this shift to the privatization of culture (cf. Wu 2002) has been aided by the adoption of IMF directives and the liberalization of the economic sector which was further enabled through the military junta of 1980 and the governments that followed it. With reference to the American model of arts funding, the call for 'arts management' as a 'more business-like' and 'professional' working mode in the artworld has been voiced increasingly in Germany as well as in Turkey where government funding for contemporary art has been extremely limited, and where private sponsors and cultural institutions have similarly appropriated the discourses of professionalization and entrepreneurship. Confronted with the pressure of implementing for-profit concepts such as 'professionalism' and 'entrepreneurship', arts organizations and individual artists have found themselves in a dilemma on how to conduct 'goal-oriented' work while reverting to for-profit, market-oriented techniques, and has led to administrative positions being filled by marketing and management majors or graduates of arts management programmes, which have sprung up at universities in both cities.

But even more importantly, these shifts have implications on how the social and economic dimension of artistic work and the societal role of artists are appraised, especially as art has been increasingly framed as a public and private investment opportunity with expectations of high financial – and symbolic – returns. This increased interest in art on the part of consumers, city officials and policy-makers has helped to reshape cultural policy into a vital tool for economic development and artists are ultimately viewed as service providers for the urban economy (cf. Lash and Urry, 1994; Karaca, 2000).

It is interesting to note that the calls for professionalization are made by quite diverse actors in each location. In Berlin, it was put forward by cultural policy-makers (mostly politicians from the opposition, not the governing coalition which was composed of Social Democrats and Socialists) and then quickly taken up by local art academies as well as some artists and artistic directors. In Istanbul, this process has taken a slightly different path as direct government funding for the arts has long been anchored within the art academy. However, private arts and cultural centres flourished especially in the 1980s. The call for the implementation of arts management techniques, institutionalization (cf. Artun, 2002) and professionalization has recently been voiced by those who administer the newly founded university arts management programmes as well as some critics and curators. This development is in part connected to the unprecedented availability of funds due to accession talks with the EU, such as *Culture 2000*, which interestingly partially funded both biennials, with foreign cultural institutions such as the Goethe Institute, and embassies located in Istanbul also being active in this regard.

A second plane on which cultural policy, economic interest and artistic practice have recently converged is that of the 'creative industries', a category which has allowed 'software producers and the major publishing and media conglomerates to construct an alliance with cultural workers and with small-scale cultural entrepreneurs, around a strengthening of copyright protection' (Garnham, 2005, p. 26). In addition, this terminology has helped to establish an affinity between certain economic sectors such as fashion, (web) design, advertising, and the automobile industry (Florida, 2006) and the arts, thus making it possible to apply cultural protectionist policies to these sectors (Hesmondalgh and Pratt, 2005). While these developments have not gone unchallenged in Istanbul and Berlin, policy-makers, corporate sponsors as well as artists and other art world actors are increasingly employing a joint vocabulary when describing their practices which take economic rationales into account (unless they choose to position themselves as clearly oppositional), making for a situation in which different sets of protagonists, generally thought of as pursuing opposing goals, are perpetuating the same discourses around the enterprising of art.

Point of Convergence: The Arts Spectacle

Throughout this chapter I have proposed that the arts spectacle is the preferred articulation of the present convergence between cultural policy, economic interests and artistic practices. This is not to suggest, however, that the 'spectacular' or the festival character *per se* impede on any critical intent that contemporary art might entail, or to propose a totalizing effect of the spectacle that turns the audience into nothing but consumers (Foster, 1985, p. 82). The art happenings of the 1960s, for instance, surely show other uses of both the parameters of the spectacle and the critique thereof (Adkins, 2004; Winkler, 2004).

While city officials tend to emphasize that art festivals succeed in engaging a broader public than 'conventional' art formats, and thus employ an originally counter-cultural argument of public 'space-claiming' (Stevenson, 1999), this presents a rather big stretch when taking current social segmentation in the urban realm into account, especially as both the Istanbul and the Berlin Biennials were instated through private rather than public (i.e. state or municipal) initiatives.

Indicative of the convergence of economic interests, cultural policy and artistic practices, the arts spectacle has come to reconcile three domains which are normally thought of as adhering to different interests and operating logics. Advanced by a different set of actors in each city, this enterprising of art is mediated by notions of 'cultural entrepreneurship' and the 'creative industries', and incorporates art in the institutions of the market, while leaving the allure of its emancipatory potential intact. This development, not least expressed in the establishment of a joint vocabulary among art world actors, puts into question long-held distinctions between cultural *politics* as oppositional to cultural *policies enacted by the state* (Stevenson, 1999), especially as the artist (and not just the artwork) – frequently against his or her own intention – becomes incorporated as part and parcel of spectacular displays. It is also against this backdrop of far-reaching incorporation that Istanbul's tenure as a European Capital of Culture, a candidacy that began under strong participatory incentives, has already been delimited.

Notes

1. Translation from the German by Levent Soysal.
2. Featuring Marlene Dietrich in her breakthrough role as the nightclub singer Lola Lola, the film tells the story of a professor's unhappy love for Dietrich's character that ultimately leads to his social and economic downfall.

3. The former tobacco warehouse (Tütün Deposu) has since become the Rodeo Gallery and Depo, a cultural centre initiated by Anadolu Kültür, an NGO pursuing diversity in art.

4. Read against this background the selection of the Feriköy Greek School as one of the locations of the Eleventh Istanbul Biennial in 2009 might be a coincidence. Its vacancy, however, is not.

5. This was all the more notable since the previous instalment of the biennial (2004), under the direction of Ute Meta Bauer, explicitly thematized the recent structural transformations of Berlin as well as relations between the arts, urban space and politics, for instance.

References

Adkins, Helen (2004) War, peace, revolution and the avant-garde, Filippo Tommaso Marinetti, Johannes Bader and Franz Jung, in Winkler, H. (ed.) *Legal/Illegal. Art beyond Law.* Stuttgart: Schmetterling Verlag, pp. 159–158.

Aksoy, Asu (2008) Istanbul's choice. *Third Text*, **22**(1), pp. 71–83.

Artun, Ali (2002) The museum that cannot be. Unpublished article based on paper presented at a seminar titled 'Exposer l'art contemporain du monde arabe et de Turquie, ici et là bas', IISMM-École des Hautes Études en Sciences Sociales, Paris, 2002.

Barragán, Paco (2008) *The Art Fair Age*. New York: Charta.

Bydler, Charlotte (2004) *The Global ArtWorld, Inc.: On the Globalization of Contemporary Art*. Uppsala: Uppsala University.

Catellan, Maurizio, Gioni, Massimiliano and Subotnik, Ali (2006) Of Mice and Men, in Catalogue of the Fourth Biennial, *Of Mice and Men*. Berlin: Haje Cantz.

Debord, Guy (1994) *The Society of the Spectacle*. New York: Zone Books.

Deutsche, Rosalyn and Ryan, Cara G. (1984) The fine art of gentrification. *October*, **31**, pp. 91–111.

Emmrich, Stuart (2010) 19. Istanbul, in 31 Places to Go in 2010. *The New York Times Online*, 10 January. Available at http://www.nytimes.com/2010/01/10/travel/10places.html. Accessed 10 January 2010.

Esche, Charles and Kortun, Vasıf (2005) The world is yours, in *Art, City and Politics in an Expanding World. Writings from the 9th International Istanbul Biennial*. Istanbul: İKSV, p. 9.

Fabian, Johannes (1983) *Time and the Other. How Anthropology Makes Its Object*. New York: Columbia University Press.

Florida, Richard (1992) *Cities and the Creative Class*. New York: Routledge.

Florida, Richard (2006) A creative crossroads. *Washington Times*, 7 May, p. B03.

Foster, Hal (1985) *Recodings. Art, Spectacle, Cultural Politics*. New York: New York Press.

Garnham, Nicholas (2005) From cultural to creative industries. *International Journal of Cultural Policy*, **11**(1), pp. 15–29.

Graw, Isabelle (2008) *Der grosse Preis Kunst zwischen Markt und Celebrity Kultur* [High Price. Art Between the Market and Celebrity Culture]. Cologne: DuMont.

Harvey, David (2001) The art of rent, in Harvey, David, *Spaces of Capital. Towards a Critical Geography*. New York: Routledge, pp. 394–411.

Herzog, Samuel (2007) Sich elegant verdauen lassen [To let oneself be elegantly digestable]. *NZZ Online*, 12 September. Available at http://www.nzz.ch/nachrichten/kultur/aktuell/sich_elegant_verdauen_lassen_1.553999.html. Accessed 12 September 2007.

Hesmondalgh, David and Pratt, Andy C. (2005) Cultural industries and cultural policy. *International Journal of Cultural Policy*, **11**(1), pp. 1–13.

Hohmann, Silke (2005) Euphorie in Moll [Euphoria in Moll]. *Frankfurter Rundschau*, 5 September, p 12.

Huyssen, Andreas (2003) *Presents Past Urban Palimpsests and the Politics of Memory*. Stanford, CA: Stanford University Press.

İçduygu, Ahmet and Kirişci, Kemal (eds.) (2009) *Land of Diverse Migrations. Challenges of Emigration and Immigration in Turkey*. Istanbul: İstanbul Bilgi Üniversitesi Yayınları.

İslam, Tolga and Behar, David (eds.) (2006) *İstanbul'da Soylulaştırma: Eski Kentin Yeni Sakinleri* [Gentrification in Istanbul. New Inhabitants of the Old City]. Istanbul: İstanbul Bilgi Üniversitesi Yayınları.

visions advanced by governmental agencies, civic organizations, and producers of culture. Remaining hopeful, but critical, Öner sets out to assess the Agency's efforts to institutionalize public-private partnerships and invent instruments of participatory policy-making in the fields of arts and culture.

The cultural representation of minorities has become the touchstone of participatory politics in the light of European accession. Marcy Brink-Danan's 'Counting as European: Jews and the Politics of Presence in Istanbul' addresses the ideal of 'cosmopolitanism', which is celebrated in official multiculturalism and public displays of difference. Istanbul's Jews, their officials, and their institutions participate in performances of difference and tolerance, which are at the heart of much public debate in Turkey, the European Union, and elsewhere. Taking an ethnographic approach, Brink-Danan shows how Jewish Istanbulites negotiate the cityscape by enacting spectacles of difference, especially in the light of European Union demands for evidence of Turkey's civilizational maturity. These and other spectacles staged by and for minorities of Istanbul as part of the 2010 ECoC programme heighten the debates over, and struggles for, diversity and recognition as a way to achieve cosmopolitanism, despite, and because of, the erasures which Brink-Danan rightly highlights.

In 'Future(s) of the City: Istanbul for the New Century', the last chapter of this Part, Levent Soysal provides a critical assessment, not only of Istanbul's 2010 venture, but also of other culture projects and spectacles undertaken to realize urban transformation under the ascriptive label of 'world city'. For Soysal, presenting Istanbul as a world city is the axiomatic starting point for imagining the future(s) of the city. This urban imaginary is tightly coupled with Turkey's 'Europeanization'. He offers a set of meta-analytical categories for mapping the cultural topography of the new metropolis as seen from Istanbul, focusing on the key cultural dimensions of what makes a contemporary city a 'world city' today. Soysal draws on Norbert Elias's notion of 'the civilizing process' in order to frame the transformations of Istanbul (and Turkey) in its quest for Europeanness. The same notion, he suggests, rightly describes and explains the making of the new Europe.

Chapter 14

The European Capital of Culture Programme and Istanbul 2010

Carola Hein

Twenty-three years after Turkey applied for full membership of the then European Economic Community[1] and on the twenty-fifth anniversary of the European Capitals of Culture Programme, Istanbul joined the ranks of cities to become a cultural capital for a year along with Essen (on behalf of the Ruhr) and Pécs.

According to the European Commission website:

> The European capital of culture is a golden opportunity to show off Europe's cultural richness and diversity, and all the ties which link us together as Europeans. The event is so attractive that Europe's cities vie with each other fiercely for the honour of bearing the title.[2]

As evident from press and other coverage, Turkey's application for full membership of the European Union has received mixed reactions. With its predominantly Muslim population, its position on the eastern edge of Europe, and most of its land in Asia, some do not see the country as sufficiently close to mainstream European culture and its perceived Christian roots. Others argue that the strategic military and economic importance of Turkey and its partial location within Europe call for its inclusion in the EU. But, Istanbul has always been seen as part of *European* history and culture.

Istanbul's selection as a European Capital of Culture (ECoC) is an example of how EU policies attempt to overcome nationalist sentiments by supporting cities, which are traditionally cosmopolitan and thus able to transcend national identity. Istanbul does not just belong to Turkey; it belongs to Europe and the world.

Granting ECoC status to Istabul allows the organizers in the city to bypass political controversies at the national or international level and to use the EU's decentralization strategy to market the city. The ECoC title is seen by cities as an opportunity; it is a case of 'opportunism' in the best sense of the word, albeit often with limited support for, or promotion of, European aspects or respect for seat of the international organizations already serving a European federation in 1877.

Indeed this was not the only proposal to locate the headquarters of international institutions in Constantinople. The nineteenth-century desire to create a single headquarters for international institutions, based on the concept of the monumental and centralized capitals of the nation-states, led to a proposal, published in 1913, by Norwegian sculptor Hendrik Christian Andersen and French architect Ernest Hébrard for a world centre (Andersen, 1913; Gresleri and Matteoni, 1982). Hébrard considered various locations for his city of 600,000 inhabitants, envisioning its placement not only in the vicinity of old European centres such as Brussels, but also on the east coast of the United States, and near Constantinople, referred to as the crossroads between Orient and Occident.

Constantinople as a site for international organizations came up again during the debates around the creation of the League of Nations to promote international collaboration and peace in 1919, when based mostly on military reasoning, the English MP Major David Davies (later Lord Davies) proposed Constantinople as the general headquarters of World Peace. He argued that due to its important geographical position, the capital had to be the property of all and not of one nation (Davies, 1919). These proposals highlight the ways in which Istanbul has been thought of as part of the larger European sphere.

The Council of Europe, of which Turkey has been a member since 1949, and the EU have greatly expanded their membership, thereby extending their reach to the borders of Turkey. Meanwhile, Europe and the EU not only interact with Turkey as a state, but also with its citizens. In the post-war period a large number of Turkish citizens and their families settled across Europe, notably in Germany, encouraged by the so-called guest worker (*Gastarbeiter*) programme, establishing a mostly low-income working-class Turkish presence within Europe which is often perceived as culturally and religiously distinct (Soysal, 2008). Poverty and religious difference are among the major arguments used within the EU against Turkish membership as observers argue for a Europe defined through its Christian heritage – an argument that is no longer defensible given the growing numbers of immigrants from diverse backgrounds in EU states.

Europe's perception of Turkey as a country with a dual heritage and partly located in Asia is reflected on maps. For example, some maps of Europe that show EU members Cyprus, Bulgaria and Romania and parts of non-members Ukraine and the Russian Federation, magically extract Anatolia, Turkey's Eastern/Asian part. One example is even to be found in the 2010 capital of culture presentation of the Ruhr – see figure 14.2. The same problem in representing Europe occurs on the Euro bills which show the geography of Europe, including European Turkey, but give the Asian side of the country a lighter shade, the same as is used for Northern Africa.

The difficulty in agreeing on the representation of the European map highlights the absence of a common cultural denominator and the difficulty

Figure 14.2. The map 'Welcome Europe' designed to highlight the Ruhr's international connections for Twins2010. The Asian part of Turkey is missing. (*Source*: 'Twins2010. Documentation of the First Preparatory Conference on the 3rd and 4th February 2006 in Duisburg, Dortmund, Essen', 2006)

among member states of making decisions on issues that involve culture and identity. Indeed, the design of the Euro bill has provoked conflict before: cities, architecture, and other structures are closely associated with national history. To avoid controversy, the EU opted for abstract representations of architectural details rather than depicting real structures that could be identified with specific nations. The selection of Istanbul as ECoC 2010 reopens the question of symbols through the design of a European map.

European Capitals of Culture: Leveraging Cities to Promote European Identity

The ECoC programme embodies the EU's pragmatic approach to the promotion of a European identity. Even though the programme has not been able to extricate itself from national pressures, its twenty-five year existence shows its inherent importance in bringing European states together. To quote again from the European Commission website:

> Cities wishing to become European capital of culture must prepare a cultural programme that meets rather specific criteria: it must reflect the European character of the event and involve the participation of the people who live there.
>
> The European flavour can be seen in the themes chosen and the artists and cultural organisers from different countries cooperating to put on the event. The programme must also have a lasting and sustainable impact on the city's long-term cultural, economic and social development.[3]

As the ECoC programme has evolved, it has adapted to emerging European priorities: featuring small cities of cultural importance first, then turning to larger metropolises, showcasing industrial cities that have used the event to re-imagine themselves as cultural centres (Glasgow) or inventing new events, such as the Zinneke Parade, a street festival in Brussels celebrating the city's multicultural character and promoting integration. Since 2000, there have been multiple simultaneous European capitals of culture, highlighting the diversity of European cities in terms of space, size, and urban form.[4]

A study of European cities of culture between 1994 and 2004 found that many could not have achieved the same 'cultural, visitor and social impacts … so quickly' and that it raised 'interest in European issues' 'offering opportunities for European countries to cooperate; reinforcing Europe as a part of each national identity' receiving media attention and recognition (Palmer-Rae Associates, 2004, pp. 162–163). The study concludes that the Capital of Culture title is a catalyst for cultural development and the transformation of a city even though the application is a financial burden, albeit one that, according to this preliminary research, may benefit the city in a wider context. Based on these findings, the European Commission is now insisting that there should be long-term benefits of the cultural development in the city and region concerned (ECOTEC, 2008). However, respondents to the study also voiced criticism concerning, for example, the designation process, the ways in which individual cities took advantage (or not) of the opportunity, the festival character of the event, or the lack of focus on Europe. Such criticism reflects wider debates on the use of festivals and other events by both the public and private sector to speed up urban transformation without fully taking into account issues of social segregation and gentrification and the needs and wishes of citizens – see, for example, Prentice and Andersen (2003) and Florida (2005).

Through European Cities of Culture and other initiatives, the EU seeks to create new 'urban imaginaries' (Huyssen, 2008) and 'European', rather than 'national', ways for citizens to perceive the cities in which they live and work. Through arts and culture, European Cities of Culture highlight and rejuvenate the chosen cities in the imagination of European citizens. The programme goes beyond national capitals to showcase cities that have long European histories, but not a major role in the nation-states, or those that were part of socialist Europe for the latter part of the twentieth century and have not yet played a part in the EU. It also includes those, such as Istanbul, which have a diverse relationship with European space due to their border location. ECoC designations also have the potential to expose the multicultural roots of European cities and to revive or create new inter-city, cross-border networks and territories of European identity.

Decentralization and weak governance comes with its share of challenges, not the least of which is the desire to take advantage of the EU presence – which I term 'opportunism'. Thus, while the EU selects the ECoC and attaches its name to ECoC events, it provides only partial funding, leaving the actual planning and organization of the year's events to the respective cities. As a result, the quality of the programmes and events depends largely on the local organizers, who approach the event as a means of place marketing and to focus on a multitude of cultural, artistic, and urban initiatives and the chance to develop new urban strategies.

Istanbul 2010 demonstrates this opportunistic approach. While the city's ECoC programme includes various specially designed European-themed activities (e.g. symposia and workshops such as 'Cultural Policies in Turkey and Europe'; 'What is European Culture?' and 'Temporary City – A Dialogue Among 2010 ECoCs'; and new art, historical renovation, and urban development projects and competitions that target problematic urban spots),[5] local authorities are combining these activities with other already existing events, such as the Istanbul Biennial 2009, the 2010 World Basketball Championship, the Encounter of Mediterranean Art Schools, and the Jazzfestival. These events would have taken place even without the ECoC, underscoring the way ECoC publicity and funds may amplify local initiatives without effectively increasing European identity, the stated purpose of the ECoC designation. Perhaps even more significant, the city is using ECoC-related art and culture projects to facilitate larger changes, such as the restoration of buildings and the gentrification of urban districts that are of particular tourist interest, for instance, the vicinity of the old city walls and the Byzantine Palace. Such interventions raise questions about the inclusiveness of the ECoC event in terms of its orientation towards selected urban groups or the city population at large.

Decentralization as Captured by the Capitals of Culture 2010

The three capitals of culture selected for 2010 present new aspects of urbanity and metropolitan form that tell us more about European space and its edges, about its capitals and centre, and about its citizenry and identity. The two EU countries slated to choose capitals of culture for 2010 were Germany and Hungary, which selected Essen (for the Ruhr) and Pécs as their representatives.[6] These cities were joined by Istanbul after a 1999 decision had opened the ECoC events from 2005 to 2019 to non-member countries and after various non-governmental actors in Istanbul had initiated a proposal to the EU in 2005 immediately after accession discussions had started with Turkey.[7]

The 2010 capitals clearly have little in common in terms of location, size or structure, or even the problems they are facing; observers may ask why three cities that are so different have together been chosen to represent European culture. As awkward as this combination seems at first sight, they represent historic and urban diversity and epitomize the slogan mentioned earlier – 'United in Diversity'. They also highlight the EU decision to decentralize its capital functions. Just as the EU started out with members of very different size (compare Luxembourg with less than 500,000 inhabitants and Germany with 80 million inhabitants) but equal decision-making power in the Council of Ministers (until the ratification of the Lisbon Treaty), the 2010 capital cities are seen as equals in their European history and in their capacity to promote European identity – as was the case for all earlier ECoC cities.

The European Capital of Culture – together with other culture, media, mobility, and creativity programmes – was intended to bring a variety of cities into the consciousness of the European public. As the European Commission states, 'Europeans share a common cultural heritage, which is the result of centuries of creativity, migratory flows and exchanges' (Commission of the European Communities, 2007). The programme thus attempts to highlight a common European background (even if in diverse ways) and the possible contributions of designated cities to the European future, to build on old exchange patterns and to create new cultural, artistic, human and administrative networks that can promote the establishment of a common European vision.

As EU policy is designed around the 'acceptance of difference as richness', matching very different and unexpected partners does not appear unusual and leads to the current combination of 2010 ECoCs:

- Essen, with about 600,000 inhabitants, a city at the core of the European space, which has associated itself with its larger region – the Ruhr (some 10 million inhabitants), a transforming industrial area, and a paradigm of the polycentric metropolis of the future;

- Pécs, a small university city of 153,000 inhabitants, located in Hungary, one of the 2004 Eastern enlargement states. It is thus a symbol of the overcoming of the iron curtain, an event captured in the title used for ECoC, The Borderless City, and for new connections with neighbouring countries in the Balkans; and

- Istanbul, with some 12 million inhabitants, a metropolis with a growing population; a long-standing cosmopolitan centre that represents a traditional part of European networks and a city with multiple cultural links to Western Europe.

The differences between the 2010 European Capitals of Culture are not only a matter of size, they are also apparent in different facets of each city's current urban planning themes. The Ruhr aims to become a paradigm of a polycentric and networked sustainable urban agglomeration, a model for a decentralized Europe that grows from multiple entities and nourishes new social links that may solidify and become physical structures in the future. By means of cultural events, including the international building exhibition (IBA) Emscher Park in the 1990s and Ruhr 2010 now, the organizers have aimed to overcome administrative splintering and create new paths, both physically and in the people's imagination. Ruhr 2010 also aims to establish links beyond the fifty-three towns and cities of the region, notably by integrating over 200 twin cities – building on the long-standing town twinning initiatives in existence today – through the programme *Twins 2010* (2006), and collaboration with the three 2010 European Capitals of Culture.[8] The Ruhr, home to many Turkish immigrants, has clear potential for collaboration with Istanbul. The MELEZ Programme, in particular, which is designed to build on the artistic and cultural variety of the area and to bring people from different cultures and immigrant backgrounds together to shape their common future, highlights the role of immigration in the Ruhr. In 2010, the MELEZ festival train will stitch the area together as it stops at regional stations.[9]

Pécs, chosen over Budapest, illustrates the importance of regional cities of long histories which have been passed over during periods of national dominance and favouritism of the national capitals. The city stands at the gate to the Balkans (beyond which we find Istanbul) and the Mediterranean. It is a multicultural city which brings together numerous historic and cultural layers including Latin, Turkish, German, and Hungarian.[10] A UNESCO World Heritage site since 2000, the city features Christian as well as Ottoman monuments and is thus a perfect link for highlighting the history and opportunities of European networking.

Istanbul adds several new facets to the ECoC 2010 programme, raising the stakes for the next set of ECoC cities. The application document presented in Brussels in 2005 advertised the concept of 'Istanbul: City of the Four Elements', a reference to the four elements – earth, water, fire and air. It was through the four elements that the pre-Socratic Greek philosophers, Thales, Anaximander and Anaximenes, who were all born in Miletus in what is now Anatolia, sought to understand the universe.[11] This reference to ancient Greece can be seen as an attempt to define Turkey's Anatolian roots beyond religious concepts. The ECoC 2010 proposals used the reference to the four elements to structure the events by making each coincide with one season. The application's focus in terms of art events was on emphasizing the ways in which the East and the West inspire Istanbul jointly. The proposal was accepted in 2006.

The application document for Istanbul 2010 presented a programme which considered the larger challenges of the city and tried to bring together numerous different constituencies. This approach was reflected in its earliest presentation, which promoted a revision of local governance structures to achieve a more broadly based participation. As one of the application's main goals, the documents argued for 'facilitating harmony between urban renewal and the transformation of daily life', ensuring the independence of cultural capital and facilitating the interaction of every class within the city.[12] Advancing democratization, through, for example, broad-based participation in the discussion of urban strategies was highlighted as a core goal in the application document and in projects such as *The Dem(art)cracy Village* project, which were aimed at the transformation of Istanbul's shanty towns into artistic centres, or bringing modern art into them. Similarly, *Forging the Future, Forging Culture* addressed children of the shantytown districts by planning to have the children participate in workshops and use art facilities. Both of these projects showcase the best of the ECoC concept, as they provide access to art and culture to people and areas which usually lack such opportunities. Concrete achievements of course will have to be evaluated after the event.

The original application for Istanbul 2010 was geared strongly towards issues of democratization and integration of various urban actors and partners. Interestingly, these aspects are no longer predominant on the current web presentation of the ECoC programme.[13] Instead, presentations of Istanbul 2010 at various tourist venues around the world are featured prominently.[14] The 2010 programme book, presented in December 2009 on the website, demonstrates the shift in focus. Şekib Avdağiç, Chairman of the Executive Board of the Istanbul 2010 European Capital of Culture Agency, states that the agency selected the motto 'Istanbul: The most inspiring city of the world' and characterized 'Istanbul, Europe's Natural Culture Capital' as 'one of the cities, which most influenced European culture and civilization' (Gültekin, 2009, p. 5). This emphasis on capital city status, which is visible also in the other two introductory statements of the project book, appears to signal a longing to overcome a lost status of capital city that is typical for many former capitals (including Philadelphia, Turin, Lagos and St Petersburg).

The shift in emphasis is also reflected in the catalogue of projects, which starts by emphasizing the restoration of major historic buildings, such as mosques, palaces, madrasas, and the city walls. These projects, as Yılmaz Kurt, Secretary General of the 2010 ECoC Agency asserts in his introduction, show that 'Istanbul is a great capital of culture and civilization' and that the projects listed in the 'program book will serve as the solid evidences of this fact' (*Ibid.*, p.7). Other initiatives developed for the ECoC event similarly feature what the description calls the forgotten history of Istanbul, including exhibitions on Greek and Armenian architects of Istanbul and the project 'Longing for Istanbul', a documentary focusing on the concerns of Greeks who left Istanbul for reasons beyond their own will and now live in Athens (*Ibid.*, pp. 39 and 127). Thus, a shift from what was promised to accompany the ECoC selection of Istanbul to what appears to be occurring in reality appears already to be taking place.

Some attempts at addressing current social, economic, and political issues of the metropolis Istanbul seem to have survived in the current programme, and 2010 Istanbul Agency's Urban Implementations Director, Korhan Gümüş, points to the need to make art accessible to Istanbulites in every part of the city (*Ibid.*, p. 8). This approach is reflected in one project entitled 'Our Home Istanbul. A Child Changes, Istanbul Changes', which is geared towards children, starting with those of the poorer districts, but appears to have been conceived with a top-down rather than bottom-up approach (*Ibid.*, p. 132). The rest of the programme is focused on art, music, literature and other cultural activities that resemble ECoC events in other cities, including a street festival entitled 'Istanbul is Having Fun on the Street' (*Ibid.*, p. 121).

Conclusion: Opportunism and the Decentralized European Capital

Owing to the absence of strong leadership from the EU, nation-states and local governments, as well as economic elites, have effectively leveraged the EU presence as an opportunity to promote their economic and political interests. As the six founding members could not agree on a capital city, in 1951, the ministers selected Luxembourg as the temporary place of the institutional headquarters together with Strasbourg as home to the parliament, adding Brussels as headquarters for the European Economic Community (EEC) and Euratom, both predecessors of the EU, in 1957. In time, a mono-functional district has grown in Brussels, mainly comprising of headquarters buildings of the European Commission, Council and Parliament, which stand out for their scale rather than for their aesthetic qualities, featuring bland architectural design that does not inspire a strong emotional attachment or make them recognizable emblems of the EU (Hein, 2004).

This development, instead of helping to build a European identity, has created strong negative reactions towards EU institutions, notably in Brussels, even though most of the policy decisions on the ground were not made by the EU directly but by national and local institutions eager to please the EU. This attitude represents one of opportunism, meaning the opportunity of the EU presence to further local goals and economy, as well as the fact that these political and economic elites have used the EU presence as the official reason to push through urban redevelopment programmes that have been detrimental to groups of citizens, particularly those living in areas where the European institutions have settled. While this attitude denotes a creative urban elite in the cities concerned, it tends to exacerbate existing social tensions and work against multicultural integration.

Zeroing in on Istanbul, one sees that over the past three decades, Istanbul has witnessed extensive neoliberal urban transformation, including the construction of office buildings, malls, and housing in the suburbs and the centre, as the result of globalization and tourism, without due public input, or participation, in the process. The ECoC programme thus enters a sphere that is already highly contested and adds a new layer to ongoing discussions about actors, goals, and methods of urban transformation, participation, and citizenship and identity in Istanbul.

The ECoC programme aims at promoting a European territorial consciousness through culture, festivals and events that are generated locally. For the EU, the decentralized and itinerant staging of ECoC events is ultimately about changing 'people's minds' and creating a new mental map of Europe, one that goes beyond the territorial definition of the nation-state. The EU has brought into existence a system that uses concurrent, decentralized spectacles and events – similar to other art, culture or sport events, but geared towards European consciousness – as tools to reform habits at the level of everyday life and consciousness. This represents a political and cultural strategy that works to convert the European space through new social networks that overcome existing national boundaries, cultural prejudices, and local policy structures. It uses festivalization as a tool not only to promote urban development, but also to nurture new European imaginaries. Istanbul is the repository of much of Turkish history; the city is the key for entry in the EU, but also the bridgehead of the East onto Europe.

In exchange for the EU brand and the marketing opportunity, creative entities in the European Capitals of Culture are called upon to stimulate dynamic processes, to establish new urban transformation procedures, to overcome the national frame, and to take advantage of the opportunities offered by the expanded European space, as well as those that help to experiment with European citizenship and bring Europe together. The programme is aimed at European goals, and the question is thus not only what Europe can do for the ECoC cities but also what these cities can do for Europe. In the case of Istanbul, the ECoC event can promote discussions on Turkish membership in the EU, highlight the existence and impact of migrants in Western Europe, and help increase artistic and cultural collaboration between Istanbulite and European institutions. The ECoC event highlights public and private policy as well as culture-making structures in Istanbul and showcases ongoing debates in the city with the hope that this process may change the way that urban renewal and governance occur.

Notes

1. Turkey was accepted as an associate member in 1963. The European Economic Community (EEC) became the European Union (EU) in 1992.

2. See http://ec.europa.eu/culture/our-programmes-and-actions/doc413_en.htm. Accessed 17 January 2010.

3. *Ibid*.

4. For scholarly evaluation of specific ECoCs see for example: Garcia, 2004*a*, *b*; Stoffen, 2004; Hitters, 2000, 2007; Gold and Gold, 2005; Miles, 2007.

5. 'Cultural Policies in Turkey and Europe' Symposium, ended. See http://www.en.istanbul2010.org/BASINODASI/BASINBULTEN/HABER/GP_587334. 'What is European Culture?' Symposium.

Available at http://www.en.istanbul2010.org/PROJE/GP_536817. Accessed 17 January 2010; ('Temporary City – A Dialogue Among 2010 ECoCs October 2009 – October 2010, Gültekin, 2009, pp. 45 and 75).

6. Ruhr.2010 and Pécs.portal. Available at: http://www.essen-fuer-das-ruhrgebiet.ruhr2010.de/en/home.html and http://en.pecs.hu/. Accessed 17 January 2010.

7. See http://europa.eu/rapid/pressReleasesAction.do?reference=IP/06/489&format=HTML&aged=0&language=EN&guiLanguage=en. Accessed 13 January 2010.

8. Twins. Ruhr 2010. Available at: http://www.twins2010.com/index.php?id=122&L=0%20%5B0%2C0%2C20953%5D. Accessed 17 January 2010.

9. MELEZ.O8 Festival der Kulturen and MELEZ.Festival. Available at: http://www.melez.de/ and http://www.essen-fuer-das-ruhrgebiet.ruhr2010.de/en/programme/moving-europe/melez/melez festival.html. Accessed 17 January 2010.

10. 'Pécs – The Borderless City. The Multicultural City'. Available at: http://en.pecs2010.hu/p/pecs/the_borderless_city/the_multicultural_city. Accessed 17 January 2010.

11. I. Istanbul: City of the Four Elements (Istanbul book1_a 2.pdf). Available at: http://www.en.istanbul2010.org/stellent/groups/public/documents/ist2010_images/gp_540557.pdf. Accessed 17 January 2010.

12. Fire. 23 September - 31 December. Forging the Future. Available at: http://www.en.istanbul2010.org/stellent/groups/public/documents/ist2010_images/gp_540583.pdf. Accessed 17 January 2010.

13. Barbara Wolbert, University of Minnesota, also noted the difference between the original application, its scope and design, and the current web presentation. She further highlighted the prominence of urban renewal projects and the selection of Greek and Armenian architecture for Istanbul 2010 during our Round Table Discussion entitled 'Orienting Europe: Capitals of Culture – from Ruhr to Bosporus' convened by Deniz Göktürk at the German Studies Association (GSA) meeting in Washington in October 2009.

14. Tourism and Promotion. Available at: http://www.en.istanbul2010.org/TURIZM/index.htm. Accessed 17 January 2010.

References

Andersen, Hendrick Christian (1913) *La conscience mondiale.* Rome: Société internationale pour favoriser la création d'un centre mondial.

Berezin, Mabel (2003) Introduction. Territory, emotion, and identity, in Berezin, M. and Schain, M.A. (eds.) *Europe Without Borders: Re-mapping Territory, Citizenship and Identity in a Transnational Age*. Baltimore, MD: Johns Hopkins University Press.

Bozdoğan, Sibel (2001) *Modernism and Nation-Building. Turkish Architectural Culture in the Early Republic*. Seattle, WA: University of Washington Press.

Bozdoğan, Sibel and Kasaba, Reşat (eds.) (1997) *Rethinking Modernity and National Identity in Turkey*. Seattle, WA: University of Washington Press.

Commission of the European Communities (2007) *Communication: A European Agenda for Culture in a Globalizing World*. COM(2007) 242 final. Brussels: Commission of the European Communities.

Davies, M.D. (1919) Constantinople as the G.H.Q. of peace. *The Architectural Review*, **46**, pp. 146–150.

ECOTEC (2008) *Final External Evaluation of the Culture 2000 Programme (2000–2006). Executive Summary*. Birmingham: EOTEC. Available at: http://ec.europa.eu/culture/archive/sources_info/evaluation/pdf_word/culture2000_final_report/executive_summary_31012008_EX_EN.pdf. Accessed 13 January 2009.

Florida, Richard (2005) *Cities and the Creative Class*. New York. Routledge.

García, B. (2004*a*) Cultural policy in European cities. *Local Economy*, **19**(4), pp. 312–326.

García, B. (2004*b*) Urban regeneration, arts programming and major events. *International Journal of Cultural Policy*, **10**(1), pp. 103–118.

Gold, John R. and Gold, Margaret M. (2005) European cities of culture, in Gold, J.R. and Gold, M.M. (eds.) *Cities of Culture: Staging International Festivals and the Urban Agenda, 1851–2000*. Aldershot: Ashgate.

Gresleri, Giuliano and Matteoni, Dario (1982) *La città mondiale*. Venice: Polis/Marsilio.

Gültekin, Ayşe Orhun (ed.) (2009) *Istanbul 2010*. Istanbul: Istanbul 2010 European Capital of Culture.

Hein, Carola (2004) *The Capital of Europe*. Westport, CT: Greenwood/Praeger.

Hitters, Erik (2000) The social and political construction of a European Cultural Capital: Rotterdam 2001. *International Journal of Cultural Policy*, **6**(2), pp. 183–199.

Hitters, Erik (2007) Porto and Rotterdam as European capitals of culture, in Richards, G. (ed.) *Cultural Tourism: Global and Local Perspectives*. Binghamton, NY: The Haworth Press.

Huyssen, Andreas (ed.) (2008) *Other Cities, Other Worlds: Urban Imaginaries in a Globalizing World*. Durham, NC: Duke University Press.

Keyder, Çağlar (1999) *Istanbul: Between the Global and the Local*. Lanham, MD: Rowman & Littlefield.

Keyder, Çağlar (2005) Globalization and social exclusion in Istanbul. *International Journal of Urban and Regional Research*, **29**(1), pp. 124–134.

Lorimer, James (1884) *The Institutes of the Law of Nations: A Treatise of the Jural Relations of Separate Political Communities*. Edinburgh: William Blackwood and Sons.

Miles, Malcolm (2007) *Cities of Culture*. New York, Routledge.

Palmer, R. (2004) *European Cities and Capitals of Culture*, Parts I and II. Brussels: Palmer-Rae Associates.

Prentice, Richard and Andersen, Vivien (2003) Festival as creative destination. *Annals of Tourism Research*, **30**(1), pp. 7–30.

Shore, Cris (2000) *Building Europe: The Cultural Politics of European Integration*. London: Routledge.

Soysal, Levent (2008) The migration story of Turks in Germany, in Kasaba, R. (ed.) *The Cambridge History of Turkey*. Volume 4. *Turkey in the Modern World*. Cambridge: Cambridge University Press.

Stoffen, Myriam (2004) *The Zinneke Parade – An Artistic Citizens' Parade?* Florence: INURA.

Chapter 15

Istanbul 2010 European Capital of Culture: Towards a Participatory Culture?

Oğuz Öner

Since it was launched in 1985, the European Capital of Culture (ECoC) programme has aimed to help cities restructure, recreate and shape their urban cultural policies. In some cities, the programme has opened up new ways for participative cultural policy-making in the long term. Istanbul's candidacy for the programme led to a collaboration between the city's *sivil toplum kurumları* (literally 'civil society organizations') or NGOs, private foundations and corporations, individuals and groups in the arts and culture sectors, as well as the municipality, and government bodies. While previously it had always been national or local governments, which took the lead in nominating their cities to the ECoC programme, Istanbul's nomination was unique in that it was led by NGOs. Their involvement became the decisive factor in the city's selection as a 2010 European Capital of Culture by the Council of Europe, ECoC Selection Panel on 11 April 2006.

As a practitioner working for the Istanbul 2010 ECoC Agency in the capacity of urban projects specialist, I will concentrate in this chapter on 'participative cultural policy', advocated by the EU in the last few decades, and evaluate the Istanbul 2010 ECoC programme in terms of the participation it has sought to achieve. The Agency's aim to facilitate participative policy and an exemplary governance model underlies the involvement of various stakeholders in the process. I argue that, in time, through the process of negotiation between civil society and government bodies, the participative objective of 2010 ECoC programme shifted from 'participation to transform' to 'participation to legitimize'. To demonstrate this shift, I analyze how and why this transformation has occurred, and how it affects the city's cultural policy-making practices.

Participative Policy-Making

'Participative policy-making' was introduced at the end of 1980s as one of the key elements of the 'learning organization' – an organization that 'learns' to transform itself (Padler *et al.*, 1988). It has had its advocates and its critics but notions and practices associated with the term such as interdependent stakeholders, consultation and involvement, participatory strategy formation, inclusion, and public contribution are taken to be critical for good governance (Şuteu, 2005).

Today, culture professionals, decision-makers, NGOs and other actors involved in making cultural policy are expected to think and plan their cities in an inclusive way. Giving local cultural groups and communities a central role in policy formation and including all civic and public stakeholders concerned in the debates and decision-making on urban cultural development are assumed to be the standard practice for all cities taking part in the ECoC programme.

In recent decades, EU policies in the field of culture management have sought to impose upon organizations the norms and practices associated with 'participative cultural policy'. Thus it can be said that 'participative policy-making' has become an EU benchmark to identify and facilitate good practice. The European Council's increasing emphasis on cultural access in funding cultural programmes suggests the centrality of the concept to policy-making and culture management.

Although there is clearly no inevitable 'truth' as to what cultural policy a given city or region should pursue, establishing a good governance model and balancing the decision-making process between the government, local administrators and other actors including the general public with an effective organizational structure could be instrumental to creating a democratic environment (Matarasso and Landry, 1998). The ideal of 'participative policy-making' anticipates the distribution of the decision-making process and governance practices so as to ensure inclusive strategies and democratic procedures. Major outcomes of this form of policy-making could also include the improvement of local autonomy, networking between different groups of society, the democratization of culture and improved social cohesion.

As such, these potentials are all ideal; in reality, participative practices can be invoked to serve two different purposes: participation to legitimize or participation to transform. In the first case, those who promote participative practices aim to strengthen the basis and justification of their policy goals and interests, in as much as they are interested in change. In the second case, the aim is not to stay 'as we are and where we are', but to strengthen the capacities of citizens to suggest and negotiate change and achieve transformation (Pindado *et al.*, 2002). When deployed to transform, participation becomes an educational process for all actors involved, including the public administration itself. This is likely to create a sustainable process for learning and sharing.

A city chosen to be a European Capital of Culture is expected to reflect all these positions, often at the same time. ECoC committees assume the task of acting as mediators between the state and the public. Indeed, the Istanbul 2010 ECoC Committee started with the goal of 'having a facilitator role between actors for transforming the ongoing system and opening up channels for participative acts in the city'.[1] Yet for a number of reasons, the Agency did not perform in the desired way and these aspirations have not been realized.

Istanbul 2010 ECoC Agency and Participative Practices

A decision taken by the European Parliament and the Council of Europe in 1999 (1419/1999/EC) enabled the title of the European Capital of Culture to be extended to include cities in countries that are not members of the EU. This created the opportunity for Istanbul to apply to the ECoC programme. A local 'Initiative Group', comprising thirteen non-governmental organizations (NGOs), was formed with the involvement of representatives from local government to work on Istanbul's nomination.[2] With further support from the Office of the Prime Minister, Ministries of Culture and Tourism, and Foreign Affairs, and the Istanbul Metropolitan Municipality, an application dossier was prepared and presented to the ECoC Selection Panel working under the Council of Europe. Istanbul's selection for 2010 was confirmed by the Selection Panel on 13 November 2006.

Figure 15.1. Istanbul 2010 ECoC Agency is located in Atlas Pasaji (Arcade) on the pedestrianized Istiklal Caddesi (Avenue) in Beyoglu close to Taksim Square. (*Source*: Istanbul 2010 Agency Archives)

Istanbul's nomination dossier emphasized participative acts whereby citizens and institutional public actors could become effective stakeholders in the city. The programme would provide platforms for establishing democratic governance in the long term and, as a result, create 'urban transformation through participation'. In unanimously selecting Istanbul as the Capital of Culture, the Selection Panel evaluated these goals positively and underlined the significant role played by this non-governmental collaboration during the application process. In 2007, a new public body was established with legal powers to conduct business under the name 'Istanbul 2010 ECoC Agency'. The Agency was assigned the task of carrying out 'a comprehensive urban development project through arts and culture, and reveal Istanbul's cultural wealth as an inspirational source for the whole world' (see the Agency's website, at www.istanbul2010.org, for more detailed information on the process and projects).

Soon after, the Agency evolved into a fully-fledged organization, with various departments responsible for managing projects in visual arts, music and opera, film, urban culture, urban transformation, cultural heritage, stage and performing arts, traditional arts and education. Among them, the Directorate of Urban Applications, the Directorate of Cultural Heritage and Museums, the Directorate of Urban Projects and the Artistic Committee, which oversees the selection of art projects, had become the most significant players in the development and production of the Istanbul 2010 programme. The criteria applied in the selection process, such as sustainability, feasibility and institutional competence and capacity, had been assigned significant weight, along with contribution to the city, artistic quality, cooperation and fairness. In other words, formal institutional capacities played significantly into the selection of projects.

What was more significant was the shape organizational structure took with the legal establishment of the Agency. The law that enabled the establishment of the Agency called for a strictly bureaucratic and hierarchical structure, directly connected to the Office of the Prime Minister (see figure 15.1). This was done in the name of facilitating speedy action without going through regular and cumbersome procedures for spending public money. The end result, however, was the creation of an internal bureaucracy which acted to the contrary, lessening the role and influence of civil initiative which was hailed as the unique aspect of Istanbul's ECoC application. Decision-making slowly shifted from the actors representing civic bodies to the governmental authorities; the decisions about which projects to fund were increasingly made by the governmental representatives; and the rift between the two groups became public as one after another civic representatives resigned from their posts or severed their ties with the Agency as advisors, consultants and directors. This is not to say that the Agency turned into a government bureau as many members of the civil initiative continue to carry out the duties and responsibilities they have assumed. They do not, however, have the final say in decision-making, and they act as advisors or coordinators under the public authorities.

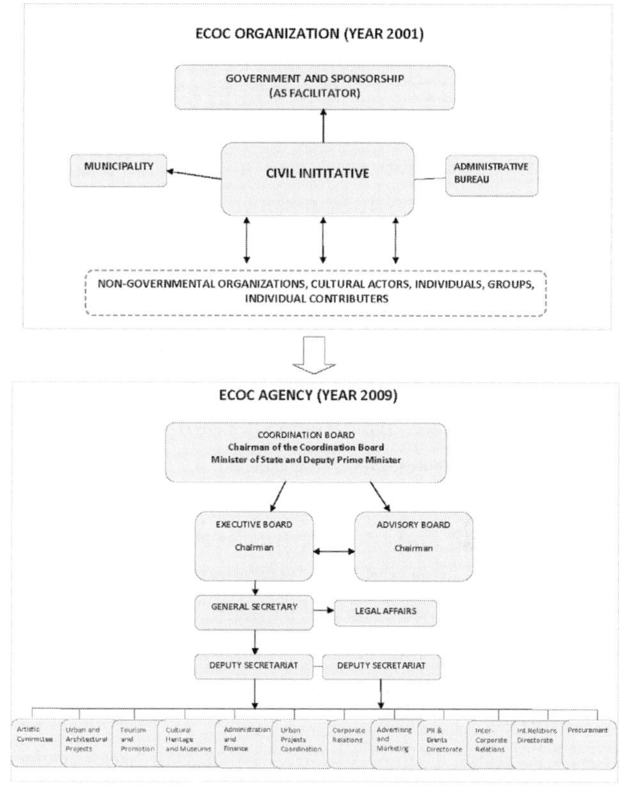

Figure 15.2. Organizational chart of the Istanbul ECoC Agency.

As the continuing resignations have exposed the rifts and conflicts among the various parties in the Agency, the existence of three categorically different groups, pursuing disparate interests, has become clear both to those working within the Agency and the public at large. At one end stand the civic activists, who, from the beginning, have considered culture as a tool for participative transformation and sought to establish transparent and decentralized governance practices. In the words of Cengiz Aktar (2006), one of the advisors to the Istanbul 2010 Agency, 'one of the most valuable benefits of the ECoC project' would be 'transforming the classical local government into good governance'. Asu Aksoy, who had been with Istanbul 2010 from the start but later resigned, stated on many occasions her expectation that Istanbul 2010 should be working as an effective interface between various public and private parties:

> [In Turkey] ... although they are chosen democratically, municipalities and local governmental bodies do not have organizing and facilitating roles in the cultural sector, and they do not create platforms for institutions and cultural actors to interact with each other. Rather, they choose to act as monolithic parties with a singular cultural vision and play central decision-making roles... Istanbul 2010 can help to solve this structural problem and create new practices for negotiation between different actors, by creating discussion platforms where citizens seek common languages to speak.[3]

At the opposite end to the civic activists stand the bureaucrats and technocrats who held governmental office prior to their involvement with the Agency. They mostly see the ECoC designation as self-evident right for Istanbul, with its magnificent history and culture, and take the Istanbul 2010 programme not as a creative but a bureaucratic tool to promote Istanbul to the world. Participation, in this context, becomes a short-term goal and a façade for legitimizing the advancement of their goals. The third category of actors, the private sector and cultural sector, appear more neutral in their expectations and seem mainly interested in facilitating means for the development of cultural industries and creating opportunities for artistic-cultural initiatives.

These categorically different perspectives on the nature and the mission of the 2010 ECoC have been in conflict with each other, made evident by the public controversies over the use of funds and management styles. Yet the reality of the task ahead – putting the show on the road, so to speak – has enabled rapprochement and coexistence. In the process, the principle of 'participation to transform' has given way to 'participation to legitimize', as the power of decision-making surely shifted to the bureaucrats and technocrats.

Istanbul 2010 ECoC Projects and Challenges

In this section, I turn my attention to the projects included in the Istanbul 2010 programme. In its final form, the programme comprises 451 projects, selected from a pool of more than 2,000 applications. The total budget for the projects approaches 300 million Turkish lira (about US$200 million), of which about 60 per cent will be spent on urban transformation projects and the rest will be distributed beween art, culture and tourism projects.[4] It is not my intention here to evaluate the content and scope of the projects, but rather to analyze them from three benchmarks used for the evaluation of ECoC programmes in Palmer (2004) – namely, 'outreach development and access development', 'community development and cultural inclusion', and 'participative governance'. These benchmarks will allow us to assess the participative claims of the Istanbul 2010 programme.[5]

Briefly, as elaborated in Palmer (2004), the objective of 'outreach development and creation of awareness' is said to lead to creation of awareness about the cultural life of the city among the city residents and to improve their access to cultural projects in the programme. The objective 'community development and cultural inclusion' is there to facilitate the extension of opportunities to those whose cultural values are marginalized by, or excluded from, the dominant cultural landscape. Primarily the aim is to enlarge the framework of cultural expression, enable new voices to be heard and, ultimately, to make the cultural space more open and participative. Finally, the objective of 'participative governance' assigns to the ECoC agencies the goal of creating a new structure for the public sphere. In such a structure, power and responsibilities can be partly transferred from centralized governmental authorities to local actors, and various stakeholders can act collaboratively in a democratic environment, sharing decision-making.

The Istanbul 2010 ECoC programme has paid specific attention to outreach development and creation of awareness. The programme prioritized the use of public spaces over private venues and prices have been kept low, if not free. Reaching the urban, economic and social periphery and facilitating artistic-cultural programmes or organizing programmes with volunteers with that aim have been taken as key to improving public awareness about the cultural life of the city. The 2010 Gönüllüleri (Istanbul 2010 Volunteers) programme was initiated in 2008 to increase the visibility of ECoC projects and the involvement of young people in the run-up to 2010. The project has worked quite successfully, albeit not at a desired pace, often because of the managerial problems inherent in the Agency. The 'Universities Committee', which was seen as a conduit to secure the involvement of forty plus universities in Istanbul and to mobilize the creative capacities of university students, failed to get off the ground after a few meetings between Agency administrators and the representatives of the universities. The Agency did not manage to secure the commitment of the universities, though many designed and presented projects, some of which were accepted.

The most successful project in this respect has been *Taşınabilir Sanat* (Portable Art), by which a portable package of artistic activities is transported to the outer districts of Istanbul, with the goal of encouraging young artists living in the margins of the city. The idea stems from the fact that contemporary art production in Istanbul is fairly concentrated in a restricted space at the centre, mainly Taksim

Figure 15.3. 'Taşinabilir Sanat' (Portable Art) project in Kartal. (*Source*: Istanbul 2010 Agency archives)

and its environs, catering for a very limited segment of art consumers. The Portable Art project set itself the goal of extending art production to all the thirty-nine districts of Istanbul using the cultural infrastructure of the municipalities. Since 2008, six portable projects have been realized in cultural centres located in the periphery of the city, such as Kartal, Ümraniye, Küçükçekmece and Tuzla.

Regarding the outcome, and evaluating the media response and public

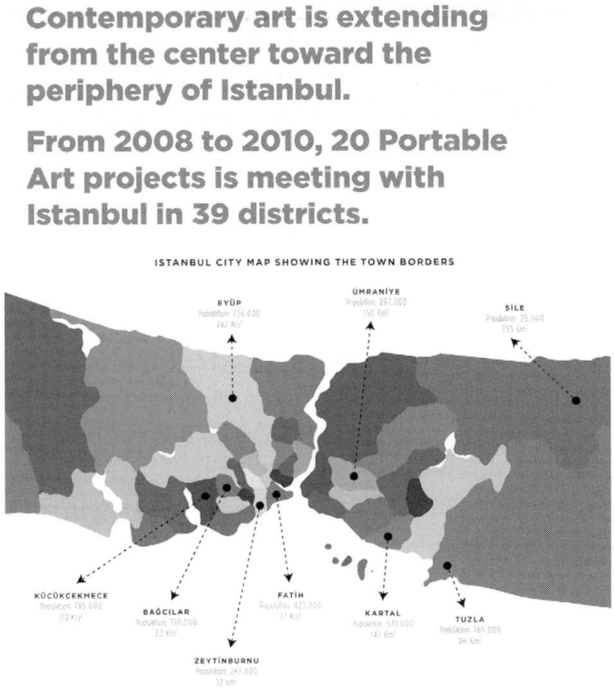

Figure 15.4. Taşinabilir Sanat (Portable Art) project aims to visit nine neighbourhoods of Istanbul. (*Source*: Istanbul 2010 Agency Archives)

participation, it could be said that the project has been moderately effective and successful. As to the overall effectiveness of outreach development and the creation of awareness, the same could not be said in the face of disconnect between the practices of the 2010 Agency and public interest in the programme – though this may improve once the programme is in full swing in the months ahead.

In terms of the objective 'community development and cultural inclusion', the Istanbul 2010 programme has sought to give visibility to socially excluded groups and encourage them to participate. One such project, 'Social Exclusion and Art', is an attempt to make visible the groups, which have remained outside the social and cultural life of the city, through works of art. Groups targeted by the project include women prisoners in correctional facilities, elderly men and women in public shelters and mental health patients in hospitals. Another project, 'Obstacle-Free Urban Projects for the Disabled', aims to conduct awareness-raising meetings organized by the disabled themselves. The project

Figure 15.5. Kentle Buluşma, Müzeyle Tanişma (Meeting the City, Getting to Know the Museum) project, which aims to visit nine Istanbul neighbourhoods. (*Source*: Istanbul 2010 Agency Archives)

Kentle Buluşma, Müzeyle Tanışma (Meeting the City, Getting to Know the Museum) project addresses women and children aged between eight and eleven who cannot participate in cultural activities in the city because of social and economic obstacles. It consists of workshops and organizing day visits to specific museums on the Historic Peninsula. Although these projects have potential to create positive outcomes and be successful in their own right, the Istanbul 2010 programme does not have a fully developed broader perspective to achieve inclusiveness. Particularly absent are issues considered to be controversial and sensitive such as those dealing with and/or targeting ethnic, religious, and socio-cultural groups and communities – Kurds, Alevites, Armenians, homosexuals, and transvestites among others.

As stated before, the objective of 'participative governance' has been central to the Istanbul 2010 programme since its inception. The programme project has aimed to develop a new kind of relationship between civil society and governmental bodies as well as building ground-up organizational structures. The goal is reflected in a number of projects, where different parties are encouraged to meet, share expertise and collaborate towards the realization of activities. In this vein, the project 'Developing and Supporting Arts and Culture Forums' called for the cooperation of local government, universities and NGOs to realize international and interdisciplinary open fora, symposia, conferences and workshops as a way to build a shared public knowledge base.

The most important project in this respect however has been the 'Project to Renovate the Port of Theodosius and Develop the Yenikapı Transfer Centre'. This project requires the expertise of architects, engineers, archaeologists, urban planners and a range of national and city administrations. Excavations at Yenikapı, the site where the *Marmaray* undersea tunnel connecting Asian and European sides of Istanbul reaches the land, has revealed the sunken remains of a Byzantine port – hence the need for joint decision-making and knowledge sharing (see Istanbul 2010 ECoC Programme Book for more information). This was seen as an opportunity to act as an interface between the actors involved and enact a model for inter-agency cooperation. There has been considerable progress to this end but there has also been strong pressure from the governmental authorities to keep the intellectual actors out of the picture and professionalize the process. It must be said that in realizing 'participative governance', the Agency has been far from achieving the mission it initially set itself.

Compared to Istanbul, Lille 2004, for instance, comes out as a very successful ECoC in achieving its set goals concerning the three benchmarks examined. Lille 2004 operated under an autonomous body and employed a decentralized approach, and the organization still continues to function under the name 'Lille 3000'. Lille 2004 set as one of its main objectives promotion of an *art de vivre* (living art), and the definition of culture was widened to include elements of ordinary activities such as cooking, sport and interior design – thus including the general public. Former factories and heritage sites were transformed into *Maisons Folies*, designed to welcome a mix of artists and locals for creative encounters and cultural production. More than 17,000 artists were mobilized during the year, half of them from the local region and one-third from foreign countries (see www.culture-routes.lu).

Lille 2004 also introduced the innovative system of *Ambassadeurs*, a very successful project, where citizens could volunteer, act as information providers, receive regular information updates, assist in the running of events and participate in special activities. More than 15,000 volunteers from all backgrounds acted as ambassadors across the territory, forming a grassroots information relay system (Sutherland and Besson, 2006). In this way, the organizers were able to get closer to the general public, develop a system of feedback on public opinion and ensure public support. In carrying out all its projects, Lille 2004 enacted a multi-partnered model that secured the partnership of public bodies, private organizations, and NGOs. Consequently, the city now has a very positive international reputation, which has made it an important cultural hub of Europe.

The projects that comprise the Istanbul 2010 programme display similarities with Lille 2004, particularly in terms of those that aim to achieve the objectives examined in this section. The single most important difference is the lack of an autonomous, decentralized organizational structure in Istanbul, which in a way underlies the success of Lille 2004. Although it began with the same goal in mind, Istanbul 2010 has not been able to enact such a model but has evolved into a more centralized body and has lost its potential to act as an effective facilitator.

Conclusion

I have so far argued that the Istanbul 2010 ECoC project has not been successful in realizing some of its initial goals, most important among them creation of a new model of public governance, bringing together a variety of institutional actors and aiming for open, transparent decision-making processes. This was, of course, an ambitious goal, one that had not been tried before, not at least with such a wide body of actors involved. From the beginning it was against the grain of the centralized administrative culture in Turkey. It could be said that culture has won again in the end and the objective of 'participation to transform' has lost in the face of deploying participation as a way to legitimize the administrative goals of bureaucracies and professionalized management techniques.

As a practitioner and officer of the Agency, however, I would like to end on a more optimistic note. As I have stated, this experience was a first for Turkey. Though arduous and full of impediments and handicaps, the process has brought together unlikely actors and competencies as participants in an exercise which has lasted for a long period. There have been lessons learned and mistakes made, and more often than not participative initiatives clashed with managerial practices. The overall experience, however, has been an archive of expertise and competency that can be mobilized towards participative policy-making in the future.

But last word on the successes, shortcomings and failures of Istanbul's 2010 European Capital of Culture venture should wait until the completion of the programme, which was set in motion on 16 January 2010.

Notes

1. Personal interview with Asu Aksoy at *santralİstanbul*, 10 March 2008.

2. Istanbul 2010 Initiative Group included representatives from the following organizations: Açık Radyo (Open Radio), İnsan Yerleşimleri Derneği (Human Settlements Association), İKSV (İstanbul Sanat ve Kültür Vakfı, Istanbul Culture and Art Foundation), İktisadi Kalkınma Vakfı (Economic Development Foundation), Tarih Vakfı (History Foundation of Turkey), Kültür Bilincini Geliştirme Vakfı (Cultural Awareness Foundation), Kültürlerarası İletişim Derneği (Association for Intercultural Communication), dDF Advertising Agency, Reklamcılar Derneği (Association of Advertising Agencies), Ulusal Ahşap Birliği (Turkish Timber Association), Marmara Belediyeler Birliği (Union of Municipalities of the Marmara Region), Türkiye Turizm Yatırımcıları Derneği (Turkish Tourism Investors Association) and İstanbul Sanat Müzesi Vakfı (Istanbul Art Museum Foundation).

3. Personal interview with Asu Aksoy at *santralİstanbul*, 10 March 2008.

4. Figures provided by the Minister of State, Hayati Yazıcı, responsible for the Agency, at the presentation of the Istanbul 2010 Programme to press and dignitaries on 10 December 2009 in Istanbul.

5. Here I take the benchmarks set by Palmer (2004), which is ordered by the European Council, as these are the benchmarks which the Council expects the ECoC cities to achieve.

References

Aktar, C. (2006) Istanbul, the European Capital of Culture in 2010. *Turkish Daily News*, 14 April.

Council of Europe (2000) *Competences and Practices in European Local and Regional Cultural Policy* (Studies and Texts, no. 69). Strasbourg: Council of Europe.

Matarasso, F. and Landry, C. (1998) *Balancing Act: 21 Strategic Dilemmas in Cultural Policy*. Strasbourg: Council of Europe, Cultural Policy and Action Department.

Palmer, R. (2004) *European Cities and Capitals of Culture*, Parts I and II. Brussels: Palmer-Rae Associates.

Pedler, M., Burgoyne, J. and Boydell, J. (1988) *The Learning Company*. New York: McGraw-Hill.

Pindado, F., Rebollo, O. and Martí, O. (2002) Eines per a la participació ciudadana: bases, mètodes i techniques. Papers de participació ciudadana [Tools for Citizen Participation: Basis, Methods and Techniques. Papers on Citizen Participation]. Barcelona: Diputació de Barcelona.

Şuteu, C. (2005) Overview on Cultural Policy in Central and Eastern Europe between 1990/2003. Ecumest Association and Romanian Academic Society. Available at: http://www.ecumest.ro/pdf/2005_suteu_overview_cultural_policy.pdf. Accessed 25 January 2010.

Sutherland, M., Besson, E., Paskaleva, K., Capp, S. and Origet du Cluzeau, C. (2006) Analysis of the Mobilising Role of the European Capital of Culture Process. Deliverable D16, EU FP6 PICTURE project. Available at: http://www.picture-project.com. Accessed 26 January 2010.

Chapter 16

Counting as European: Jews and the Politics of Presence in Istanbul

Marcy Brink-Danan

On a recent stroll down İstiklal Caddesi, a popular shopping thoroughfare on Istanbul's European bank where Jewish, Greek and Armenian businesses once lined the streets,[1] I noticed a shirt displayed in the window of a clothing store. By replacing letters in the city's name with a Christian crucifix, a Jewish star and an Islamic crescent, this cultural artefact captured a moment in which the symbols of cosmopolitanism seem increasingly fashionable in Istanbul.

Figure 16.1. Shirt for sale, Galata, Istanbul. (*Photo*: Marcy Brink-Danan, 2007)

Scholars note a growing nostalgia for Ottoman cosmopolitanism (Mills, 2007; Örs, 2006; Özyürek, 2006, 2007; Potuoğlu-Cook, 2006) or what is often called the Ottoman 'mosaic'.[2] Festivals, exhibitions and concerts celebrating religious tolerance increasingly punctuate Istanbul's landscape (Yardımcı, 2007). In addition to these public displays of cosmopolitanism in Istanbul, nostalgia for religious difference has been observed in particular neighbourhoods, or *mahalleler*. Kuzguncuk (Houston, 2001) especially, but also '… Jengelköy, Beylerbeyi and Arnavutkoy, and other parts of the city known for minority history like Fener, Samatya and most especially Beyoğlu also locate nostalgia for the lost Istanbullus, the Greeks, Jews and Armenians who took the character of the city with them when they departed' (Mills, 2006. p. 371). These areas, once inhabited by Jews and other minorities, only became quaint when emptied of them: 'As attested to by its cemeteries, Kuzguncuk is a multireligious and multicultural suburb. But mainly in the solidarity of death' (Houston, 2001, p. 17).

How do Istanbul's *living* religious minorities, those persons whose beliefs, rituals and languages constitute the difference celebrated, relate to this kind of geographic nostalgia? Are they critics? Symbols? Moving beyond an analysis of nostalgia, this chapter attends to the staging of cosmopolitanism in Galata (encompassing, also, Beyoğlu and Karaköy) by Jewish Istanbulites in conjunction with the local municipality, national politicians and European organizations and the ensuing re-signification of the area's architectural and historical landscape. First, I describe this neighbourhood's importance in Jewish Istanbul's past and present. I then discuss the role of the Jewish museum in a historic synagogue in this neighbourhood and the use of this particular site in Istanbul as an effective stage to announce Turkish tolerance, *vis à vis* the Jewish population, to the world. I then turn to a particularly fraught site for the performance of Judaism, the Neve Shalom synagogue in Karaköy, thrice bombed over the past two decades but also the site of regularly scheduled weddings, Bar Mitzvahs (a coming of age ritual) and other Jewish celebrations, as well as the scene of the investiture of the Turkish Chief Rabbi in 2002, whose audience included not only community members but also Turkish government officials, and whose rhetoric cast Istanbul's Jews as players in European Union accession.

Given their public face as model minorities, one might expect Jewish Istanbulites to be colourful peacocks, parading their multicultural feathers along İstiklal Caddesi. However, this story of Jewish Istanbul is more complicated, involving in equal parts presence and erasure. Despite the idyllic official discourse of multiculturalism, Istanbul's Jews today gather in homes, schools, and synagogues unmarked as Jewish spaces. Visitors to Jewish sites in Istanbul march through metal detectors; security cameras record the scenes, perched like watchful birds above the barbed wire and cement barricades that secure Istanbul's Jewish institutions.

In light of these obvious tensions between public performance of difference and private fears of displaying diversity, how might we understand the role that

Istanbul's Jews, their officials and their institutions play in the ongoing debate about cosmopolitanism, difference and tolerance that today is the focus of so much public debate in Turkey, the European Union and elsewhere (Blommaert and Verschuren, 1998)? I take an ethnographic approach to show how Jews in Istanbul today negotiate the cityscape by defining various stages for the erasure and performance of difference, especially in light of European Union demands for evidence of Turkey's civilizational maturity (Levent Soysal, Chapter 17).

As Soysal *et al.* (2005, p. 27) have argued, public efforts towards 'recognition of diversity', however imperfectly matched with celebrations of national and pan-European identity, have become a pillar of European self-definition. Jews, particularly, occupy a central role in European claims to cosmopolitanism, especially as a foil against which to counter cries of intolerance made by other differentiated citizens (Peck, 2006, pp. 154–174). Istanbul's re-signification, as invoked by the multi-religious iconography of the shirt in the shop window, echoes trends across Europe's urban landscapes, in places such as Berlin (*Ibid.*), Krakow (Kugelmass and Orla-Bukowska, 1998) and Vienna (Bunzl, 2003), in which Jewish spectacles (museums, memorials, cafes, etc.) are key sites through which cities enact their tolerance of diversity. Gruber (2002, p. 4) notes, 'More than half a century after the Holocaust, an apparent longing for lost Jews – or for what Jews are seen to represent – is also evident. In a trend that developed with powerful momentum in the 1980s and accrued particular force after the fall of communism in 1989–90, Europeans … have stretched open their arms to embrace a Jewish component back into the social, political, historical, and cultural mainstream'. To be a European city, it seems, is to 'have Jews'.

As capital of the Ottoman Empire, Constantinople was known for its cosmopolitan array of Muslims of varying sects, Armenians, Greeks and Jews. Although Jewish life in Asia Minor dates back to the fourth century BCE, the majority of Jews in Istanbul are 'Sephardic', that is, they trace their ancestry from Spain. Sephardic Jews sought refuge in the Ottoman Empire during the fifteenth century when Sultan Beyazıt II 'welcomed' Jewish exiles from the Iberian Peninsula, populating his territory with skilled and grateful subjects and purportedly uttering the following critique of Christian monarchs: 'Spain's loss is my gain'. Even before the expulsion, Rabbi Yitzhak Sarfati, himself a German refugee living in Ottoman lands, issued a call to Jews throughout Europe to '… leave the torments they were experiencing in Christendom and to seek safety and security in Turkey' (Braude and Lewis, 1982, pp. 135–136). Although historians suggest that the Rabbi may have been coerced by the Sultan to write the letter (Levy, 2002, p. 5), these origin myths are widely recounted today.

During Ottoman rule, Jews and Christians enjoyed collective rights under the Islamic 'pact of toleration' accorded them as 'People of the Book'. After four centuries in which an Islamic code for minorities recognized Ottoman Jews as a corporate identity, in 1923 Jewish community officials exchanged special status for universal citizenship in the nascent Republic of Turkey. Local anti-minority episodes, such as the population exchanges, the Thrace Pogroms (1934), devastating Property Taxes levied on minorities (1942), riots in Istanbul (1955 and 1963), coupled with the out-migration of over 30 per cent of Turkey's Jews to Israel post-1948, increasingly transformed Istanbul into what has been called a 'depopulated cosmopolis' (Komins, 2002).

Although the relocation of Turkey's capital to Ankara, the out-migration and exile of huge numbers of religious minorities and the Turkification process deeply altered Istanbul's character, Jews never abandoned the city; if anything, they increasingly abandoned the rest of Turkey to relocate in the former capital of the Ottoman Empire. The shift in settlement patterns of the Turkish-Jewish communities reflects their increased urbanization: over 99 per cent of Turkish Jews live in urban centres – probably over 90 per cent (Tuval, 2004, p. xxxiii), with communities remaining in Izmir and a handful of other cities (Weiss, 1975, p.169).

While there is no official census of Jews in Turkey, unofficial counts estimate their numbers to be between 20,000 and 25,000 (Toktaş, 2006, p. 123), thus a negligible fraction of Turkey's overall population of approximately seventy million. '"Jews?", Mete Tapan, Istanbul city planner, had to think for a minute. "One knows they are there but they don't even comprise 1% of the population. No one knows how they vote or what their interests would be. They don't count"' (Fleminger, 2003). While, as suggested by the city planner, Jews might not 'count' for much in the polling booth, they count in symbolic ways not measurable by census or survey.

Jews in Istanbul have recently taken on an increasingly public role, brokering Turkish diplomatic ties with Israel and advocating for the Republic as Turkey vies for European Union accession. As Turkey continues its half-century march toward the European Union, its Jews have likewise been called upon as living proof of Turkey's fulfilment of the 'recognition of diversity' criterion. Playing the part of the 'good minority' in international arenas, Jews regularly proclaim Turkey's eternal hospitality and tolerance for difference to a global audience as a counterpoint to European politicians' regular criticisms of Turkey's treatment of Armenians, Kurds and Islamists. This shift marks a change in the way Turkish Jews represent themselves in the urban landscape and is but one reflection of the myriad ways in which Turkey's European Union overtures have set the stage for Jews to stand symbolically for the tolerated 'other'. Istanbul, home to the vast majority of Turkey's Jews, is the obvious theatre for the Jewish community to perform this role.

Jews today overwhelmingly live in neighbourhoods such as Şişli, Nişantaşı, Gayrettepe, Göztepe or Etiler (Tuval, 2004, pp. 85–87). These demographic shifts map the changing socio-economic status of Istanbul's Jews and reflect larger social changes which have altered the overall pattern of residence in the city (Varol, 1989; Kastoryano, 1992). However, Jews largely *perform* cosmopolitanism not where they eat, sleep or work, but, rather, in a handful of buildings and streets in what was once Pera, but today is called Galata, Beyoğlu or Karaköy. During my ethnographic fieldwork conducted between 2001 and 2003, I often came to this area to visit the Jewish Museum, for a meeting at the Offices of the Chief Rabbinate or to attend a function at Neve Shalom synagogue. 'Pera' – a nineteenth-century nexus for Muslims, foreigners and non-Muslim Ottoman subjects – served as a test-bed for European-style modernization: banks, widened streets and other architectural innovations marked the area as progressive (Karmi, 1996; Bartu, 1999).

Early twentieth-century Jewish elites tied their fate to this neighbourhood's Westernizing elite as a claim and 'path to modernization' (Kastoryano, 1992, p. 258). Evidence of this era survives in the architectural formation called the 'Kamondo Steps',[3] named after its sponsor, Abraham Kamondo, a Jewish-Ottoman banker whose alliances with major European Jewish figures, such as the Rothschilds and Baron de Hirsch, opened a path to the West for Ottoman Jewry. Seni (1994, p. 664) has noted the Kamondos' impact on the 'urban fabric of Istanbul … where the family business began. Their influence derived principally, although not solely, from a real-estate empire centred mainly in Galata, the European section of the city. The family also played an important role in transforming the city in other ways, for example, in the development of a system of transportation'.

Figure 16.2. Kamondo Steps, Galata. (*Photo*: Henri Cartier-Bresson/ Magnum Photos, 1964)

Made famous by French photographer Cartier-Bresson in a 1965 photo, the Kamondo Steps signal a 'Europeanist' moment in a changing architectural and human landscape of Ottoman minorities, but the tale of Istanbul's Jews does not end there. The pirouetting curves of the stairs connect Constantinople to Istanbul, recalling how Turkish Jews have remained continually connected to the rest of what many now simply call 'Europe'. Highlights of this deep relationship include: the presence of 'Greek' Jews in the Byzantine Empire; the influx of Jews to the Ottoman Empire after their expulsion from the Iberian Peninsula; the maintenance of Judeo-Spanish as this community's lingua franca until the mid-nineteenth century; the introduction of French as an elite Jewish language with the arrival of the Alliance Israélite Universelle (AIU) schools in Constantinople and throughout the Ottoman Empire (a process precipitated by Kamondo and his sympathizers). With genealogical roots in Spain and a legacy of French education, European influences are observable in their linguistic practices, travel itineraries and musical styles (Dorn, 1997).

Given its history, it is not surprising that Galata is today the staging ground for performances of European-ness upon which Istanbul's Jews – and other Turks – act out their role as cosmopolitans. However, I would argue that this role is not shaped by nostalgia. Rather, Jews (or their officials) are motivated by a very practical goal of using Istanbul's landscape as a site through which to argue for Turkey's – and their own – deep ties to Europe. One indication of this is visible in the recent marking of the Kamondo Steps with a plaque acknowledging their creator and the Kamondo Mausoleum Rehabilitation Project, supported by the

Figure 16.3. Plaque marking the Kamondo Steps. (*Photo*: Marcy Brink-Danan, 2007)

Istanbul 2010 European Capital of Culture Agency (Gültekin, 2009, p. 38). Among Istanbul's Jews, self-nomination in the public sphere, in this case, in Galata, is a relatively new phenomenon.

During the early years of the Republic, becoming Turkish, and the fear of not being perceived as Turkish enough, engendered a profusion of effacing social practices among Jews and other minorities in Istanbul (Bali, 2001). Despite the loyalist attitude of many Jews (Benbassa and Rodrigue, 2000), including ideologues and founders of Turkism such as Abraham Galante, Tekin Alp (who was born Marcel Samuel Raphael Cohen) and others, the early years of the Republic of Turkey saw an increase in xenophobia, in which minority languages were banned and devastating riots occurred. Further, the 'Capital Tax' instituted during World War II penalized and pilfered small Jewish (and other minority) businesses to the point of bankruptcy (*Ibid.*, p. 182).

In spite of these hardships, thousands of non-Muslims remained in Turkey. However, their traces of difference were erased from the national narrative when the Republic of Turkey redefined the status of its minorities as full citizens. After centuries in which Jews in the Ottoman Empire had protected status, the Jewish community exchanged inequality for universal citizenship under the new Republic. Just over half a century later, European Union overtures set the stage for Jews to stand symbolically (and publicly) for the tolerated 'other'. This role was consolidated in 1992 with a Turkish-led international celebration of the 500-year anniversary of Jews finding refuge from the Spanish Inquisition in the lands of the Ottoman Empire. Taken from the statement of purpose of the Quincentennial Foundation, an organization led by Jewish and Muslim Turkish elites, the commemoration:

> not only celebrates the 500th anniversary of the arrival of the Sephardic [Spanish] Jews on Turkish soil in 1492, but also the remarkable spirit of tolerance and acceptance which has characterized the entire Jewish experience in Turkey. This spirit is not an insulated [*sic.*] instance of humanitarianism; throughout its history Turkey has welcomed people of different creeds, cultures and backgrounds. The Jewish community of Turkey is a part of this tradition.[4]

This description set the agenda for a public awareness campaign to improve Turkey's international image. If, due to their small numbers and historically low-profile, Turkish Jews were not once considered meaningful players in the Turkish political scene, this campaign, with its museums, academic treatises and heritage tours, offered them a public platform from which to speak as quintessentially, and quincentennially, tolerated minorities. Quincentennial celebrations, attended by Turkish and Israeli politicians, gathered momentum at the same time that Turkey's relations with Israel were warming. In parallel, the 1990s witnessed an increase in Jewish contributions to Turkish efforts to court Europeans, who made Turkish officials' recognition of their own heterogeneous populations a central condition for Turkey's acceptance into their club.

My first visit to the Jewish Museum in Istanbul, created by the Quincentennial Foundation in the defunct Zulfaris Synagogue, took me through the back streets of Beyoğlu and Karaköy. Given the lack of signage at all major Jewish community buildings in Istanbul, one needs prior knowledge of the location of the space, acquire permission, and announce one's arrival beforehand even to find the places where Jews congregate. In 2003, I observed a group of young Turkish-Jewish adults during an organized tour of the museum. The museum's curator introduced the goals of the museum to an audience which he perceived as already familiar with the exhibition's historical narratives; he explained he was compelled to create the museum after the 1986 synagogue bombing, when the general population in Turkey (as well as the international press) recorded its shock at the revelation, through a news-worthy event, that Jews *still lived* in Istanbul. The museum's director also explained that while the museum indeed serves the local Jewish population, it is really intended to 'present the face of Judaism to the outside' as a platform for Jews to have a political presence through which they advocate, in no uncertain terms, for Turkey to be seen as civilized, tolerant and modern; in short, European.

Foreign museum guests respond to the tolerance trope much in the way that the museum's curators intended. They engaged in its 'civilizational discourse' (Brown, 2006) equating tolerance of Ottoman and Turkish Jews with modernity and civilization. Museum visitors remarked, 'It is amazing to see how tolerant the Turks are; America can learn something from them'; and, 'I was surprised to learn how tolerant the Turks were even in medieval times; *even in* Europe the Jews were treated much worse'. As a public relations campaign, the museum projects a civilized society which American and European visitors identified as similar to their own, as reflected in comments in the visitor guestbook. Examples included a US ambassador to Turkey, who, in 2001, wrote, 'The United States and the American people are proud and honored to stand with Turkey in fostering peace and tolerance among all … peoples' (Kamhi and Ojalvo, 1997, p. 47). A British Ambassador expressed similar sentiments, noting that the museum 'Is a symbol of the significance of religious tolerance for us all' (*Ibid.*). By staging the museum's civilizational message, Turkey's Jews perform their Europeanness and prescribe to themselves an ongoing and central role in Turkish affairs, offering the republic 'a people to be tolerated' at a time when tolerance is on the minds of so many (especially lawmakers in Europe).

The museum's location in Karaköy brings diversity to an area of Istanbul where few Jews still live, but where their visibility can be carefully managed and presented to the outside. One need not hypothesize about the parallel rhetorics of the museum narrative and the goal of Jews' public performances as both Europeans and tolerated subjects. The neighbourhood itself is home to the offices of the European Capital of Culture Agency, the planning centre for activities held in honour of Istanbul's designation for 2010, which through its multiple projects imagines and transforms the urban landscape as a stage for Turkey's European-

Figure 16.4. 'Mono-love, multi-culture' (ECOC poster displayed at the Jewish Museum). (*Photo*: Marcy Brink-Danan, 2007)

ness. This was brought into sharp relief when I noticed, hanging on the wall next to the museum guest book, a poster commissioned by the Istanbul Metropolitan Municipality, the Istanbul Tourism Atelier, and Istanbul 2010: European Capital of Culture, with the slogan 'mono-love, multi-culture' emblazoned above the word 'Istanbul' with the city's historic names testifying to its mythical cosmopolitanism.

If the tolerance trope is increasingly familiar to observers of Turkish-Jewish life, so too is the community's impenetrability to visitors and other non-intimates. When confronted with the tension between the pervasive tolerance trope and constant attention to 'security', Turkish Jews claimed that security measures began as a response to the 1986 attack in which twenty-one people were massacred by Palestinian sympathizers at Istanbul's largest synagogue. Built in the 1950s, the ironically named Neve Shalom (Hebrew: Oasis of Peace) is situated in what is now a lighting wholesale district along Büyük Hendek Caddesi in Karaköy. Before I began fieldwork, I was warned, 'If you want to learn about Jewish life in Istanbul, don't look in the synagogue', suggesting that Turkish Jews, known for their secularism, would not regularly be found at prayer. What I found, however, was that the Neve Shalom synagogue was a key symbol through which Jews in Istanbul understand their presence in the city, not through prayer, but through a politics of performance and a negotiation of sometimes dangerous cosmopolitanism. While most members of the community say that they retreated to a more private Jewish space since the first Neve Shalom attack in the 1980s, one educator in Istanbul's Jewish community explained that 'the attack made us open up, we were suddenly visible'.

I visited Neve Shalom many times for various functions, including friends' weddings and the visit of an Israeli dignitary. I also attended a Jewish community *iftar* – a break-fast meal for Ramadan (in Turkey, *Ramazan*) – held in the Neve Shalom synagogue in 2002. The main social hall was converted into a dining room, filled with hundreds of attendees seated at tables set with a typical Turkish *iftar sofrası*, including the *Ramazan* staple – fresh *pide* (flatbread) – prepared in the kosher kitchen of the Jewish home for the aged. As if to signal the 'spatialization' of the event (see Soysal, 2001, p. 23), the *iftar* menu, placed conspicuously on each table in the social hall, displayed an image of the Galata Tower, a short walk from Neve Shalom, and itself an architectural memory of fourteenth-century Genoese habitation of the quarter. For at least the past decade the Jewish community has regularly hosted an *iftar* (last year it was held in tandem with Istanbul's 'European Day of Jewish Culture') that provides a stage from which to proclaim the similarities between Jews and Muslims in Turkey and in the world.

Perhaps the most memorable event I attended at Neve Shalom was the investiture of the Turkish Chief Rabbi in 2002. The ceremony took place in the synagogue, whose inconspicuous entrance, fortified with a cement wall, metal detectors, and security cameras, gives way to a resplendent interior. Wooden pews line the main floor in a semicircular pattern around the ark which holds the Torah scrolls, above which an 8 ton chandelier dangles precariously. Above the main section, the women's balcony is accessible by two main stairways. The space holds 1,000 people and it was overfull with excited community members and officials.

The men sat in the synagogue's main hall, their heads covered with blue skullcaps emblazoned with the golden image of the breastplate worn by the chief rabbi. From above in the women's section, where I sat, a sea of sapphire satin seemed to cover the congregation. Male Turkish governmental leaders, lay leaders of the Jewish community, and a handful of Israeli politicians sat in the front pews in the company of at least one American Rabbi, bookended on each end by attentive security guards.

The upper floor of the sanctuary overflowed with women, the scent of perfume overwhelming, well-coiffed hair and polished nails ubiquitous. Packed tightly together in the balcony, women angled for a better view. Because the community attendees included at least as many women as men, a projection screen was set up in the upper halls of the synagogue with a broadcast of the event. Simultaneity did not compensate for these women's distance from the main hall, about which many of them quietly grumbled.

The newly elected Turkish Chief Rabbi, İshak Haleva, entered the hall dressed in regal purple with a silky headdress. While the chief rabbi wears a royal purple gown on formal occasions, the 2002 investiture at Neve Shalom was cloaked in the red and white hues of the Turkish flag as well as the blue and yellow of the European Union. From the permissions sought by the community from the Ministry of Internal Affairs and the Governorship of Istanbul to the invocation of the European Union during the Rabbi's investiture, the community ensured that a ritual which might have been private became a public, indeed, a national and international endeavour.

When community officials envisioned the staging of the investiture, there was little doubt that the event would take place in Istanbul. The Rabbi's investiture served as a stage from which to praise Turkey, best represented, perhaps, in the symbolic, if not demographic, evidence of tolerated difference in Istanbul, as his speech proclaimed:

> It is my pleasure and my responsibility to announce to the world that we are indebted to the Turkish nation's dignified nobility and its government's recognition of rights for standing here today as a Jewish community with a size and a quality enough to elect a chief rabbi by free will [*özgür iradesiyle*].

Here, the new rabbi highlighted the value and 'countability' of the community. The election and investiture were invoked as not only impacting Turkish Jews but as having potential ramifications beyond the community, as indicated in speeches given by representatives of Turkish political parties, such as the Justice and Development Party (AKP) Deputy, Mercan, who spoke for Prime Minister Erdoğan: 'Mr. Haleva's cultural and philosophical approach will not only benefit our Jewish Community but will bring serenity and peace to our country as a whole' (Yannier, 2002).

Jewish Community President Bensiyon Pinto used the installation of the Chief Rabbi as a platform to praise the Turkish Republic and to be explicit about the role of Jews as the positive face of difference. His words spoke of the ritual as a politically opportune moment to represent the community as the 'good minority' to an audience that included not only Turkish Jews, but also diplomats and other foreigners:

> Our community has contributed generously to various stages of European Union accession processes which the majority of Turkish people regard positively. We especially try to invite delegations who have the power to shape public opinion in their home countries so they can personally get to know our country and its people. We are happy to observe how effective and credible this strategy has proved so far. Our guests are impressed with what they see and experience. On their return they promote Turkey in various aspects. The common wish of people living in this country is to reach the level of modern civilizations as glorious Atatürk had himself signalled. I believe that cooperative efforts spent in Turkey's name toward attainment of equal member status in the European Union will be rewarded sooner than expected. We, the Turkish Jews, will continue to work towards this goal to the best of our ability.

These remarks champion Turkish Jews' cultural citizenship (Rosaldo, 1994), arguing for Jews in Istanbul to count as Turks while retaining their particular difference as minorities. Partridge (2008, p. 668) calls this 'exclusionary incorporation' by which '... previously abject beings become subjects, but in a way that preserves and even depends on their position as outsiders'. However, in a complicated twist, at Neve Shalom in 2002, Jews counted as Turks because of their exclusions, and by extension, Turkey counted as European, because of the ability – *vis à vis* the Jewish community's efforts – to recognize diversity in Istanbul. This site- or city-specific tolerance was reiterated during the April 2009 visit of US President Barack Obama who vowed to support Turkey's European Union admission bid, meeting with the Turkish Chief Rabbi and other religious leaders not in Ankara but in Istanbul.

If, historically, Istanbul's minorities kept a low profile through *kayadez* (Judeo-Spanish for 'quiet' (Bali, 2007)), Jews (or, more correctly, their officials) are increasingly changing tactics. They have successfully carried out semi-public appearances, such as that at Neve Shalom, and are moving onto increasingly visible stages in Galata (and elsewhere in the city). Galata offers Turkish Jews not a nostalgic return to the 'old neighbourhood', but, rather, a political platform they might otherwise be denied due to their small numbers. Political scientists distinguish between a 'politics of ideas' and a 'politics of presence', emphasizing how the former does not always address the needs of the marginalized: 'Political exclusion is increasingly ... viewed in terms that can only be met by political presence' (Phillips, 1994, pp. 77–78). The value of understanding a politics of presence is key to understanding the desire of collectives, such as Turkish Jewry, to claim cultural citizenship in the face of historical exclusion and demographic decline.

In 2003, I attended a rehearsal of Jewish schoolchildren who were busy learning songs in Judeo-Spanish, the language their Sephardic (Spanish), which many elderly Jews in Istanbul still speak today. The children, and other musicians, intellectuals, craftspeople and volunteers were engaged in preparations for Istanbul's first celebration of European Day of Jewish Culture, inaugurated in Europe 1996 and in which twenty-one cities now participate. I followed news about the transformation of Galata and its environs into an outdoor stage which would come alive with Jewish music, food, exhibitions and displays, with no small amount of disbelief.

How would the community that I had found so concerned with security open its doors wide open to the public? How would it – how could it – perform Jewishness on the streets of Galata? In fact, reports of the event mark its great success, and so it came as an ironic surprise when, just two months later, Neve Shalom synagogue was again attacked, this time by Turkish perpetrators of Islamist orientation.

Seemingly without irony, given the intolerant and even genocidal treatment of Jews in various episodes throughout European history and virulent anti-

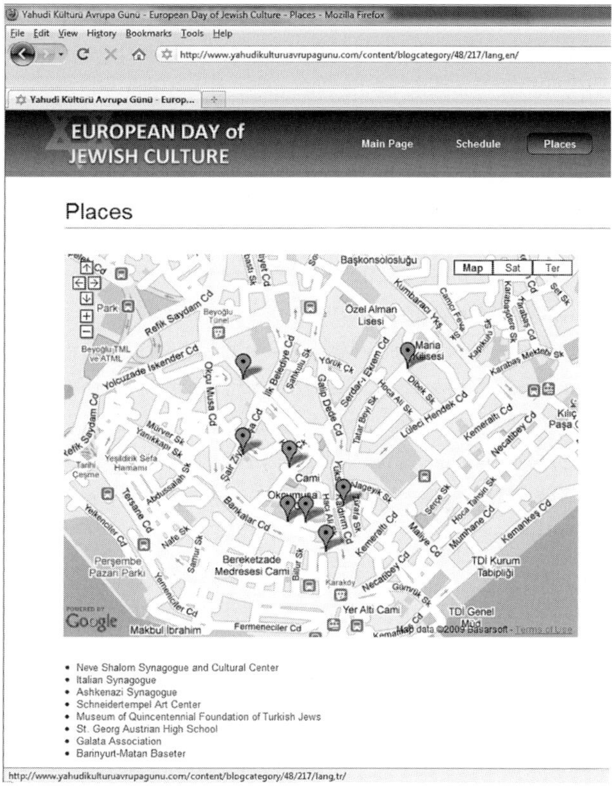

Figure 16.5. Map of Events for Istanbul's European Day of Culture (*Source*: http://www.yahudikulturuavrupagunu.com/)

Semitism in European member countries, European Union officials framed the attack in Istanbul as incompatible with European norms. The MidEast Dispatch Archive summarized the European Union and world reaction to the 2003 events in Istanbul: then European Union High Representative for Common Foreign and Security Policy, Javier Solana responded thus: 'The attacks close to the two synagogues are an unacceptable expression of intolerance and rejection that has to be eradicated'. In a similar vein, then-European Commission President Roman Prodi remarked: 'There is no Europe without tolerance, this episode is incompatible with our very idea of Europe'. In the streets of Istanbul attacks were likewise repudiated, as, just two days after the attack, the Beyoğlu Municipality staged a 'Tolerance for Peace' march to condemn terrorism and to celebrate an 'International Day for Tolerance'.[5] Thus, Istanbul, and this neighbourhood specifically, remains a symbolic centre of European imaginations of Turkey (and Turkey's imaginings of its own European-ness) *vis à vis* recognition – and tolerance – of diversity.

Following the 2003 bombing, political scientist (and Turkish Jew) Şeyla Benhabib invoked another Istanbul neighbourhood as a model for coexistence:

> In Istanbul this past summer, an Israeli friend and I visited a synagogue in Ortakoy (*sic*), a neighborhood on the shores of the Bosporus. Ortakoy is home to a mosque, a synagogue and a Greek Orthodox church, which huddle together between the sea and gently sloping hills…

> During my childhood in the 1950's, the number of Jews in Istanbul alone was 80,000. With settlements in other cities, the size of the Turkish Jewish community was well above 100,000. With the founding of the state of Israel and then the growing political instability of Turkey in the 1970's and 80's, many Jews started leaving. The Jewish population in Turkey now numbers about 30,000.
>
> Yet the presence of the Jews in Turkey cannot be measured in numbers alone. They are a testament to the peaceful coexistence of Jews and Muslims… This is something the murderous forces of Islamic terrorism, who attacked synagogues in other areas of Istanbul on Saturday, would like to obliterate.
>
> To be sure, some of Turkey's other minorities – the Greeks, Armenians and Kurds, who unlike the Jews, have had territorial claims on Turkish lands – have fared far less well. What matters now, though, is which historical model Turkey will be encouraged to embrace… Ortakoy remains a beacon of hope for those … fighting against the nihilistic message being spread by terrorists. The synagogues of Istanbul will be built again. No one should doubt it. (Benhabib, 2003)

The synagogues of Istanbul have been rebuilt, with even more stringent security. In August 2009, I attended a luncheon at the Ortaköy synagogue in honour of a friend about to marry in Neve Shalom. Where the Ortaköy synagogue once had a lovely open courtyard, now it is capped with industrial-strength glass and metal to prevent the possibility of attacks from above. Invitations to the wedding, a lovely design based on an historic image of the Galata Tower, exhorted guests to bring proper identification in order to pass through security at the synagogue. The bride recounted a story about wanting to take photographs in front of the synagogue but of quickly being reprimanded by security guards for 'loitering'. Instead, she and her fiancé posed in front of the Galata Tower. As such, the actual lived space of Istanbul Jewry seems far removed from, or at least more complicated than, visions of Istanbul as a 'beacon of hope'. Nonetheless, the cultural displays I have described in Galata need not actually be '… representative of the cultural orientations and political aspirations…' of the minorities who perform them, rather, they are a way of 'articulating utopias' (Soysal, 2001, p. 22) that speak to the larger society in which they hope to participate, despite their marginality. For example, the Beyoğlu Jewish Rabbinate Foundation recently conducted a survey of Turkish opinions about minorities with funds from the European Commission. Despite the fact that the survey results indicated that 42 per cent of Turks 'do not want to have Jewish neighbours' (with an increase to 61 per cent among those who identify themselves as Muslims)[6] – or perhaps because of it – evidence of Jews engaging in the politics of presence in Istanbul underscores the (perhaps utopian) aspirations of this tiny community to count for something and to stand 'for Europe'.

The creation of the Jewish Museum, the Chief Rabbi's investiture, and Istanbul's participation in the European Day of Jewish Culture have created stages upon which Turkish Jews perform a politics of presence in Galata and its environs. Despite the bombs that threaten to force Istanbul's Jews back in the 'closet' just as they were finally coming out, the community has celebrated the European Day of Jewish Culture every year since its inception in 2003. Last year's celebration fell on the anniversary of the 1986 attack on Neve Shalom and organizers of the 2009 Day of Jewish Culture; on their promotional website, acknowledged this tragic coincidence with the following text: 'We commemorate with respect, the victims of this incident that coincides this year with European Day of Jewish Culture that is organized simultaneously all over Europe'. Rather than making a retreat, making alliances with the European Union and European Jewish communities has allowed Istanbul's Jews to represent themselves as counting, symbolically, beyond what their numbers might predict.

Notes

1. İstiklal Caddesi, once known as Rue de Pera, connects Taksim, a hub of contemporary transportation, with Tünel, a short mid-nineteenth century era funicular, built by a French engineer, and the sloping streets that lead to the Galata Tower and Karaköy.

2. The 'mosaic' model is possibly being replaced by the metaphor of '*ebru*' – Turkish paper marbling – in which identity is seen as porous and mixed (Altınay, 2007).

3. 'Kamondo' is the Turkish spelling; the name is also frequently written as 'Camondo'.

4. GCI 'General Agenda' press release; 1991: The Records of the Quincentennial Foundation; Box 1; American Sephardi Federation at the Center for Jewish History.

5. http://www.anadoluajansi.com.tr/ 17 Kasım 2003 Pazartesi. Accessed 14 January 2010.

6. See http://www.turkyahudileri.com/images/stories/dokumanlar/perception%20of%20different%20identities%20and%20jews%20in%20turkey%202009.pdf. Accessed 21 January 2010.

References

Altınay, A. (2007) Ebru: reflections on water, in Durak, A. (ed.) *Ebru: Reflections of Cultural Diversity in Turkey.* Istanbul: Metis.

Bali, R. (2001) *Les relations entre Turcs et Juifs dans la Turquie moderne* [The Relations between Turks and Jews in Modern Turkey]. Isis: Istanbul.

Bali, R. (2007) The alternative way to come to terms with the past/Those who try to forget: Turkey's Jewish minority (unpublished lecture). Istanbul.

Bartu, A. (1999) Who owns the old quarters? in Keyder, Ç (ed.) *Istanbul: Between the Global and the Local.* Lanham, MD: Rowman and Littlefield.

Benbassa, E. and Rodrigue, A. (2000) *Sephardi Jewry: A History of the Judeo-Spanish Community, 14th–20th Centuries.* Berkeley, CA: University of California Press.

Benhabib, S. (2003) In Turkey, a history lesson in peace. *New York Times* (online), 18 November. Available at: http://www.nytimes.com/2003/11/18/opinion/18BENH.html?pagewanted=1. Accessed 21 January 2010.

Blommaert, J. and Verschueren, J. (1998) *Debating Diversity.* London: Routledge.

Braude, B. and Lewis, B. (eds.) (1982) *Christians and Jews in the Ottoman Empire: The Functioning of a Plural Society.* New York: Holmes & Meier.

Brown, W. (2006) *Regulating Aversion: Tolerance in the Age of Identity and Empire.* Princeton, NJ: Princeton University Press.

Bunzl, M. (2003) Austrian Zionism and the Jews of the new Europe. *Jewish Social Studies*, **9**(2), pp. 154–173.

Dorn, P.J. (1997) Change and Ideology: The Ethnomusicology of Turkish Jewry. PhD Dissertation. Indiana University.

Fleminger, M. (2003) Istanbul: The Jewish municipality in the Turkish million-metropolis. *Juedische Allgemeine*, 26 March. Available at: http://kehaberler.blogspot.com/2003/05/bu-satn.html. Accessed 24 January 2010.

Gruber, R. (2002) *Virtually Jewish: Reinventing Jewish Culture in Europe*. Berkeley, CA: University of California Press.

Gültekin, A. (2009) *Istanbul 2010 European Capital of Culture Program*. Istanbul: ECOC Agency.

Houston, C. (2001) *Islam, Kurds and the Turkish Nation State*. Oxford: Berg.

Kamhi, J.V. and Ojalvo, H. (1997) *Exhibit of the Quincentennial Foundation*. Istanbul: The Quincentennial Foundation.

Karmi, I. (1996) *The Jewish Community of Istanbul in the Nineteenth Century: Social, Legal and Administrative Transformations*. Istanbul: Isis.

Kastoryano, R. (1992) Du millet à la communauté: les Juifs de Turquie. [From millet (minority religious group) to community: the Jews of Turkey]. *Pardes*, **15**, pp. 136–157.

Komins, B.J. (2002) Cosmopolitanism depopulated: the cultures of integration, concealment, and evacuation in Istanbul. *Comparative Literature Studies*, **39**, pp. 360–385.

Kugelmass, J. and Orla–Bukowska, A. (1998) 'If you build it they will come': Recreating an historic Jewish district in post-Communist Kraków. *City & Society*, **10**(1), pp. 315–353.

Levy, A. (2002) *Jews, Turks, Ottomans: A Shared History, Fifteenth through the Twentieth Century*. Syracuse, NY: Syracuse University Press.

MidEast Dispatch Archive (2003) World reaction to Istanbul bombings. 19 November. Available at: www.tomgrossmedia.com/mideastdispatches/archives/000155.html. Accessed 26 January 2010.

Mills, A. (2006) Boundaries of the nation in the space of the urban: landscape and social memory in Istanbul. *Cultural Geographies*, **13**, (3), pp. 367–394.

Mills, A. (2007) Gender and Mahalle (Neighborhood) Space in Istanbul. *Gender, Place & Culture,* 14, no. 3, pp. 335–354.

Örs, İ. (2006) Beyond the Greek and Turkish dichotomy: the rum polites of Istanbul and Athens. *South European Society & Politics*, **11**(1), pp. 79–94.

Özyürek, E. (2006) *Nostalgia for the Modern: State Secularism and Everyday Politics in Turkey*. Durham, NC: Duke University Press.

Özyürek, E. (2007) *The Politics of Public Memory in Turkey*. Syracuse, NY: Syracuse University Press.

Partridge, D. (2008) We were dancing in the club, not on the Berlin Wall: black bodies, street bureaucrats, and exclusionary incorporation into the new Europe. *Cultural Anthropology*, **23**(4), pp. 660–687.

Peck, J.M. (2006) *Being Jewish in the new Germany*. New Brunswick, NJ: Rutgers University Press.

Phillips, A. (1994) Dealing with difference: a politics of ideas, or a politics of presence? *Constellations*, **1**, pp. 74–91.

Potuoğlu-Cook, Ö. (2006) Beyond the glitter: belly dance and neoliberal gentrification in Istanbul. *Cultural Anthropology*, **21**(4), pp. 633–660.

Rosaldo, R. (1994) Cultural citizenship and educational democracy. *Cultural Anthropology*, **9**(3), pp. 402–411.

Seni, N. (1994) The Camondos and their imprint on 19th-century Istanbul. *International Journal of Middle East Studies*, **26**, pp. 663–675.

Soysal, L. (2001) Diversity of experience, experience of diversity: Turkish migrant youth culture in Berlin. *Cultural Dynamics*, **13**(1), pp. 5–28.

Soysal, Y., Bertilotti, T. and Mannitz, S. (2005) Projections of identity in French and German history and civics textbooks, in Schissler, Hanna and Soysal, Y.N. (eds.) *The Nation, Europe, and the World: Textbooks and Curricula in Transition*. New York: Berghahn.

Toktaş, Ş. (2006) The conduct of citizenship in the case of Turkey's Jewish minority. *Comparative Studies of South Asia, Africa and the Middle East*, **26**(1), pp. 121–133.

Tuval, S. (2004) *The Jewish Community in Istanbul 1948–1992*. Jerusalem: WZO.

Varol, M. (1989) *Balat-Faubourg juif d'Istanbul*. Istanbul: Isis.

Weiss, A. (1975) *The Jewish Community of Turkey*. Jerusalem: Center For Jewish Community Studies.

Yannier, E. (2002) Hahambaşı Rav İzak Haleva'nın İsad Töreni'nden... [From the Instalment Ceremony of the Chief Rabbi, Rav İzak Haleva ...]. *Shalom*, 25 December, p. 2.

Yardımcı, S. (2007) Festivalising Difference: Privatisation of Culture and Symbolic Exclusion in Istanbul. *EUI Working Papers, Mediterranean Programme Series*, pp. 1–26.

Chapter 17

Future(s) of the City: Istanbul for the New Century

Levent Soysal

ISTANBUL
*i*rresistible
*s*upersized
*t*itillating
*a*dmirable
*n*oisy and
*b*eautifully
*u*gly
*l*abyrinth
(t-shirt epigram, *Mavi Jeans*)

'Nothing here is perfect; everything, even failure, is magnificent'.
(Herzfeld, 2009, p. 7)

'[Petersburg] differs from all other European towns by being like them all'.
(Alexander Herzen, cited in *Natasha's Dance*, Figes, 2002, p. 9)

Introduction to the Future(s) of the City[1]

There are numerous potentially productive ways to enter a discussion on the future(s) of Istanbul. One could begin with a demographic and spatial expansion of the city, the population now standing at an astronomic fifteen million (officially twelve million; nearly one fifth of Turkey's population) and the urban sprawl literally covering the entire land between the two adjacent cities of Kocaeli and Tekirdağ, and with no end to growth in sight. One could speak about the city's economy as a prelude and possibly ponder on what it means for the city and Turkey that Istanbul's contribution to *gayrı safi katma değer* (national gross value added) is nearing 30 per cent, not only greater than any other city but any region of Turkey. One could also engage in a deliberation on the political impact of a mega-city with seventy (out of five hundred and fifty) representatives in the assembly and home to the largest number of political, cultural, and recreational associations, headquarters of all the major banks, corporations and media conglomerates.

All these points of entry provide justifiable preambles for inquiry into the future of Istanbul and allow for predictions on what the future holds for the economies, politics and societal prospects of the city, leading to fruitful projections that outline trends and make prognoses for the usual urban predicaments such as traffic jams, housing shortages, financial and democratic deficits, poverty and homelessness. They all concern consequential challenges posed by the contemporary condition of urbanity in Istanbul. They are, however, limited in their scope and close the future of Istanbul onto itself; they carry the danger of confining 'problems' and 'solutions' to the national domain. In fact, this has been the first and foremost handicap to most domestic debate on Istanbul, though the signs that indicate a move away from narrow definitions of 'urban problems' and 'national solutions' are in evidence – as chapters in this volume clearly highlight.

My intention is to locate Istanbul, as well as the discussion of its future, in a framework that takes the world as its analytical starting point. Since Saskia Sassen's *Global City* (2001) was first published in 1991, we are more attentive to the fact that the global character of metropolitan centres intensifies at the expense of national configurations. Major urban centres increasingly project themselves beyond the nation-state within which they are located. Cities come to prominence no longer by being the capital or industrial hub of their nation-states and seek their fortunes and fame elsewhere. Increasingly, they fashion themselves as 'world' or 'global' cities and, under tutelage of marketing companies, they strive to develop themselves as 'brands', as cities of culture, art, fashion, science, sports and various other forms of entertainment. This is true not only for cities such as New York, Los Angeles, London and Paris, which have always been perceived as detached from their nation-states, but also for Bilbao, Rotterdam, Berlin and Istanbul.

Specifically in Europe, the brand 'European Capital of Culture' is one instance of being and projecting an image of a city beyond the conventional geography of local or national import and achieving global stature. As its counterparts have done before, Istanbul braced itself for the title and the status at hand, with a budget of 300 million Turkish lira (about US$200 million) and potpourri of spectacles and urban projects – 451 in all.[2] The project is embraced – and presented – as a momentous feat, envisioned as a glorious comeback; it opened with seven popular concerts and artistic performances in seven districts of Istanbul (for Istanbul is known as a city built on seven hills) and a towering fireworks display on the shore of the Golden Horn.[3] On the other hand, the Europeanness of Istanbul necessarily rehashes the prolonged debate on the commensurability of civilizations (that is, compatibility of Turkey – or more precisely of Islam – with European culture and values).[4] Though a cliché, the idea that Istanbul is a bridge between the East and the West, Asia and Europe is dangerously fashionable and conveniently marketable, as Fatih Akın's film *Crossing the Bridge* (2005) made unmistakably apparent – it inevitably invites engagement (see Deniz Göktürk, Chapter 10).

I begin my discussion where Çağlar Keyder (Chapter 1) stopped, 'embark[ing] on the path to globalization' or 'informal globalization' (Keyder, 1999) in the post-1980s, particularly in the last decade of the past century: the city's 'success in this path' is 'undeniable in the new century' and 'grounded on solid institutional foundation'. Entering the twenty-first century, or the new millennium, Istanbul has been successfully elevated to the league of global cities.

In his account, Keyder succinctly outlines the political economy of the process that relies heavily on real-estate development and city-marketing, as well as the fragilities of being a global city in the crisis-ridden world economy (see also Introduction to this volume). My goal here is to map the cultural topography of Istanbul, the new Metropolis, and its future(s), and introduce what I consider the key features of what makes a contemporary city a City today – be it called Metropolis, Global City, or World City. In order to do so, I first locate the latest transformations of Istanbul in the framework of the process known as Europeanization and the changes at the world level.

For that I turn to Norbert Elias's notion of the 'civilizing process' (1939 [1982]). I argue that Istanbul is undergoing a civilizing process, discovering and establishing ways of doing things in a civilized manner, adopting and enacting at institutional levels the codes, standards and norms of being civilized and furnishing the urban landscape of the city with monuments, buildings, bridges and emblems, logos and signs that imply the contemporary state of civilization. If one source for what is being adopted and enacted is the inventory and models provided by what Lechner and Boli (2005) call 'world culture',[5] the other is the repertory of codes, conventions and norms that comprise the covenants of the European Union, as well as supranational organizations such as the United Nations and UNESCO.

This civilizing process, however, only seemingly resembles the linear models of change prescribed by modernization or Westernization, in which the West always remains the West, the periphery is caught up in a non-ending performance of catching up with the centre, and the relationship between periphery, say Istanbul, and the West is never 'coeval' (Fabian, 1983). The same civilizing process that is overwhelming Istanbul, I suggest, is simultaneously facilitating the making of the new Europe, the West, so to speak. In that, Istanbul's transformative venture is co-terminus with those of its counterparts in Europe and the world at large.

The second task I undertake is to identify the elemental aspects of the change that Istanbul is undergoing within this civilizing process. What are the signs and signatures (buildings, monuments, spectacles) that mark the 'new' metropolitan topography? What are the formative visual elements and aesthetic traits of public events that saturate the city space? What do the new spectacles say as to the globalization of everyday lives and the salient movements between the margins and the core of Europe? What are the visions that accompany this new role for a city with a long historical memory, as well as a claim to be located between East and West, past and present? I present my responses to these questions under such headings as Gentrification, Heritage Displays, DesignCity, SpectacleCity, Culture Incorporated and Sociality Amplified. These I propose as meta-analytical categories that confer a comparative basis for capturing the similitude and difference, as well as locating presences and absences, in the contemporary processes that shape cities East and West. They serve to delimit – and accentuate – the elemental facets of what it means to be a world city; and, the changes captured within these categories are symptomatic of the civilizing process underway in Europe and beyond.

Europeanness, Civilizing Process and Future(s)

Writing for Associated Press on the Picasso in Istanbul exhibition, Benjamin Harvey quotes 'the patroness of the Sakıp Sabancı Museum (SSM), the head of a multibillion-dollar Turkish conglomerate', Güler Sabancı, saying that 'she hopes the show will accelerate Turkey's cultural shift from the ancient, Islamic and Oriental to the cutting edge, contemporary and European' (Associated Press, 24 November 2005). Admittedly, the opinion on what this civilizing project involves is varied. However, as seen from within and from without, it is definitely a major shift in the scale of the change that took place at the turn of the twentieth century and the goal – once again – is to catch up with the West, with Modernity as such. Europeanization, as spelled out by the 'patroness' of SSM, or is implicit in Associated Press wire, becomes the new brand name (or new disguise) for the familiar, and timeless, prescription of modernization.

This paradigm finds its most succinct form in Bernard Lewis's *The Emergence of Modern Turkey* (1968). Roughly premised upon the proposition that the Ottoman Empire had started to 'decline' after its failed attempt to conquer Vienna in the seventeenth century, Lewis argues that the reformer sultans of the Ottoman Empire, Selim the Third and Mahmud the Second, were the first to embark on a mission to modernize the institutions of the Empire.[6] After the collapse of the Empire, the newly founded Republic, so the story goes, took on the task of modernization, with zeal, through a series of radical reforms, involving not only the institutions but also 'culture'.

This version of 'Ottoman/Turkish' history, deeply colours both the scholarly and popular imagination within and without Turkey. Lately scholarly efforts are under way to break from the rudimentary story of modernization/Westernization and explore the process of change that shaped contemporary Turkey not as an isolated case, but a nation-state located in the history, politics and economies of the world. One such effort is relevant for my account here. In a special issue of *Toplum ve Bilim* (Society and Science, no. 84, 2000), a social studies journal published in Turkey, an interdisciplinary group of scholars makes a preliminary but concerted effort in approaching the so-called process of modernization through the lens of Elias's concept, the 'civilizing process'. Mostly focusing on the Republican period, the essays deal with topics ranging from rationalization of home economics to civilizing the body, from the question of violence to national character and memory.

Rather than cataloguing the issue in detail, I draw attention to two essays that make arguments pertinent to my own. In his comparative treatment of Turkish and Japanese modernization, Selçuk Esenbel (2000, p. 19) argues against problematizing Turkey (and Japan) within the East-West paradigm and felicitously states that 'in the complex world of today, we are all, in different ways, the founders and uninvited guests of the same modernity', placing the experiments and experiences of modernity in the rest of the world on the same analytical level with the West. Ayşe Öncü (2000, p. 15), on the other hand, asserts that 'the orientalisms and primitiveness produced and served for consumption by the culture industries today' are embedded in globally situated power relations and forcefully suggests the newness of today's civilizing process.[7] In bringing in Elias and the 'civilizing process' into the discussion of 'Europeanization' I proceed in the same spirit.

As Elias puts it, a civilizing process attests to changes in 'a wide variety of' things: being civilized may compel changes in 'the type of manners' (there have been repeated calls by prominent politicians, journalists and TV personalities for Turkish people to act in a manner appropriate for Europe). It may necessitate the transfiguration of the extant 'religious ideas and customs' (the new emphasis on the equality of religions and inter-faith dialogue is a case in point) or it may facilitate the transformation of 'the manner in which men and women live together' (lately there has been renewed attention to the schooling of girls and condemnation of arranged marriages and honour killings). The revamping of 'the form[s] of judicial punishment' may be yet another manifestation of being civilized (Turkey abolished capital punishment and is debating the revision of penal laws and constitution for compatibility with European codes). Last but not least, mundane everyday practices, such as 'the way in which food is prepared', may need to be abandoned (European health codes require changes in the preparation and sale of *kokoreç*, a popular fast-food made of grilled intestines, and popular ways of displaying fish, meat and other perishable foodstuffs in open markets) (Elias, 1982, p. 5). These all point to standardization of norms, codes and procedures across Europe in order to create a coherent (but not necessarily without conflict) civilized space.

'Strictly speaking', as Elias states (1982, p. 5), 'there is almost nothing which cannot be done in a "civilised" or an "uncivilised" way'. In this sense, one could, as the officials of the European Union and their local informants do, find a variety of ways in which Turkey (and the Turks) need to civilize themselves. Again citing Elias, one could say, at a more macro level, the concept of civilization 'expresses the self-consciousness of the West', [summing] up 'everything in which Western society of the last two or three centuries believes itself superior to earlier societies or "more primitive" contemporary ones' (*Ibid.*).

What is more significant is the connection Elias sets up between the civilizing process and state formation. According to him, 'mechanisms of integration' which come with state formation (in Elias' case the absolutist state) – that is 'the whole reorganization of human relationships' – '[have] direct significance for the change in human habitus, the provisional result of which is our form of "civilized" conduct and feelings' (*Ibid.*, p. 366). 'Only if we see [this] compelling force with which a particular social structure, a particular form of social interweaving, is pushed through its tensions [and conflicts] to a specific change' (*Ibid.*, p. 367) can we grasp the symptoms and procedures of the civilizing process Turkey – and for that matter the West itself – is undergoing. Europeanization of the new EU member states is one of the forms that this new civilizing process assumes. The transformation of Europe from a collection of well-established national welfare states (*Sozialstaat* or *soziale Marktwirtschaft*, as Germans call it) into a competitive free-market economic zone is another (the struggles over the European budget and constitution are signs of the difficulties and obstacles to civilizing missions). Finally, the dismantling and re-formation of Iraq, Afghanistan, Somalia, and many other Middle Eastern, ex-Soviet, and African nation-states is another.

Here I want to reiterate two elemental features of state formation which underwrite the contemporary civilizing process: one is the 'shift to entrepreneurialism' (Harvey, 2001) and the related 'privatization of the executive branch' (Sassen, 2006), both of which carry momentous repercussions on how cities and states are governed. The second is the universalized discourses of diversity, human rights and democracy, which are expounded by supranational organizations (UN, EU, UNESCO and others) and facilitate the new regimes of rights and the reorganization of state-citizen relations (Beck and Grande, 1997; Soysal, 2002). These developments provide the models to be approximated on the way to become 'civilized'.

The civilizing process undertaken worldwide is not of course without its hierarchies, the West waving the flag of civilization ahead of the rest. Nonetheless, the civilizing process of the Turkish kind is not particular but can be generalized to the West. In the words of Elias (1982, p. 380) 'the "trend" of the movement of civilisation is everywhere the same', however 'differences may arise in particular cases'.

The next section captures the dimensions that enable the rise of Istanbul to the global stage and maps out its new cultural topography. The changes that mark Istanbul's architectural and cultural landscape reflect upon the state of its civilizational progress and carry the imprint of (now normalized and taken-for-granted) models and trends that have expedited the ascent of cities to the world stage.

Istanbul is Now a Metropolis!

'Istanbul is now a Metropolis', confirmed the banner that had adorned the city's celebrated Taksim Square for about a year after the passing of the millennium. Then came a series of mega events, all celebrated as milestones in Istanbul's march to world prominence: NATO Summit, World Congress of Architects, Formula One Races and the Tenth Istanbul Biennial. 'Istanbul Won' was the public sentiment and the standard response by jubilant officials, organizers and the media after every event. Istanbul became the winner by offering its hospitality, rich culture and diversity, and showing 'the world' that 'we [the Turks or Turkey]' have the capacity to stage such grand events. Then Istanbul made it to the cover of *Newsweek* (29 August 2005) as the coolest city on earth. This series of events more or less recapitulates Istanbul's advance from being Turkey's cultural and economic capital to becoming a world-order city. As the banner prophetically declared, Istanbul has become a Metropolis – in the eyes of its admirers, foreign and native.

If compared with New York, Tokyo and London, one might see the label Metropolis – or World City – attached to Istanbul as a mere intention or a desire to be in a different league. The *prescriptive* models elaborated by scholars of globalization reinforce such conclusions by setting benchmarks for being proper global/world cities – benchmarks that emphasize their control and command functions in the world economy (Friedmann, 1986, 2002; Sassen, 2001).[8] The models that shape our understanding assume hierarchies and levels to be achieved as a city fulfils one or the other criteria on the road to being a global city.

Istanbul's label, however, is not a mere aspiration. Istanbul is not a financial centre on a global or regional level but is a World City, *par excellence* (though the government is intent on making it a financial centre by bringing Turkey's central bank from the national capital Ankara to Istanbul). For World City is now an *ascriptive* label and simply denotes an extant 'global present' (Flusty, 2004) in which we are all active participants in making and enacting in its myriad diverse and conflictual forms. Rather than measuring cities by the prescriptions of global city models, it is more productive to accept their ascriptive claims and explore them 'in the fullness of their particular linkages with the worlds outside their boundaries', as Michael Peter Smith (2001, p. 71) suggests.

Of course, a world city as such is specific to the neoliberal milieu and has its distinguishing and emblematic features. I now map out the newly emerging iconography and the transformation of the lived space in Istanbul since the 1980s – the imaginary timeline that marks the incorporation of Turkey into the 'globalizing' world of culture and economy. The categories I use – gentrification, heritage displays, design city, spectacle city, culture incorporated and sociality amplified – are elemental components that constitute the cultural content of the world city today. Developments in each of the fields delimited by these categories are taken for *bona fide* steps towards a civilized state of affairs; and they are closely coupled with institutionalized practices promoted and legitimated by the EU and the Council of Europe and the UN and UNESCO.

These categories are intertwined in complicated ways but I treat them as discrete features as a way to capture their distinctive contributions to the totality of the world city. The order in which I list these key features does not necessarily correspond to an order of importance; and in the examples below the emphasis is on scope and scale rather than the specificity of individual events and enactments.

Gentrification and Development, or as it is expressed in Turkish, *Kentsel Dönüşüm* (literally, urban transformation) – a much more poetic and seemingly less loaded phrase than gentrification – is the name of the most fundamental and lucrative game in town. Since 1980, the urban *terra firma* of the city, especially the public land, is rapidly turning into real estate. What is being built is quite varied and ranges from luxury high-rise residences, office towers, gated communities, shopping and entertainment centres and five-star hotels to parks, roads, urban infrastructure and public housing complexes. The construction frenzy is readily visible, resulting in ever-rising real-estate prices and exceptional traffic congestions.

Sunday newspaper real-estate supplements are full of advertisements for a peaceful and comfortable life in newly built housing estates or residences, with names like *Mashattan*, from Maslak and Manhattan, and *Bahçeşehir* (Garden City) and its Ottoman sounding version, *Şehr-i Bahçe*. The Mayor of Istanbul Metropolitan Municipality holds frequent press conferences and announces one grand project after another. One day the residents of Istanbul are given the happy news that their city will host the second tallest skyscraper in the world, called Dubai Towers Istanbul, with the compliments of Gulf investment. Another day the Mayor's office announces plans for building a safari park; an aquarium and research centre; a statue of a whirling dervish on one of the Princess Islands, visible from everywhere and with enough space under its skirt for shopping, entertainment and fast food joints; the 'rehabilitation' of Sulukule, a historic Roma neighbourhood, moving its inhabitants from the centre of the city to a rather faraway place and away from the entertainment districts; and wish lists for turning historic buildings into luxury hotels. Not all these projects, of course, were realized. The Courts, for instance, blocked the building of Dubai Towers but Istanbul now has Sapphire, a residence and office complex built with domestic capital, competing to be the tallest building in town.

Law number 5366, known for short as *Kentsel Dönüşüm Yasası* (Urban Transformation Act, adopted in 2005), has been instrumental in accelerating the speed with which the gentrification is taking place.[9] Cihangir and Galata, two neighbourhoods overlooking the Bosphorus and next to Beyoğlu, the entertainment and culture centre of Istanbul, have already undergone gentrification in the style of East Village in New York or Prenzlauerberg in Berlin. It is almost impossible to find property at reasonable prices in the district of Tarlabaşı, another area neighbouring Beyoğlu, next in line for gentrification. Fener and Balat, historically multi-ethic and multireligious neighbourhoods of the city on the shores of the Golden Horn, will substantively change once the plans for the 'rehabilitation' are implemented. Two major ports of the city, Karaköy and Haydarpaşa, are up for redevelopment as entertainment centres, shopping malls, hotels, office towers and residences. The agreements for the first, which will be called Galataport, were signed between the municipality and a consortium of Turkish and Israeli companies but turned down by the courts. Public debate regarding the second is ongoing but nobody expects the projects to be abandoned.

In addition to private enterprises and municipal governments, supranational entities are heavily implicated in this process. For instance, the EU, European Commission and UNESCO, all partake in the development and execution of the programme for the 'rehabilitation' of the Fener-Balat district, along with the local (Fatih) municipality, central government and neighbourhood NGOs. The EU has allocated a budget of seven million euros for implementation of the programme.[10] As such, the gentrification of Istanbul is not simply a local matter; it involves European and other supranational actors and is closely connected to processes of Europeanization and globalization.

Public opinion on gentrification is not oppositional but mainly affirmative. Some among the intelligentsia and professional associations of architects and civil engineers have always raised their voices against policies of unrestrained gentrification and drawn attention to its social consequences, but these voices have remained marginal and ineffective against the potency of developmental dogmas. Only recently, particularly with the onslaught of mega projects, continuing privatization of public land and the heavy-handedness of government-led gentrification, have contestation and dissent come to play a role in making and implementing public policy (Özlem Ünsal and Tuna Kuyucu, Chapter 3).

Heritage Displays are a corollary of gentrification. In Istanbul today every 'historic' building, that is, every building that could be attributed with some historical significance, is illuminated at night and opened to the gaze and use of tourists during the day. Galata Tower and Kız Kulesi, two major historic landmarks, are now accessible and house fancy restaurants. Eminönü and Sultanahmet, the two major tourist districts since the 1960s and home to Topkapı Palace, the Blue Mosque, Hagia Sophia, Covered Bazaar and Egyptian Bazaar, and churches and synagogues towards the Golden Horn, have now come to be known as *Tarihi Yarımada* (Historic Peninsula). On and off, newspapers report of 'promising' projects to turn this part of the city into a history-museum park.

This historical refurbishing of the city began in the early 1980s with the conversion of some dilapidated palaces into cafés with winter gardens, painted in pastel pink and yellow and continued with the excavation of the city's Ottoman-

Greek-Jewish past and offering them for public consumption as nostalgia (see Marcy Brink-Danan, Chapter 16), and reached its height with the construction of Miniaturk, a theme park with exact replicas of palaces, castles, buildings and bridges from all periods of Ottoman and Turkish history and from the west of the empire in the Balkans to the east of the Republic in Anatolia. With Miniaturk, heritage and history are now brought home (see İpek Türeli, Chapter 6).

In addition to these popular instances of displaying heritage, there are numerous tangible historic preservation projects actively pursued and executed over the wide expanse of Istanbul. Under the tutelage of the Istanbul 2010 Agency, a substantive number of projects are being implemented, ranging from conventional restitution and restoration projects (e.g., restoration of sections of Topkapı Palace, Hagia Sophia and various other hitherto forgotten or unused historic buildings) to the creation of archaeological parks, conversion of old industrial buildings into cultural centres (e.g., Hasanpaşa Gashouse Centre) and multicultural heritage projects (e.g., exhibitions on the works and biographies of Greek and Armenian architects of Istanbul), as well as the opening of several new museums (see Oğuz Öner, Chapter 15).[11] As with gentrification, it is necessary to note the expediency of the EU, Council of Europe and UNESCO, and the documents and discourses they provide on tangible and intangible cultural heritage, and on the legitimation of heritage preservation programmes.[12]

DesignCity is another corollary of gentrification. If heritage displays are an attempt to locate nostalgia for consumption, DesignCity is a means to capture the present and race for the future, propelled by the increased salience of neoliberal and European frameworks for creativity, innovation and skills, which finds its expression in the increasingly influential talk of 'creative industries'. In the juxtaposition of design and city, there lie the implications of the recent premium placed on design (as in designer everything) for the life, culture, and physical organization and planning of the city. In this sense, the 'city' in question, Istanbul in our case, is both the place with its streets, bridges, and buildings, and the concept realized in designs of all sorts (social theory, film, advertisement, photograph, dance, music, web, consumer products, and urban planning).

In Istanbul, as in every corner of the globe, designer products abound and designer places are in the making. The buzz is that almost all architects of world renown are most eager to place one of their trademark buildings there. The names of Zaha Hadid (designed the regeneration plan for Kartal, an industrial district on Anatolian side), Frank Gehry (designed a towering museum-cultural centre complex to be built overlooking the Golden Horn), Santiago Calatrava (a bridge imitating his style is planned across the Golden Horn), Ken Yeang (designed an ecological park between two lakes of Istanbul, Büyük and Küçük Çekmece) and other royalties of architecture hover in the air when the topic is a new monumental construction project, such the aforementioned Galataport or a museum, a bridge, or a hotel.

One does not need to wait for the coming of celebrated architects to encounter designer spaces in Istanbul. As the city is seeking its future in becoming a brand, more and more designer places happen to be part of everyday life. Many, like the newly built *Fransız Sokağı* (French Street) with its high end cafés, restaurants and boutiques, are listed in *Style City Istanbul*, a recent addition to the family of Istanbul guides devoted to paying attention to style and design. The trend among the new private universities is to convert old industrial buildings into designer spaces. Kadir Has University is now located in an elegantly renovated old tobacco factory. Bahçeşehir University is the proud occupant of a refurbished textile warehouse. Bilgi University restored an old power station into a centre for visual arts and education, *à la* Tate Modern, under the brand *santral*Istanbul. The fashion for design is sure to stay here for a while.

SpectacleCity underwrites the re-organization of public spaces and the proliferation of public spectacles in metropolitan centres. On the one hand it suggests the emergence of a new performance genre located in the city. Large and small, the totality of spectacular events constitutes the cultural fabric of the contemporary metropolis, provides it with an aura of cultural creativity and finesse, and often aims to facilitate formation of community and solidarity. With the intensification of media attention and global travel, public spectacles, particularly those that meet dimensions of grandeur, have become indispensable and identifying features of the contemporary metropolitan order. They provide the raw material for creating a brand name for the city by catering for contemporary 'lifestyles' and for the demands of the new economies of consumption.

On the other hand, SpectacleCity suggests that the city has become a novel spectacle in its own right. This may not be considered a new phenomenon. After all, Baudelaire and Walter Benjamin's *flâneur* testify to the centuries old tradition of experiencing the city as a spectacle and Harvey (2006, p. 210), speaking of Paris at the turn of the nineteenth century, states that 'spectacle, even that of the city itself, has always been fundamental to urban life'. However there is something substantially particular about the new metropolitan order: not only 'capitals of modernity' (Harvey, 2006) like Paris, but every city emulates spectacle and competes with the spectacular performances taking place on its terrain. Today the ascription of the label world city necessitates the stupendous architectural monuments designed by brand name architects, diversity displays that duly commemorate metropolitan plurality and renovated (and scaled up) living spaces and designer lifestyles (see Soysal (2005) for further elaboration of spectacle and city). Furthermore, European Capital of Culture programmes and other European funding schemes contribute substantially to the spectacularization of urban spaces by facilitating the construction of SpectacleCities across Europe as a way to create an aura of Eurpeanness (Sassatelli, 2009).

In this respect, Istanbul, a city not famous for its carnivals, is no exception. In the last two decades, the city has become a year-round stage for film, music, art and street festivals of all sorts. The moment one of them ends, another begins. İstanbul Kültür Sandt Vakfı (Istanbul Culture and Arts Foundation) (İKSV), for instance, sponsors a series of festivals starting with the Istanbul Film Festival and followed by a Jazz Festival, Theatre and Classical Music Festivals, which span the spring and summer of culture in Istanbul. Istanbul Biennial, also underwritten by İKSV, along with many young entrepreneurs with new money, turns Istanbul into an internationally acclaimed art event and market. Pop and rock concerts and festivals, such as Rock'n Coke, carrying the unmistaken brand of Coca Cola, attract large crowds. Every municipality has its own festival and festive events; hotels in the main tourism districts play a leading role in staging street fairs and festivals. Simply put, counting (or accounting for) festivals is an exercise in futility.

The most spectacular of the events is undeniably Istanbul 2010, during which Istanbul presents its cosmopolitan ambience and cultural riches to its blasé natives and curious visitors. Billboards and television advertisements announce that the stage belongs to Istanbul (*Sahne Senin İstanbul*) and call everyone to rediscover Istanbul, its historic sites, natural beauty and everyday life. Istanbul is simultaneously the stage and the act, the city and the spectacle.[13]

Culture is *Incorporated*[14] in Istanbul today, as in other world cities; in other words, Istanbul is no exception to the worldwide trends of states' diminishing role in cultural production (of national culture, so to speak) and of the private takeover of culture. High and low, every other event is a sponsored event – to the extent that sponsorship comes as a prerequisite for legitimacy, both for the organizers and sponsors. Being and having a sponsor has come to be the benchmark for social value and cultural significance. From the point of the event maker, not having a sponsor's logo on a poster lands an event in the nowhere of marginality from the start. From the lens of the corporation/institution, having one's logo on an event poster is the mark of social responsibility and corporate citizenship.

Among the major sponsors are financial corporations, media conglomerates, producers of consumer goods and retailers, cultural offices of foreign emissaries (such as the British Council, Goethe Institut, l'Institut Français), and various native and foreign NGOs. Three of Turkey's major financial institutions, Akbank, Garanti Bankası and Yapı Kredi Bankası, not only sponsor cultural events but are also the major producers in the field of culture. They all have galleries, stage musical events and film festivals and publish books. Sony, Turkish cell phone giant Turkcell, and METRO Group, the German-based hypermarket chain, are the most noticeable corporate sponsors among the producers of consumer goods and retailers.

İstanbul Kültür Sandt Vakfı, one of the oldest private institutions in the culture arena with the backing of Eczacıbaşı Group, is the producer of a number of high-profile festivals exemplified above (see Yardımcı, 2005) and behind the launching of the first private modern art museum in Istanbul, Istanbul Modern.

The two most important corporate families, Sabancı and Koç, the Rockefellers of Turkey, fund the other private art museums – namely the Sakıp Sabancı Museum and Pera Museum. The increase in number of private museums (others are on the way) only confirms the new status of art as private/corporate business.

Finally, an *amplified* sense and experience of *sociality* derives from the body politic of late twentieth-century cities, characterized by the intensification of public spectacles, recreational activities and entertainment. That is, sociality has increasingly moved to spaces outside the home – to cafés, streets, shopping malls, stadia and city squares. In the early 1990s, Istanbul's urban landscape was not as saturated with cafés, restaurants, large and small eateries and global and local fast food chains as it is now. Eating out was not a routine activity as it is now; going out to a restaurant was meant to be a celebration and required a special occasion. Istanbul's streets and squares did not host so many fests, revelries and celebrations. Since the end of the 1990s, it is almost impossible for a passer-by not to encounter some sort of festivity in Taksim Square: one day a commemoration with parades and fireworks, another day a fest with traditional arts and songs, yet on another day a party with rap, techno and pop music. The Istanbul 2010 project promises to contribute extensively to the intensification of recreation and entertainment by facilitating the staging of more events per day than ever happened in the recent history of the city. After 2010, the city's annual event inventory will certainly be more extensive. It is not inaccurate to suggest that nowadays people spend more time out, in public, in the open, social spaces of the city. It is as if the outside, the city spaces, is the new inside, the new space where people socialize and engender 'public intimacies' (Soysal, forthcoming).

Conclusion

In this chapter, I strived to achieve two seemingly disparate tasks: providing an account of the transformations taking place in Istanbul under intertwined meta-analytical categories (Gentrification, Heritage Displays, DesignCity, City-Spectacle, Culture Incorporated and Sociality Amplified) and locating Istanbul in a civilizing process, colloquially called Europeanization. To achieve the state of civilization dictated by Europeanization, as a candidate country, Turkey adopts and implements the formal frameworks provided by the EU, which range from the revisions that expedite the transformations in the areas of law, economy, education and civic life to the imposition of new codes that affect the everyday life of the city – how traffic signs are arranged, how street sellers are dressed and how restaurants serve food. Not all models enacted on the road to civilization come via Europe. World cities elsewhere afford blueprints for reorganizing cityscapes and public space; and supranational organizations (UN, UNESCO, Council of Europe) make available covenants that legitimate the changes. In this respect, the dizzying expanse of gentrification, spectacle and sociality I have attempted to capture here is necessarily related to the globalizing economies of imitation, circulation and mobility (Appadurai, 1996; Inda and Rosaldo 2002; Lechner and Boli 2005; Tsing 2002; and Yudice 2003); and the social/cultural worlds we encounter in Istanbul approximate the global as we know it. In that, they are not incontrovertibly particular to Istanbul but conjure and contribute to imaginaries beyond the confines of the national.

The future(s) of Istanbul are profoundly shaped by the developments envisaged and narrated in the scripts of Europeanization and globalization. In turn, as a world city proper, Istanbul's future(s) in culture industries, real estate development, heritage production and expansive sociality will have an unmistakable mark on the shape the 'civilized' future holds for the city and Europe at large.

Here a cautionary remark is in order. As Anna Tsing (2002, p. 454) warns us, when speaking of city and globalization, it is not helpful to 'throw ourselves into endorsements of globalization [and transformation of city] as a multilayered evolution, drawing us into the future' but '[s]ometimes our critical distance [as social scientists] is less useful than our participation'. That is to say, like Tsing, I am writing about globalization and the city, in this case Istanbul, as a 'participant' not only in the scholarly debate but also in larger 'definitional struggles' towards 'a more nuanced and surprising appreciation of the making and remaking of geography' (*Ibid.*, p. 477).

Let me now conclude by referring to the publicity blurb on the cover of the 1999 edition of Julian Barnes's novel, *England, England*. It reads as follows:

> As every schoolboy knows you can fit the whole of England on the Isle of Wright. In Julian Barnes' new novel, the grotesque visionary tycoon Sir Jack Pitman takes the saying literally and does exactly that. He constructs on the island 'the Project', a vast heritage centre containing everything 'English', from Buck House to Stonehenge, from Manchester United to the White Cliffs of Dover. The project is monstrous, risky, and vastly successful. Indeed it gradually begins to rival 'Old' England and threatens to supersede it. (Picador, 1999)

This I believe is a hilarious but suitable end. As the cultural and social topography changes under the duress of neoliberal rebuilding and modernizing/globalizing desires, World City Istanbul, like its counterparts Berlin and London in the west of Europe, gets detached from the imaginary of the 'national' and attaches itself to the visions and icons of the 'global'. This is a grand transformation on the road to the new and the civilized. As envisioned, the project is simply monstrous and extremely risky but promises to be vastly successful. Like the 'New' England of grotesque visionary tycoon Sir Jack Pitman, it threatens to supersede the Old Istanbul, give or take the matter of a decade. The same I believe goes for Old Turkey and Old Europe.

Notes

1. This chapter draws on a larger project, which I have been pursuing since 2001 in Berlin and Istanbul. That project, called WorldCity Berlin and the Spectacles of Identity, concerns the changing meaning and constitution of public events and the reconfiguration of the cultural landscape in the contemporary metropolitan centres of new Europe, in particular in Berlin, partially funded by the Social Science Research Council, New York, and the Berlin Program for Advanced German and European Studies, Free University, Berlin, and Mirekoç, Migration Research Program at Koç University, Istanbul. An earlier version of this chapter was presented to an audience at a colloquium at the Institute for European Ethnology, Humboldt University-Berlin. I would like to thank my friends and colleagues at the Institute, Wolfgang Kaschuba, Michi Knecht, and Sefan Beck, for intellectual sustenance and hospitality over the years. I also thank Deniz Göktürk and İpek Türeli for their valuable comments and critques.

2. Minister of State, Hayati Yazıcı, responsible for the 2010 Agency, provided the figures during his speech at the presentation of the Istanbul 2010 Programme to press and dignitaries on 10 December 2009 in Istanbul at which I was present.

3. The selection of performers and acts on seven hills were telling. In Kadıköy, a district home to intellectuals, professionals and radical youths and with a reputation to vote *en masse* for the left and social democrats, *Mor ve Ötesi* (Purple and Beyond), a rock band with a progressive record took stage. In Pendik and Bağcılar, peripheral districts with working and lower middle class residents, on stage were two prominent folk singers, Kıraç and Zara. In Sultanahmet, on a stage in between Hagia Sophia and the Blue Mosque, the performer was Mercan Dede, a musician of international fame, who is known for his performances with a whirling dervish on stage. Nil Karaibrahimgil, a pop singer with a vast teenage fan base and famous for her songs *à la Girls wana have Fun*, entertained the residents of Beylikdüzü, where new housing complexes are being built. The centre, Taksim Square, was reserved for Tarkan, the most popular pop icon of all, as if to bring the city – and the nation – together for a celebration on the way to becoming, again, a European capital. For the videos of these shows, see http://www.istanbul2010.org/HABER/GP_618537. Accessed 18 April 2010.

4. During the panel discussion that was held in the European Parliament on 2 February 2010, to promote Istanbul's reign as Cultural Capital, the participants, were asked to deliberate on the question whether the 'idea of Istanbul 2010' was 'an irony, a confrontation (with Europe) or a natural definition', reports journalist Nilgün Cerrahoğlu (*Cumhuriyet*, 6 February 2010, p. 11). As it is explicit in the question, Istanbul's location in Europe and Turkey's Europeanness is open to debate but not easily deniable. Such debates, though raged in Europe, do not make it to the Turkish press, as if not to spoil the show.

5. According to Lechner and Boli (2005, p. 6), 'world culture' comprises 'norms and knowledge shared across state boundaries'. Such norms and knowledge is 'promoted by nongovernmental organizations as well as for-profit corporations, enacted on particular occasions that generate global awareness'. World culture as such is 'riven by tension and contradiction' but facilitates 'the multiple ways particular groups relate to universal ideals'.

6. For an excellent analysis of what is called the 'decline' paradigm, see Kafadar, 1997/1998.

7. Translations are mine.

8. The literature on the global/world city is vast and it is not my intention to review and criticize this here, but I should underline that Saskia Sassen's work was informative on my thinking on global city; so were Abu-Lughod (1995, 2000), Harvey (2001), Herzfeld (1991, 2009), Zukin (1995, 2010). Of course this list is in no way exhaustive.

9. The full name of Law No. 5366 is 'Yıpranan Tarihi ve Kültürel Taşınmaz Varlıkların Yenilenerek Korunması ve Yaşatılarak Kullanılması Hakkında Kanun' (roughly translated, Law for the Protection, Renovation, and Sustenace of Historic and Cultural Real Estate under Erosion). The Law is heavily criticized for giving licence to municipalities for wholesale gentrification and lack of protective measures (see Ayşegül Baykan *et al.*, Chapter 4).

10. For more on Fener-Balat project, see http://www.fenerbalat.org/. Accessed 10 April 2010.

11. For details of some of these programmes, see http://www.en.istanbul2010.org/KENTSEL PROJELER/index.htm and http://www.en.istanbul2010.org/KULTURELMIRASMUZELER/index.htm. Accessed 19 April 2010. For complete listing, see Istanbul 2010 European Capital of Culture Program (published and distributed by Istanbul 2010 Agency).

12. See the special theme section of *Focaal: Journal of Global and Historical Anthropology* (No. 55, 2009) on 'The persistence of "culture talk" in the making of Europe', edited by Mary Taylor, with contributions from Levent Soysal, Katharina Bodirsky, Banu Karaca and Mary Taylor.

13. Istanbul is not alone in staging itself. Every summer Berlin stages a show called *Schaustelle Berlin* (Stage Berlin) and presents itself to its visitors and residents.

14. I borrow the subheading from Rectanus (2002).

References

Abu-Lughod, Janet (1995) *From Urban Village to East Village: The Battle for New York's Lower East Side*. Cambridge, MA: Blackwell.

Abu-Lughod, Janet (2000) *New York, Chicago, Los Angeles: America's Global Cities*. Minneapolis, MN: University of Minnesota Press.

Appadurai, Arjun (1996) *Modernity at Large: Cultural Dimensions of Globalization*. Minneapolis, MN: University of Minnesota Press.

Barnes, Julian (1999) *England, England*. London: Picador.

Beck, Ulrich, and Grande, Edgar (2007) *Cosmopolitan Europe*. Cambridge: Polity Press.

Elias, Norbert (1939, 1982) *The Civilizing Process (Über den Prozess der Zivilisation)*. Oxford: Basil Blackwell.

Esenbel, Selçuk (2000) Türk ve Japon modernleşmesi: 'Uygarlık süreci' kavramı açısından bir mukayese [Turkish and Japanese modernization: a comparison in terms of the concept of 'civilizing process']. *Toplum ve Bilim* [Society and Science], No. 84, pp. 18–36.

Fabian, Johannes (1983) *Time and the Other: How Anthropology Makes Its Object*. New York: Columbia University Press.

Flusty, Steven (2004) *De-Coca-colonization: Making the Globe from the Inside Out*. New York: Routledge.

Friedmann, John (1986) The world city hypothesis. *Development and Change*, **17**(1), pp. 69–83.

Friedmann, John (2002) *The Prospect of Cities*. Minneapolis, MN: University of Minnesota Press.

Harvey, David (2001) *Spaces of Capital: Towards a Critical Geography*. New York: Routledge.

Harvey, David (2006) *Paris, Capital of Modernity*. New York: Routledge.

Herzfeld, Michael (1991) *A Place in History: Social and Monumental Time in a Cretan Town*. Princeton, NJ: University of Princeton Press.

Herzfeld, Michael (2009) *Evicted from Eternity: The Restructuring of Modern Rome*. Chicago: University of Chicago Press.

Inda, Jonathan Xavier and Rosaldo, Renato (2002) Introduction: a world in motion, in Inda, Jonathan Xavier and Rosaldo, Renato (eds.) *The Anthropology of Globalization: A Reader*. Malden, MA: Blackwell, pp. 1–34.

Kafadar, Cemal (1997/1998) The question of Ottoman decline. *Harvard Middle Eastern and Islamic Review*, **4**(1/2), pp. 30–75.

Keyder, Çağlar (1999) *Istanbul: Between the Global and the Local*. Lanham, MD: Rowman and Littlefield.

Lechner, Frank J. and Boli, John (2005) *World Culture: Origins and Consequences*. Malden, MA: Blackwell.

Lewis, Bernard (1968) *The Emergence of Modern Turkey*. Oxford: Oxford University Press.

Peter Smith, Michael (2001) *Transnational Urbanism: Locating Globalization*. Malden, MA: Blackwell.

Öncü, Ayşe (2000) Elias ve medeniyetin öyküsü [Elias and the story of civilization]. *Toplum ve Bilim* [Society and Science], **84**, pp. 8–17.

Rectanus, Mark W. (2002) *Culture Incorporated: Museums, Artists, and Corporate Sponsorships*. Minneapolis, MN: University of Minnesota Press.

Sassatelli, Monica (2009) *Becoming Europeans: Cultural Identity and Cultural Policies*. Basingstoke: Palgrave Macmillan.

Sassen, Saskia (2001) *The Global City: London, New York, Tokyo*, 2nd ed. Princeton, NJ: Princeton University Press.

Sassen, Saskia (2006) *Territory-Authority-Rights: From Medieval to Global Assemblages*. Princeton: Princeton University Press.

Soysal, Levent (2005) Karneval als Spektakel. Plädoyer für eine aktualisierte Perspective [Carnival as Spectacle. Propositions for a revised perspective], in Knecht, M. and Soysal, L. (eds.) *Plausible Vielfalt. Wie der Karneval der Kulturen denkt, lernt und Kultur macht* [Plausible Diversity. How the Carnival of Cultures Thinks, Learns and Makes Culture]. Berlin: Panama Verlag, pp. 260–274.

Soysal, Levent (forthcoming) Intimate engagements of the public kind. *Anthropological Quarterly*.

Soysal, Yasemin N. (2002) Locating Europe. *European Societies*, **4**(3), pp. 265–284.

Tolum ve Bilim [Society and Science] (2000) Uygarlık Süreci [Civilization Process], No. 84, Spring.

Tsing, Anna (2002) Conclusion: the global situation, in Inda, Jonathan Xavier and Rosaldo, Renato (eds.) *The Anthropology of Globalization: A Reader*. Malden, MA: Blackwell, pp. 453–485.

Yardımcı, Sibel (2005) *Kentsel Değişim ve Festivalizm: Küreselleşen İstanbul'da Bienal* [Urban Change and Festivalism: The biennial in a globalizing Istanbul] Istanbul: İletişim Yayınları.

Yudice, George (2003) *The Expediency of Culture: Uses of Culture in the Global Era*. Durham, NC: Duke University Press.

Zukin, Sharon (1995) *The Cultures of Cities*. Oxford: Blackwell.

Zukin, Sharon (2010) *Naked City: The Death and Life of Authentic Urban Places*. Oxford: Oxford University Press.

Epilogue

Istanbul: Cultural Politics in the Kaleidoscope

Michael Herzfeld

The kaleidoscopic array of this book reproduces the frenetic transformation of what those on the other side of the Aegean, the Greeks, call '*the* City' – a transformation so raucous, so contemptuous of mere space and time, that it swamps the poor and the local in a deafening, global blare of speculation and spectacle; and so much in hock to a neoliberal market that, as Jale Erzen demonstrates in these pages, local artists and their sometimes terminally marginal subjects risk total banishment from the scene they once animated – so much so, indeed, that the galloping rediscovery of ethnic enclaves and Ottoman traces virtually depends on the disappearance of those populations that animated such identifications in the first place. The best the newly marginalized artists and ethnic minorities can hope for may be their preservation as picturesque and suitably detoxified traces of a past that only survives as a mark of what the present does not want to see in itself. Here we have, not the simulacrum of a socially experienced past (see Baudrillard, 1994), but a sanitized redrawing of the past that labours mightily to destroy and replace its presence. The more that past is turned into something that never existed, the more it seems to invade the present, a nostalgia industry fuelled by the technology and economic power of a historically oblivious modernity.

The complex new reality that incorporates such a reconfigured past is composed of what Levent Soysal, with an ironic sensibility, here invokes Norbert Elias to call a 'civilizing process': a neoliberal recreation of the aesthetic and ethical values that in the nineteenth century sustained the European project of colonial domination, providing it with a rhetoric of justification that today hides somewhat disingenuously behind labels such as cosmopolitanism, globalism, and the free market. In some ways, however, it is still the same project of Europeanization (rather than of globalization, I would argue), its core the paradoxical necessity of demonstrating that all this noisy difference actually bears witness to a transcendent unity, a genius that can only be understood through the infinite variety of its forms and achievements.[1]

If for the Europeans of high modernism this claim to a unifying identity invoked diverse incarnations of the common ideal of national genius (thereby exemplifying the unity-in-diversity paradox at an even more inclusive level and so anticipating the global straitjacketing of national traditions by such world organizations as UNESCO as well as the more narrowly continental articulations of the European Union), for Istanbulites today it means, rather, embracing the signs and symbols of 'diversity' – perhaps best exemplified here by the increasingly self-assertive but also perhaps increasingly beleaguered Jewish community described by Marcy Brink-Danan. That community's emergent ebullience appeals to a cosmopolitanism of clearly the same kind of European lineage as the newly gentrified Greek quarters, the exhibition of Armenian architects, and the quietist or moderate neo-Ottoman Islamic activism. Neo-Ottoman activism poses a far more subtle challenge to the crude twentieth-century Euro-nationalism of the military establishment than any headscarf, particular because, as Jeremy F. Walton's essay shows, spatial expressions of a reversion to officially prohibited cultural identities are much harder to contest or suppress – or even to identify persuasively – than outright fundamentalism or revolt.

Space, in Walton's account, reorders time, by implying the possibility of reversion to an older order and thus to a rejection of state modernism. This is a striking iteration of the clash between the monumental time of the state, symbolized in Turkey by the Atatürk Mausoleum in Ankara, and competing social rhythms some of which for the moment can only assert themselves through their tactical insubordination of an increasingly powerful world economic order. But the state is arguably no longer the most powerful purveyor of monumental time; UNESCO, with its list of World Heritage sites and its curiously self-contradictory rhetoric about 'intangible heritage', represents an extraordinarily far-reaching amalgamation of what had hitherto been single national interests into a huge machinery of recognition and reification.

In that context, Istanbul's status as a non-capital gains peculiar significance, begging for comparison with Sydney and New York and shedding interesting light on the frequently expressed desire to de-capitalize Rome by some of its more traditionally-minded civic leaders. Istanbul now – again? – occupies a disproportionate space in the national imaginary, materially captured – as İpek Türeli demonstrates – in the organization of the Miniaturk theme park, with its evocation of the Ottoman past. Today the city's dominance may owe less to its having been the capital of an empire than to its *not* being the capital of what, in other respects, still looks to international observers like a rather parochial and inward-looking nation-state.

Is Istanbul now, as Çağlar Keyder argues here, becoming a true capitalist city (and will it, in that case, come down with the collapse of the neoliberal self-deception of infinite resources)? Is it engaged on a convergent process of post-national civilization, as Soysal's delicately non-evolutionist invocation of Elias suggests? Or is this little more than the next phase of a process of European domination initiated on the other side of the Aegean in the elevation of Venetian

('Western' and Christian) architecture over its Ottoman ('oriental' and therefore diabolical) rival (see Herzfeld, 1991), the erasure of that oriental past through processes of gentrification rhetorically mitigated by the framing of the ghosts of minorities dwindling and dead as a monument to the equally Eurocentric ideal of liberal tolerance (as in the use of a historic synagogue as the site of official interactions between the Jewish community and the Turkish state; see also Rabinow, 1989, p. 299)?

That ideal is today reconfigured as the no less Eurocentric model of 'diversity' embraced and controlled, but we should not forget that what has locked it into place was an earlier erasure of minorities in celebration of the new city, as Türeli shows in an elegant discussion of the film *Birds of Exile*. Is the evolutionism of which Soysal tries to strip the Elias model simply the discursive face of Eurocentric globalism, now detemporalized in its pursuit of total domination although perhaps also internally redirected as other global players such as China and Japan begin to impose their own aesthetics and ethics? Are the processes described in this book as irreversible as the vicariously fatalistic language of the market – itself the reincarnation of earlier, colonial forms of implanted fatalism – insistently claims? The minorities will presumably never return in their earlier form, but a process of 'Turkification' that opens the gates to a cosmopolitan reconfiguration of the entire city through the educated citizens' growing predominance undermines its own narrowly nationalistic intentions while at the same time rearranging the class hierarchy that characterized the old Istanbul.

Much of what we read in this book can certainly be replicated elsewhere. If Turkey was once the site of particularly crass invocations of the European *mission civilisatrice*, with an impassioned dualism pitting *alla franka* cultural modes against the *alla turka* traditions of an earlier age in exactly the same way as the Europeanist and neoclassical aesthetic smothered a nativist and partially Ottoman-derived alternative on the other side of the Aegean (Dorn, 1991), other countries that are still coming to terms with the conditional nature of their political independence perhaps illustrate those roots more clearly today. In Thailand, the desire to be *siwilai* – the English derivation of that word is no less significant than the Latinate roots of *alla franka* – was expressed through crypto-colonial projects of cultural Westernization not unlike those we can observe in Greece and Turkey (see Harrison and Jackson, 2009; Herzfeld, 2002; Thongchai, 2000); moves to redesign the central artery of Bangkok's old royal centre as 'the Champs-Elysées of Asia' show a powerful desire to have one's Eurocentric cake but to eat it in the name of independence as well. Are Istanbul's rapid gentrification, spectacularization, and engagement in a market-dominated manipulation of artistic activity as well as of land use simply signs of a similar reaching for respectability on the world stage, or do they bespeak the particular conditions of a local power struggle?[2]

I prefer not to engage in such restrictively binary choices, but, instead, to ask what larger processes are indexed by these geographically localized manifestations. Even the forms of protest – the variable tactics of those resisting eviction from newly prestigious inner-city zones, for example (see Özlem Ünsal and Tuna Kuyucu, this volume) – are connected to similar phenomena elsewhere on the planet, but are also refracted through local particularities of power and social organization. Just as many forms of resistance to globalization have little choice but to use global tactics, for example, so the techniques of resistance to eviction (and thus also gentrification) benefit substantially from a rapidly intensifying information network that would have been inconceivable in pre-electronic days.

The emergent forces of resistance are not imagined communities in Benedict Anderson's (1983) sense, but new coalitions that, like minorities beleaguered by powerful nation-states and thereby compelled to adopt their rhetorics and symbolisms, increasingly adopt the world-encompassing discourse and methods that neoliberal consolidation makes the only viable alternative to extinction. Those who adapt well to such global trends can expect to be more successful in protecting their local interests; they engage in a struggle in which they are already partially complicit, as one always must be in any successful battle, with the dominant economic ideology; that complicity here takes the form of convergent rules of engagement.

To that extent, we may indeed be watching a unilinear process – for the moment. But such developments carry within themselves the seeds of other possibilities. The vibrancy of street-life alternatives to the official demographic imaginary – the sheer cheek of the *keyif* celebrated by Engin F. Işın – suggests that Istanbul also exemplifies the emergence of perhaps mutually competing versions of an increasingly global but resolutely counter-hegemonic insubordination. The pessimism against which Işın directs such sharp criticism represents the hegemonic fatalism and Eurocentric nostalgia and orientalism wished upon this radically disobedient city even by some of its own intellectuals, serving, whether they acknowledge it or not, as agents of the current forces of economic and political domination.[3] The authors of *this* book certainly do not play *that* game, although they do sometimes ride its powerful waves.

The old Eurocentric state sought to perpetuate itself, Ozymandias-like, in the temporal idiom that I have called 'monumental time' (Herzfeld, 1991). But monumental time was always just one among many modalities of 'social time', although most nation-states made strenuous attempts to deny this: the spectre of Ozymandias was too corrosive of their pretensions to be allowed to hover in the foreground. There have been few exceptions, and those mostly in nation-states whose nationalism has been too weak to restrain local exuberance. Italy is one such exception. The restorers of Rome, at least until recently, were among the few who recognized the value of celebrating human fragility by maintaining the marks of time's corrupting passage in buildings allowed to remain stained by weather and time and flaunting their mongrel beauty in all its gloriously dilapidated indiscipline. That visible mortality, the mark of a doctrinally recognizable 'original sin' – the very 'imperfection' of which Ayşegül Baykan, Zerrin İren Boynudelik, Belkıs Uluoğlu and Burak Sevingen speak here, but in an explicitly Christian idiom – also permits official complicity with the insubordinate local social attitudes that insist on compromise with zoning and conservation laws. The city is a palimpsest of such pragmatic complicities.

Yet these arrangements today are threatened by an accelerated sense of time of which the first traces appeared long ago. In Istanbul, whose 'fall' to the Turks in 1453 marks for Greeks a secular reproduction of the religious myth of the creation of original sin through the fall of Adam and Eve and an excuse for all the 'Turkish' imperfections in their 'European' polity, the role of cinema in aligning an existing temporality with European and even global events and in allowing the city's residents to feel that they were taking part in events on the world stage, as Nezih Erdoğan shows, fuelled the motor of the Europeanizing process and prepared Istanbul both for its relative detachment from the rest of Republican Turkey and for the city's new role as one of several cosmopolitan centres in the world as a whole. In its rush towards a desired modernity, it also prepared the way for today's frenetic rebuilding and reconfiguration of once leisurely spaces.

The marks of restoration in Istanbul do not, as they do in Rome, signal a reluctance to leave behind the damage wrought by time, to forget the temporality made manifest by the corrosion of the architectural fabric. Even that Roman acknowledgment of this temporally grounded imperfection – and Rome, again, represents the exception rather than the rule – can also serve as an attempt to co-opt the signs of impermanence in the cause of 'conservation'. But all forms of conservation are also destructive of some memory traces. It is usually easier to expropriate the marks of a complex past, as the Turkish authorities did in Istanbul with the properties of departed Greeks and Armenians, than to celebrate the embarrassments of an intolerant history. There are partial parallels here with the old Jewish quarter of Le Marais in Paris and with the Ghetto of Rome. In Rome, the Jewish community has long exercised a cultural influence, and indeed been considered an emblematic metonym for the city as a whole, but apparently without the need for 'making alliances with the European Union and European Jewish communities [that] has allowed Istanbul's Jews to represent themselves as counting, symbolically, beyond what their numbers might predict' (Brink-Danan, this volume).

Like post-Franco Spain embracing a Jewish and Muslim past erased by a violent church and its supine monarchy, Istanbul tries physically to co-opt the signs of a diversity that was no less violently uprooted and turn these traces into a politically correct and therefore internationally marketable resource. Here, perhaps, the spectacle of the Roman popes, successors to one of the last European states to practice public executions, now fulminating against capital punishment, with a compliant municipal authority using the bloodstained Colosseum to protest each execution in other countries, represents a closer approximation to the Istanbul experience, at least in terms of an apparent unawareness of the ironies that all such revisionist self-representations entail. Or, for that matter, we might instance the papal treatment of Rome's Jewish community, once taken under the popes' 'protection' and humiliatingly exploited and mistreated, only to find itself reconstituted by John Paul II as the Catholic church's 'elder brothers'. Whatever the motives that underlie such apparent changes of heart, they lend themselves to subsequent manipulation. In an age when political correctness itself sometimes seems to substitute simulacra for sincerity, skilled social actors and powerful institutions can turn the new courtesies into extremely effective instruments for erasing the remnants of an embarrassing past, and nowhere can they do this more effectively than in the so-called 'restoration' of 'historic space' (as in the *newly* named 'Historic Peninsula' of Istanbul, an appellation that recalls the Italian term *centro storico* or historic centre, but with a geographical twist that also emphasizes what makes Istanbul different).

There is another interesting difference between Istanbul and Rome that deserves comment, and that is the fact that in Istanbul the Jewish presence has been evolving against a background of Islamic religious practice in a determinedly secular and modernist nation-state. This creates an interesting dynamic in Istanbul's relationship with notions of European identity. The tutelary attitude of the European Union became especially clear in its leaders' condemnation of the 2003 Istanbul synagogue bombing as incompatible with European values – indeed, we might say, with the global hierarchy of value in which 'old Europe' still claims primacy. The presence of Jews in Istanbul remains disproportionately significant, and seems to have become ever more so. As Brink-Danan so succinctly puts it, '[t]o be a European city, it seems, is to "have Jews"'. At the same time, however, this presence *also* bespeaks a difference between at least some versions of Istanbul's Islamic heritage and that of Arab and some other Islamic states hostile to Israel (a stance towards which prime minister and former Istanbul mayor Recep Tayyip Erdoğan, who not coincidentally has also been disappointed in his attempts to court the European Union, may yet steer the Turkish state).

The long history of Ottoman tolerance toward Jews is compatible with the kind of neo-Ottoman Islam that Walton describes here, and allows Istanbul to persist in its claims of being both Islamic and European – and thus also to challenge the smugly offensive insistence of some European Union leaders that Europe is by definition Christian. But this rhetoric also disguises the fact that the actual size of the Jewish community is dwindling in proportion to the city as a whole. The gentrification and re-creation of the old non-Muslim districts, like the masking of slums, makes for an increasingly tendentious reconstruction of what was a complex, bloodstained, and infinitely troubled past.

All these changes are written into a monumental claim on permanence and, increasingly, an electronically powered lifestyle that does not even countenance any sustained reflection on temporality or the changes that it encapsulates. Modernity creates its own 'machines for the suppression of time' (Lévi-Strauss, 1964, p. 24), as it has always done at least since the invention of nationalistic historiographies. In their account of the film *The Magic Carpet Ride*, Deniz Bayrakdar and Elif

Akçalı very effectively capture this rejigged and internationalized experience of time in Istanbul, with its breathless excess and its exploitation of mediatic techniques that carry the impact of such changes far beyond the confines of mere geography. In related vein, Deniz Göktürk provides a compelling picture of the ways in which placing the new technology in the hands of the disenfranchised – nicely emblematized by the circumstance that these were in fact children – can transcend even the destruction of a beloved neighbourhood such as the Rom settlement of Sulukule (or, in a more cynical reading that we might consider either as an alternative or as a supplement, can serve as a palliative that ultimately supports the developers' interests by substituting for substantive opposition).

As the authority of the nineteenth-century nation-state wanes, and cities like Istanbul become more globally than nationally oriented and dispossess themselves of the messy sights that once offended European sensibilities,[4] the temptation to embrace an ever more superficial rhetoric of permanence increases and the visibility of the alternatives becomes easier to deny – although, by the same token, those alternatives also, and paradoxically, become much harder to eradicate. The city authorities' efforts notwithstanding, these other visions insistently mark the smooth surfaces of officially sanctioned architecture with their own insistent messages, whether those in power care to read them or not. Even here, however, economic reasons may make the difference between immediate extinction and at least some prospect of survival; the foundrymen who can make small objects in small quantities cater to a 'niche market', but that term suggestively points up the dependent and fundamentally economic terms of the sufferance that saves their livelihood for the moment. More often, the only path to survival is self-reification through the establishment of zones of artisanal activity. Here, too, the ironic dismissal that such moves towards the creation of artisans' neighbourhoods encountered in Rome may point up the ways in which Istanbul may be significantly more representative of global trends; I have observed similar proposals in Bangkok, where they appeal to (in this regard, at least) a more convergent relationship between popular and official aspirations. Rome, seat of an ancient empire and a still powerful if damaged church, nevertheless still partially revels in its *campanilismo*,[5] its intense localism, with its widely despised local dialect and its national reputation as a cultural backwater and the seat of a recalcitrant and antiquated bureaucracy; Istanbul, by contrast, extracts even from the swelling population of recent peasants a sense not only of being connected to global trends but actually of representing them to the world. That, in turn, entails a greater collective willingness to freeze and sanitize the presence of heritage and to deploy the current population in accordance with that project.

Relocation of artisans in a single neighbourhood entails a 'museumification', not only of the articles they produce, but of the artisans themselves. This is a process of reification and de-humanization that reproduces the repressive logic of the intangible heritage concept. It also speaks to the way in which the reordering of urban space entails, and reflects, changes in the uses to which urban spaces are put. When Hou Hanru disarmingly tells Nilgün Bayraktar, 'I am not interested in authenticity or turning a city into a museum' but then goes on to acknowledge that 'these buildings are also part of a certain heritage', we may be persuaded that a liberal imagination desires the inclusion of populations representing the city's past in its dynamic present; but we might well be equally convinced that a *neoliberal* imagination is more concerned with *co-opting* that past in the interests of speculative economic futures, and that Hou Hanru's expressed optimism, however sincere it may be, ingenuously facilitates surrender to the market. It is no coincidence that at least two of the mega-architects (called *archistar* in Italian!) mentioned in this volume, Zaha Hadid and Santiago Calatrava, were invited by the right-wing mayor of Rome in 2010 to participate in a rethinking of the city's future that seems increasingly to involve rejecting the old policy of acknowledging the past's glorious imperfections as integral to the city's identity; that other, more critical voices were simply excluded may be equally significant.[6]

Here, Istanbul's membership in the club of cities able and willing to host Biennale events sheds some light on the global dynamics in question. As Banu Karaca shows here, this is a growth industry; just as such events may actually marginalize the more original forms of street artistry, so they increasingly enmesh Istanbul in what, in a very different context, I have called 'the global hierarchy of value' – the worldwide articulation of taste, which serves, not unlike aesthetic judgment on a much smaller scale (Bourdieu, 1984), to confirm and reinforce a political hierarchy that largely reproduces the colonial order through a ranked 'entextualization' (Silverstein and Urban, 1996) of cultural value. Entextualization may not always literally be textual; here it is also, and to an at least equal degree, spatial and architectural, redeploying the geographical accident of straddling two continents to claim, in the restoration and conservation of a rich array of architectural styles and resuscitated memories, a glowing transcendence of the radical difference that each of these two continents represents ideologically in the global hierarchy of value.

That hierarchy is largely but not exclusively of Western origin, which means that it takes on a peculiarly poignant force in a city struggling to position itself in relationship both to an ideologically European present and to a historically Ottoman past. Driven today by the forces of the neoliberal economy, which indeed justifies itself in the moral terms of that same hierarchy (as when it calls gentrification 'urban renewal', for example), the new Istanbul aesthetic represents a valiant claim on superior ranking. As Carola Hein points out, moreover, Istanbul's recognition as a European Cultural Capital for 2010 elides the embarrassment of Turkey's failure to gain rapid admission to the European Union as well as the hostility of some European leaders to the idea of a predominantly Muslim nation's potential membership in what has hitherto been a 'Christians only' club.

But this success is only possible because it is granted by others, and because it carries with it an unspoken obligation to reproduce top-down cultural authority internally. It does not even carry comprehensive funding; the humiliating privilege of emulating 'the West' can be expensive. By inducing countries like Turkey to scramble for this kind of recognition, and by recognizing Istanbul for a cosmopolitanism that in its earlier incarnations similarly served to articulate its veiled subjection to the colonial order, the more powerful of the West European nation-states thus still succeed in confirming their own hegemonic advantage: their collective culture, however vaguely defined, remains the yardstick by which all others are judged. Claiming to be the birthplace of ancient Greek philosophers is, as Hein remarks, 'an attempt to define Turkey's Anatolian roots beyond religious concepts'; but it is also, and significantly, the result of Turkey's subjection to the oppressive adulation of a West European version of the Hellenic past that the Greek nation-state itself has had to bear. Such straining to escape orientalist stereotypes can result all too easily in re-orientalization, or at least in humiliation and renewed cultural subjection; and it also emerges internally as official hostility towards supposedly un-European practices in everyday life.

Indeed, Europeanization – political as well as cultural – becomes an excuse for the suppression of much of what is most familiar in everyday life. While Hein's and Oğuz Öner's contributions suggest that this is facilitated by the top-down authority structure favoured by the Cultural Capital bureaucracy, however, other essays in this volume suggest that the undisciplined side of Turkish life will not so easily be silenced. Indeed, across the Aegean in Greece, the recent emergence of enthusiasm for some (admittedly highly sanitized) features of the Ottoman past suggests that one consequence of this monolithic process of Westernization in Turkey may likewise eventually prove to be a growing rejection of its own success.

For that to happen, however, Turkish intellectuals will need to rethink their enthusiasm for the orientalizing models of Western observers. Soysal suggests that this has already begun to happen, as, for example, in the weakening grip of Bernard Lewis's vision of Turkish history on the Turkish intellectual imagination. Such reversals are important; I would argue, for example, that the continuing marginalization of Italy's southern areas in economic, social, and cultural terms owes much to the equally persistent enthusiasm in circles of power and intellectual leadership for the deflating ideas of Edward Banfield (1958) and Robert Putnam (1993).[7] Patterns of domination, thus locally endorsed by influential thinkers from powerful countries, have a way of becoming self-fulfilling prophecies.

There is some risk that Istanbul might follow other big cities down this path, encouraged in that direction by the hierarchical identity politics underlying the European Cultural Capital project. But the lively sense of criticism that the present collection exudes holds out the promise of a far richer if largely unpredictable cultural future (as my preferred metaphor of the kaleidoscope is intended to imply). This collection of essays, too, participates in kaleidoscopic Istanbul, as I indicated at the start of my remarks; and, as such, it deserves to be read both critically and as a richly expressive outcome of the very processes its authors describe. It represents an act of insouciant refusal in the face of authoritarianism, whether intellectual or political. It will – productively, one hopes! – infuriate those who are committed to one or another of the competing essentialisms that inevitably, in a city that makes so much of its geographical ambiguity between stereotyped renditions of orient and occident, rush to the service of every faction embroiled in the unceasing refashioning of a hyperactive urban space. In short, like Lévi-Strauss's Sigmund Freud analyzing the Oedipus myth, it will serve as a new incarnation of the processes its authors have so astutely analyzed, while also opening up new vistas of intellectual adventure. 'The City' demands no less of those who would write about that conceptual space between orient and occident that continues to tease the global imagination.

Notes

1. Some years ago, I advanced a very similar argument in relation to the European claims of Greece in the nineteenth century (Herzfeld, 1987).

2. The official preference for the term 'urban transformation' should not disguise the fact that it is a process that serves particular class interests rather than simply architectural 'renewal' (see Smith, 2006). This may be linked to the phenomenon noted here by Banu Karaca, in which the sponsorship of public art can actually create what she calls 'exclusionary practices' – a sage and timely warning relevant to many other cities as well.

3. For an important discussion (albeit in a very different context) of how minority interests may adopt the rhetoric of a dominant institutional structure as well as of the ethical issues that this raises for analysts, see Jackson (1995).

4. Göktürk mentions one dramatic example of this physical erasure of eyesores on the urban landscape, a phenomenon that I would put in the same category as the Thai authorities' attempts to conceal slums from visiting dignitaries' eyes (see, for example, Klima, 2002, p. 40).

5. Technically, this term means identification with a neighbourhood within earshot of a particular church tower (*campanile*). See Herzfeld (2009) for a more amplified discussion of the aspects of Roman culture discussed here.

6. See especially the reportage in the Rome sections of the newspapers *La Repubblica* (3 April 2010, p. XVI; 6 April, p. VI) and *Il Messaggero* (3 April, p. 36; 6 April, p. 35; 7 April 2010, p. 36).

7. Putnam's ideas are considerably more subtle, and are based on a much more sophisticated model, than Banfield's, to which Putnam nevertheless acknowledges a degree of indebtedness. I am not sure whether this is why Banfield's ideas are cited more frequently than Putnam's in Italian political life, or whether that (admittedly impressionistic) observation simply reflects the chronological fact that Banfield's key notion of 'amoral familism' has been sedimented over a longer period. For further discussion, see Herzfeld, 2009, pp. 77–79; for critiques of these stances by Italian scholars, see especially Lai, 1992; Sabetti, 2000.

References

Anderson, Benedict (1983) *Imagined Communities: Reflections on the Origin and Spread of Nationalism*. London: Verso.

Banfield, Edward C. (1958) *The Moral Basis of a Backward Society*. Glencoe, IL: Free Press.

Baudrillard, Jean (1994) *Simulacra and Simulation*. Ann Arbor, MI: University of Michigan Press.

Bourdieu, Pierre (1984) *Distinction: A Social Critique of the Judgement of Taste*. London: Routledge & Kegan Paul.

Dorn, Paméla (1991) Change and Ideology: The Ethnomusicology of Turkish Jewry. Ph.D. dissertation, Indiana University, Bloomington, IN.

Harrison, Rachel V. and Jackson, Peter A. (eds.) (2009) *The Ambiguous Allure of the West: Traces of the Colonial in Thailand*. Hong Kong: Hong Kong University Press.

Herzfeld, Michael (1987) *Anthropology through the Looking-Glass: Critical Ethnography in the Margins of Europe*. Cambridge: Cambridge University Press.

Herzfeld, Michael (1991) *A Place in History: Social and Monumental Time in a Cretan Town*. Princeton, NJ: Princeton University Press.

Herzfeld, Michael (2002) The absent presence: discourses of crypto-colonialism. *South Atlantic Quarterly*, **101,** pp. 899--926.

Herzfeld, Michael (2009) *Evicted from Eternity: The Restructuring of Modern Rome*. Chicago, IL: University of Chicago Press.

Jackson, Jean E. (1995) Culture, genuine and spurious: the politics of Indianness in the Vaupés, Colombia. *American Ethnologist*, **22**, pp. 3–27.

Klima, Alan (2002) *The Funeral Casino: Meditation, Massacre, and Exchange with the Dead in Thailand*. Princeton, NJ: Princeton University Press.

Lai, Franco (1992) Imprenditori e contesto culturale: Il dibattito sull'invidia come vincolo all'attività imprenditoriale nella ricerca antropologica, in Siniscalchi, Valeria (ed.) *Frammenti di Economie: Ricerche di antropologia economica in Italia*. Cosenza: Pellegrini.

Lévi-Strauss, Claude (1964) *Le cru et le cuit* [The raw and the cooked] (Mythologiques I). Paris: Plon.

Putnam, Robert D. (1993) *Making Democracy Work: Civic Traditions in Modern Italy*. Princeton, NJ: Princeton University Press.

Rabinow, Paul (1989) *French Modern: Norms and Forms of the Social Environment*. Cambridge, MA: MIT Press.

Sabetti, Filippo (2000) *The Search for Good Government: Understanding the Paradox of Italian Democracy*. Montreal & Kingston: McGill-Queen's University Press.

Silverstein, Michael and Urban, Greg (eds.) (1996) *Natural Histories of Discourse*. Chicago, IL: University of Chicago Press.

Smith, Neil (2006) Gentrification generalized: from local anomaly to urban 'regeneration' as global urban strategy, Fisher, Melissa S. and Downey, Greg (eds.) *Frontiers of Capital: Ethnographic Reflections on the New Economy*. Durham, NC: Duke University Press.

Thongchai Winichakul (2000) The quest for '*siwilai*': a geographical discourse of civilizational thinking in the late-nineteenth-century and early-twentieth-century Siam. *Journal of Asian Studies*, **59**, pp. 528–549.

Index

Note: illustrations are indicated by *italic page numbers*, notes by suffix 'n' (e.g. '123n[2]' means page 123, note 2); 'ECoC' = 'European Capital of Culture'